CORNELII TACITI
DE VITA AGRICOLAE

EDITED BY
R. M. OGILVIE
AND
The Late SIR IAN RICHMOND

OXFORD
AT THE CLARENDON PRESS

OXFORD
UNIVERSITY PRESS

Great Clarendon Street, Oxford OX2 6DP

Oxford University Press is a department of the University of Oxford.
It furthers the University's objective of excellence in research, scholarship,
and education by publishing worldwide in

Oxford New York

Auckland Bangkok Buenos Aires Cape Town Chennai
Dar es Salaam Delhi Hong Kong Istanbul Karachi Kolkata
Kuala Lumpur Madrid Melbourne Mexico City Mumbai Nairobi
São Paulo Shanghai Singapore Taipei Tokyo Toronto

and an associated company in Berlin

Published in the United States
by Oxford University Press Inc., New York

Special edition for Oxbow Books, Oxford
(and their American partner Powell's Books, Chicago)

British Library Cataloguing in Publication Data
Data available

ISBN 0-19-814438-5

1 3 5 7 9 10 8 6 4 2

Printed in Great Britain
on acid-free paper by
Biddles Ltd.,
Guildford and King's Lynn

I. A. R.

PREFACE

ALL too often has the work of British scholars on the *Agricola* been cut short by death. Furneaux's edition was first published in 1898. Professor Haverfield had embarked on a thorough revision of it but did not live to complete the task. Much of his material was incorporated in the new edition which Professor Anderson brought out in 1922. 'Furneaux–Anderson' for long held its place as the best commentary on the *Agricola*, but the progress of knowledge, particularly in the archaeology of Roman Britain, made parts of it increasingly outdated. In 1964 Sir Ian Richmond, who had for many years entertained the project, invited me to collaborate with him in revising it anew. The original plan was that Richmond was to be responsible for Sections 1, 5, 6, and 7 (as now numbered) of the Introduction and for the Appendixes, while I was to undertake the rest of the Introduction and the Text; we were each to revise such parts of the commentary as seemed necessary and then to collate and combine our suggestions. Before his untimely death in 1965 Richmond had completed drafts of his Sections of the Introduction and of the Appendixes. We had gone over these together and I had shown him drafts of Sections 2 and 3, but he had not gone further than making a few marginal annotations on the Commentary. We had, however, met every week for several months and discussed major topics of text and interpretation, so that I was encouraged by the generous assistance of so many of his friends to complete the edition alone. In the event, the text and commentary are my sole responsibility and Sections 5 and 6 of the Introduction and Appendix 3 have been substantially rewritten in the light of further study. Except where it is explicitly stated, I would not claim that Richmond would necessarily have agreed with my conclusions.

We planned it as a revision because 'Furneaux–Anderson' had proved itself a commentary of outstanding usefulness. We have, therefore, kept the form and the language of the original notes wherever practicable but it was impossible not to make radical and far-reaching changes amounting in fact to about three-quarters of the whole. In addition, the text and *apparatus criticus* have been entirely recast. We also decided to alter the numeration of the sections of the *Agricola* (and of Tacitus' other works) to bring it into line with the numeration which is now generally used on the Continent, e.g. in the Teubner and Budé editions. It is to be hoped that this will become standard practice for citing passages of Tacitus, since the present variety of reference-systems only leads to confusion. As this edition is also intended for use in schools, we have tried to pay particular attention to elucidating and translating Tacitus' meaning. At the same time we felt that the value of the edition would be impaired if we did not include discussions of controversial matters and some bibliographical documentation. Where these are not of primary importance for understanding the text we have enclosed them in square brackets ([ ]) and they may be passed over.

The numbered sites on the maps (Figs. 2, 3, 4, 5, 7) are identified by the key, pp. 337 ff.

Richmond had an unrivalled knowledge of Roman Britain which he would have used to illuminate much of the Commentary. I am very conscious of my own deficiency in this field and, although I have tried to visit all the relevant sites in Scotland, limitations of time and finance have handicapped me from making as extensive an investigation of the sites in England and Wales as I should have liked; but I hope that, however inadequate, this edition may be a memorial to a great scholar and a dear friend.

Errachd R. M. OGILVIE

CONTENTS

ACKNOWLEDGEMENTS

IT would have been impossible to have completed this work without the assistance of Miss Helen Waugh, who not only compiled the maps and secured the illustrations but also made throughout valuable improvements on archaeological and historical matters. The curators of the Ashmolean Library gave me permission to use Sir Ian Richmond's papers and Lady Richmond has throughout taken the deepest interest in its progress. The greater part of it was written during the academic session of 1965–6 when I was privileged to hold a visiting appointment at University College, University of Toronto, and to be a Senior Resident at Massey College in the same University. It would be difficult to imagine more agreeable or stimulating surroundings in which to work. I owe a great debt of gratitude to Professor S. S. Frere and Mr. R. H. Martin, who read most of the typescript, and to Mr. Jasper Griffin, who read the proofs. Sir Ian Richmond's many friends rallied to my aid and, in particular, I should like to thank Dr. J. K. St Joseph, who also conducted me over the sites in Strathmore as only he can do, Mr. D. F. Allen, Dr. J. B. Tait, Mr. J. W. Brailsford, Professor K. H. Jackson, Dr. A. Graham, Mr. B. R. Hartley, and Mr. T. G. E. Powell for advice on detailed points. κοινὰ τὰ φίλων. I pestered my colleagues for help and, in acknowledging their generosity and forbearance, I must single out Professor F. R. D. Goodyear, Professor G. Goold, Professor Hugh Lloyd-Jones, Mr. R. Meiggs, Mr. A. M. Snodgrass, Mr. J. S. Morrison, Mr. D. R. Wilson, Mr. C. H. V. Sutherland, Mr. G. C. Boon, and Mr. K. Wellesley. Mr. Michael Burton proved a disarming critic of my wilder ideas. I have to thank the following bodies for permission to reproduce illustrations: Fig. 1 the Ordnance Survey; Fig. 6 Dr. St Joseph and the Society for the Promotion of Roman Studies; Fig. 9 Dr. St Joseph; Fig. 11 Professor S. S. Frere; Figs. 12, 14, 15, 16 the Society of Antiquaries; Fig. 13 the Prehistoric Society; Figs. 17 and 18 Glasgow Archaeological Society and Dr. Anne Robertson, F.S.A.; Plate I (*b*) Chester Corporation and Mr. D. F. Petch, the Curator of the Grosvenor Museum, Chester; Plate I (*a*) Professor S. S. Frere; Plate II the Warburg Institute and the Trustees of the British Museum; Plate III the Trustees of the British Museum and Mr. K. S. Painter; Plate IV the American Numismatic Society; Plate VI *Antiquity* and Dr. I. M. Stead; Plate VII the National Museum of Antiquities of Scotland and Mr. R. B. K. Stevenson; Plate VIII the Committee for Aerial Photography in the University of Cambridge and Dr. St Joseph.

PLATES

between pp. 336 *and* 337.

FIGURES

INTRODUCTION

1 · AGRICOLA

OUTSIDE his biography by Tacitus, which would seem to have been little read in the Roman world at large, Agricola is hardly known.[1] The other works of Tacitus which must have mentioned him do not survive at the appropriate points and the only other writer to name him is the third-century historian Cassius Dio, who refers to him twice.[2] Two inscriptions contain his name as governor of Britain, one from the *forum* at St. Albans (*Verulamium*), the other from the legionary fortress at Chester (*Deva*).[3] Yet his biographer wrote truly, *narratus et traditus superstes erit*: for in the *Agricola* the general and administrator is nobly and sensitively portrayed, in a fashion vouchsafed neither by fortune nor accident of literary survival to any other comparable figure in the Roman world. Thus, while in ancient Britain his name rather than his fame survived in early Welsh genealogy, modern Britain, thanks to a lively interest in early history and a traditional attachment to the classics, knows him well. In the latter days the avowed object of the biography has been achieved.

Gnaeus Iulius Agricola was born on 13 June A.D. 40, at Forum Iulii (Fréjus), in the province of Gallia Narbonensis, where, a trifle more than eighty years previously, Julius Caesar had planted a settlement, which Augustus refounded

[1] For discussions of his life see I. A. Richmond, *J.R.S.* 34 (1944), 34 ff.; A. R. Burn, *Agricola and Roman Britain* (Teach Yourself History Library, 1953); Syme, *Tacitus*, 1. 19 ff.

[2] 39. 50, 4; 66. 20, 1. Dio is probably dependent on Tacitus: see Appendix 2 and notes on c. 28; c. 43, 2; c. 44, 4.

[3] Quoted in the note on c. 46, 4. See Plate I.

as a *colonia*,[1] peopling it with time-expired men from his Eighth Legion—men who, as legionaries, held the full Roman franchise and were not Gauls but, in the main at least, were born of Italian stock.

Agricola's *nomen* Iulius, however, suggests that his paternal ancestor had been enfranchised either as an army officer or as a wealthy *incola* of the *colonia*, and therefore bore the *nomen* which referred to either Caesar or Augustus. In any case, Agricola was by legal status a full Roman citizen, and had behind him three generations of the sentiment implied by such citizenship. His father had sat in the Roman Senate; both his grandfathers had been Roman *equites*. On his mother's side he may well have inherited more ancient provincial sentiment. Her *cognomen*, Procilla, is frequent in Gaul[2] and rare outside it, even in Cisalpine Gaul. This would support a family connexion with the Romanized Gallic landed proprietors and a pride of local ancestry not uncommon among them.

The family entered Roman administrative circles by the usual ladder, of wealth allied to ambition. In the earliest generation known to us, Agricola's grandfathers were procurators of equestrian rank. His father, Iulius Graecinus, son of a procurator, married another procurator's daughter, and, apparently by favour of Tiberius, received senatorial rank and had reached the praetorship before his career was cut short. He was one of the many southern Gauls who were fully Romanized in the early Empire: he had literary tastes, and wrote on the cultivation of the vine—and is said, indeed, to have written with wit and learning.[3] The elder Pliny quotes him. Seneca, too, refers to him as *vir egregius*,[4] recounting how he had been unwilling to accept political

[1] *CIL*. xii, p. 38. Its full name was *Colonia Octavanorum Pacensis* (or *Pacata*) *Classica Forum Iuli* (c. 4, 1 n.). See also *A*. 4. 5, 1.

[2] Cf., e.g., *ILS* 7383.

[3] Columella 1. 1, 14. The *cognomen* Agricola given to his son probably reflects the father's interest in agriculture; cf. note on c. 4, 1.

[4] *Ep*. 29, 6; *De Benef*. 2. 21, 5.

favours from public men of notoriety. Attachment to principle took him in the end into dangerous waters: as praetor he declined the request of the Emperor Gaius to prosecute Marcus Silanus, and, having thus become a marked man, was executed in unknown circumstances in A.D. 40, soon after the birth of Agricola.[1] A fragmentary inscription, apparently from a tomb in altar form, commemorating this Graecinus and erected either by a brother or, less probably, by a son, was found on the Esquiline in Rome in 1940.[2]

When Agricola reached school age, Procilla took him to the Graeco-Roman city of Massilia (Marseilles), the centre of culture and the university of south Gaul. Here Agricola started his education and completed it by those courses in rhetoric and philosophy that were the normal training for young men belonging to the upper class of Roman society.

A senator's son with ambitions usually entered the senatorial career of higher office and Imperial administration.[3] His birth gave him the initial qualification for admission to that career, the right to assume with the dress of manhood the *latus clavus*, or broad purple stripe on the tunic, which men of lower degree could acquire only by favour of the emperor. The first step in a senatorial career of this kind was the tenure of a military commission, as *tribunus laticlavius* in a legion; this commission, which was no sinecure, Agricola held in Britain under the governor and commander-in-chief Suetonius Paulinus (58–61), getting to know the province during the peaceful years of that governorship, and serving on headquarters staff during the grim years of the Boudiccan revolt. Then he passed through the usual round of offices in Rome, becoming quaestor in 64, tribune in 66, and praetor in 68. The Civil Wars brought him two extraordinary duties, under Galba the discreet recovery of stolen temple goods and under Vespasian the

[1] See note on c. 4, 1.

[2] *AE*, 1946, 94.

[3] On the senatorial *cursus* see Syme, *Tacitus*, App. 17.

charge of legionary conscription in Italy. He was then appointed legate of the turbulent Twentieth Legion in Britain under Petillius Cerialis, and rewarded by elevation to the patrician aristocracy and by the governorship of Aquitania. In A.D. 77 he held a *suffect* consulship for some months, at a time not specifically determined, during which he betrothed his daughter to Tacitus. Thereafter came the marriage and a pontificate, followed by the governorship of Britain in A.D. 78,[1] during which, with the normal title of *legatus Augusti pro praetore*, he is recorded on two inscriptions.

In Britain for seven years, a longer term than that of any known governor, he fought with success and administered with efficiency. He first rapidly completed the subjugation of Wales substantially accomplished by his predecessors, and rounded off the annexation of the Brigantes begun by Cerialis and himself eight years before. He then embarked upon the conquest of north Britain, defeating first the Lowlanders and then the Caledonian tribes of the Highlands and devising arrangements for their permanent blockade. The fact that after his recall these dispositions had to be at once substantially curtailed, and a generation later completely abandoned, was due to no fault of the British command. It was governed by the urgent need for troops in central Europe and the lack of means to supply them except by reducing the armies of other frontiers: and among the frontiers that of Britain counted for least in Imperial policy. If Agricola's undeveloped ideas on the conquest of Ireland have seemed, on the basis of the few words reported, to be unduly optimistic, his actual campaigning in Scotland, as well as the subsequent dispositions of permanent design, evince a highly talented and astute appreciation of the task and a judicious approach to its problems. The mention by Tacitus of the admiration expressed by contemporary experts for Agricola's skill in

[1] On the date see Appendix 1.

choice of sites has been decried as stock praise: but a topographical study of the sites concerned does at least demonstrate that the statement was in fact true, however valueless it may have been in other contexts or however conventionally bestowed in this one.

Besides military gifts, Agricola had other virtues well suited to a provincial governor. He was simple in manner, hard-working, balanced in judgement, honest in money matters, kindly towards the native population. The portrait by Tacitus bears every mark of truth. These were the characteristic traits of the new aristocracy that rose to prominence under Vespasian, himself the embodiment of the new spirit.[1] Agricola's provincial origin and education gave him sympathy with the people whom he ruled. Like many men of his age, and perhaps with less ambivalence than his son-in-law,[2] he saw the advantage of diffusing Roman civilization in the provinces, and, though Tacitus does not claim that Agricola initiated this policy in Britain,[3] he plainly recognized its merit and zealously forwarded it. It is from his time that the clearest beginnings are seen in Britain in development of orderly, civilized life, the growth of towns, and the diffusion among higher provincial society of the Roman language and ideas.

These achievements and, in particular, the victory of Mons Graupius were appropriately and fairly rewarded by triumphal honours, a public statue, and a long and honourable citation: no general could then expect more and none received it. But after that, nothing. The times were indeed difficult: the Emperor, faced by invasion, unsuccessful or half-successful countermeasures on the Danube, and a *pronunciamento* in Germany, gave no further commission to Agricola; perhaps because the problems of the theatre of war were different, perhaps because rival generals were better courtiers or pressed their claims harder, perhaps

[1] *A.* 3. 55, 4. See Syme, *Tacitus*, 2. 585 ff. [2] c. 24, 3.
[3] Cf. c. 21, 1–2.

again, for personal reasons. Tacitus hints at the two last possibilities, but does not admit the first, nor could he do so in view of the slur which it might seem to imply. On the other hand, the circumstances provided rich material for innuendo, and the culmination was reached when the governorship of Asia became a possibility, after the execution of Civica, and Agricola excused himself from it. Here was a golden opportunity for a biting description of the Imperial entourage in action. Finally, Agricola's last illness could be made a vehicle for insinuations of poisoning, and Imperial solicitude could be adroitly misrepresented as a desire for his death. The sinister impression was indeed conveyed so well that it tricked Cassius Dio into an affirmation which Tacitus deliberately did not make, even when all could be written with impunity in the years that followed Domitian's fall.

Was there then a moral, or did Tacitus intend to point one? In one passage the intention seems clear. Just before the tense description of Agricola's death, in significant context, there is a firm and striking vindication of his tranquil acceptance of events as they were.[1] 'Let the customary admirers of forbidden conduct know that great men can exist under bad Emperors; that obedience and modesty, combined with hard work and energy, can attain to the same heights as others have glorified by ostentatious deaths, bent upon danger but not upon aught of profit to the community at large.' This enlightened and humane attitude was the creed not only of Agricola but of Tacitus himself, and amid all the irony and innuendo it rings true, the *forma mentis aeterna*, the *fama rerum*.

2 · THE LIFE OF TACITUS AND THE DATE OF THE *AGRICOLA*

1. C. (or P.)[1] Cornelius Tacitus was born in A.D. 56 or 57;[2] his *patria* was perhaps Vasio, the capital of the Vocontii, in Gallia Narbonensis.[3] He was thus a provincial.[4] His father was an *eques*, probably the procurator of Gallia Belgica mentioned by Pliny the Elder (*N.H.* 7. 76) who served as an officer with the Rhine armies from about A.D. 46 to 58.[5] If so, the close friendship between the historian and the

[1] Sidonius Apollinaris calls him Gaius (*Epist.* 4. 14, 1; 4. 22, 2) but the *subscriptio* of the Codex Mediceus of *Annals* 1 and 3 records *P. Corneli.*

[2] This is to be inferred from the dramatic date of the *Dialogus* (17, 3: A.D. 75) in which Tacitus describes himself as *iuvenis admodum* (1, 2: i.e. about 18 years of age; cf. *Agr.* 3, 2) and from the regular sequence of promotion in the senatorial *cursus honorum.*

[3] As persuasively argued by Syme (*Tacitus*, 2. 622 ff.). Transpadane Italy and Gallia Narbonensis have been favoured candidates. Gallia Narbonensis, and Vasio in particular, is suggested by Tacitus' allusions to Afranius Burrus who came from there and by his knowledge of the province (Syme, op. cit., App. 95). Tacitus is a rare *cognomen* but occurs four times on inscriptions from Narbonensis, including one from Vasio (*ILS* 4841 *Marti et Vasioni Tacitus*). Cornelius is, of course, a common *nomen* and does not help to elucidate the historian's origin but it may be noted that at Vasio several Cornelii make dedications to Mars (cf. e.g. *ILS* 4542). See also note on c. 21, 2. E. Koestermann (*Athenaeum* 43 (1965), 167 ff.) has re-argued the case for the Transpadane. He bases his argument on the peculiar prominence given in the *Agricola* to the philosophical circle of Paetus, whose family was Paduan, and to Tacitus' sense of guilt in respect of the members of that circle (c. 2, 1; 45, 1). He suggests that Tacitus was originally an adherent of the circle, as would be natural if he too came from near Padua. He sees confirmation of his view in the reference to Livy (c. 10, 3), another Paduan, and in the presence of several Transpadanes, as well as Tacitus, in the will of Dasumius (*ILS* 8379*a*: see below).

[4] Cf. his ambiguous reply recorded in Pliny, *Epist.* 9. 23, 2.

[5] Pliny (*N.H.*, loc. cit.) implies that the historian had a brother who died in infancy.

younger Pliny (born A.D. 61 or 62) can be traced back to the previous generation.[1] Both men came from similar equestrian backgrounds and enjoyed similar careers. Tacitus was granted the *latus clavus*, the right to wear the broad purple stripe on the tunic which was the symbol of the senatorial order, by Vespasian (*H.* I. I, 3) and he devoted the last years of his education to the study of oratory, as was normal for a youth who was aspiring to a career of public service. To judge by allusions in the *Dialogus* he was perhaps a pupil of Quintilian,[2] the leading exponent of rhetoric in that age. He will have held one of the minor offices of the vigintivirate before serving at the age of 20 or so his initial military service as a *tribunus laticlavius*. We do not know under whom or where he served but at this same time, A.D. 77, he married the daughter of Julius Agricola (*Agr.* 9, 6). His acquaintance with Agricola and his family may well have been of long standing: Agricola himself came from Forum Iulii in Gallia Narbonensis and had married the daughter of Domitius Decidius, a Narbonensian senator (*Agr.* 6, I; cf. *ILS* 966). He was quaestor in A.D. 81 or 82, and tribune of the plebs or aedile some three years later. He held the praetorship in A.D. 88, which was the year in which Domitian celebrated the Secular Games, and he was also a member of the priestly college, *quindecimviri sacris faciundis*, which was responsible for organizing the celebrations (*A.* II. II, I). These positions are significant. They show that Tacitus' career continued uninterrupted at least through the early years of Domitian's reign, and his membership of one of the four senior priesthoods at such an early age indicates the high reputation which his oratorical talents had already won for

[1] Pliny addresses eleven letters to Tacitus, which bear witness to their friendship and community of tastes. They may be paired together in the will of Dasumius (*ILS* 8379*a* (A.D. 108):] Secundo, Cornelio [).

[2] R. Güngerich, *Class. Phil.* 46 (1951), 159 ff.

him.[1] Among his colleagues in the quindecimvirate were the poet Valerius Flaccus and the statesman Fabricius Veiento (*cos. II* 80, *cos. III* 83). It was normal after the praetorship to hold a three-year command of a legion and not uncommon thereafter to be appointed for a year as proconsul of one of the minor provinces. Tacitus states that he was abroad when Agricola died (23 August A.D. 93) and had been abroad for a *quadriennium* (*Agr.* 45, 5). He does not specify what his employment was but we may infer that it was of the normal kind.[2] He was elected consul in A.D. 97, holding a *suffect* consulate in the second half of the year. His name is as likely to have been selected for the consulate by Domitian before his assassination as by Nerva. Tacitus was 40 or 41 when he was elected. For the son of an *eques* this was fast promotion. The official *cursus honorum* would bring a man to the consulate in his 43rd year (*suo anno*) but patricians and nobles were exempted from certain intermediate offices and could hope to become consul in their thirties. As consul he delivered the funeral oration of the aged Verginius Rufus (Pliny, *Epist.* 2. 1, 6) but he must also have participated, whether as consul or as ex-consul, in the political crisis that resulted in the adoption of Trajan in the autumn of that year. His standing as an orator was supreme. With Pliny he prosecuted Marius Priscus who had been proconsul of Africa under Nerva. Priscus was condemned in January A.D. 100 and exiled from Rome (Pliny, *Epist.* 2. 11, 2 ff.; Juvenal 1. 49–50). Syme conjectures[3] that Tacitus was absent from Rome for the three years 101–4, perhaps as governor of one of the Germanies, but a letter written to Tacitus by Pliny in 104 or 105 (*Epist.*

[1] Cf. also the testimony of Pliny (*Epist.* 7. 20, 4; 9. 23, 2; 2. 1, 6; 2. 11, 2, 17).

[2] It is certain that he was not praetorian legate of Gallia Belgica as has sometimes been maintained (Borghesi, *Œuvres* 7 (1872), 323) since Glitius Agricola was legate there immediately before his own consulate which was also in A.D. 97 (*ILS* 1021).

[3] *Tacitus*, 1. 71 f.

4. 13) implies that he was still actively engaged as an advocate, and it is more likely that he spent those years in Rome.[1] But the scope and rewards of oratory were severely reduced under the principate, as the *Dialogus* makes clear, and much of Tacitus' time and energy must have been devoted to the composition of the *Histories* on which he was certainly engaged in A.D. 106 (Pliny, *Epist.* 6. 16; 6. 20) and, later, of the *Annals*. He did not, however, like Sallust retire from political life, for a chance inscription (*O.G.I.S.* 487) reveals that he reached the summit of the senatorial career, the proconsulate of Asia, which he held in 112/13. The date of his death is uncertain but must have fallen in the reign of Hadrian.

2. The *Agricola* is one of the three lesser and earlier works of Tacitus. Of the other two, the *Germania* was completed in A.D. 98; the date of the *Dialogus de Oratoribus* is still disputed but is probably contemporary with the *Histories* and published about A.D. 107.[2] That the *Agricola* was written shortly after Domitian's death is plain from its whole character. A more exact date can be obtained only from references in it to Nerva and Trajan. In c. 3, 1 Nerva is mentioned as Nerva Caesar without being called *divus* while Trajan, called Nerva Traianus, is described as 'daily increasing the happiness of the times'; whence it has been inferred that Tacitus was writing when Nerva was still alive but after he had adopted Trajan as his son and made him *socius imperii*, i.e. between October A.D. 97 and 28 January A.D. 98, when Nerva died. On the other hand, in c. 44, 5 Trajan is spoken of as *princeps*, which shows that

[1] See Sherwin-White, *The Letters of Pliny*, p. 287.

[2] The date is discussed by Syme, *Tacitus*, I. 112 ff. and App. 28. The *Dialogus* is dedicated to L. Fabius Justus (*cos. suff.* 102) and may be the work which Pliny mentions on two occasions, about the year 107, as having been sent to him by Tacitus (*Epist.*7. 20, 1; 8. 7), unless that is part of the *Histories*. The alleged allusion to a passage of the *Dialogus* (9, 6 *in nemora et lucos*) in a slightly later letter (*Epist.* 9. 10, 2) is not certain: see Sherwin–White ad loc.

he was already emperor and that Nerva was dead. It is possible that in the former passage Tacitus has merely omitted to style Nerva *divus*[1] and that the inference that Nerva was still alive is false, but it is perhaps more likely that the work was begun in the late autumn of A.D. 97 and not completed until the spring of A.D. 98. Frontinus' contemporary treatise on aqueducts discloses the same phenomenon: in the earlier chapters Nerva is referred to as *Augustus* (1, 1) and *imperator* (88, 1), that is, as still living, whereas later he is called *divus* (102, 4; 118, 3) and Trajan is introduced as Imperator Caesar Augustus (93, 4). The introduction of the *Agricola* (cc. 1–3) makes it clear that it was Tacitus' first historical undertaking, and we may therefore conclude that it was published in the same year as the *Germania*, A.D. 98, but before it.[2] The date is of importance for the understanding of the work. Tacitus had just held the highest office in the land: he must also have played an active part in the stirring events of the recent months—the death of Domitian, the accession of Nerva, and the adoption of Trajan. The experience was a vivid lesson in the realities of politics.

3 · THE LITERARY CHARACTER AND PURPOSE OF THE WORK

The *Agricola* is a biography.[3] Its full title is *de Vita Iulii Agricolae*, and Tacitus makes it plain that he has set out to

[1] Similarly Pliny (*Paneg.* 7–10) calls Nerva *divus* once but on five other occasions omits the epithet.

[2] If an external *terminus ante quem* were needed, it is afforded by allusions to the *Agricola* which Pliny made in his *Panegyricus*, delivered on 1 September A.D. 100; cf. especially c. 6, 1 and *P.* 83, 6; c. 30, 4 and *P.* 48, 5; c. 46, 3 and *P.* 55, 10–11, and see R. T. Bruère, *Class. Phil.* 49 (1954), 161 ff. Tacitus may have contemplated writing a biography soon after his father-in-law's death (see n. on c. 1, 3 *opus fuit*) but it is evident that he did not undertake it as long as Domitian was alive.

[3] So also Büchner, Till, Forni, and most modern editors. See W. Steidle, *Mus. Helv.* 22 (1965), 96 ff.

write a life of his father-in-law in the accustomed manner (c. 1, 4 *narraturo mihi vitam defuncti hominis*; c. 46, 4 *Agricola posteritati narratus*)—*narrare vitam* is the phrase which Nepos used to distinguish the character of his *Lives* from other historical works.[1] There was at Rome a long tradition of biography: Cicero had written a life of Cato, Augustus a *Vita Drusi* (Suetonius, *Claudius* 1), Julius Secundus a life of Julius Africanus, and several other such biographies are known;[2] but fortune has preserved, apart from the *Agricola*, the works of only two classical biographers, neither of whom can be regarded as typical of the genre—Nepos' brief summaries and Suetonius' formless and gossipy *Lives of the Caesars*.[3] In origin biography was a Greek literary form and Roman authors looked to Greek models, in particular to the *Agesilaos* of Xenophon and the *Euagoras* of Isocrates, which were regarded as blueprints of the art by later theoreticians.[4] The distinctive feature of such works, as, for example, Polybius makes clear in reference to his own *Philopoemen*,[5] is that they relate the achievements of a man as illustrations of his character and not as part of a wider, more complex historical framework. It was also natural that when the life of a good man was being told the resulting biography should be laudatory: it is the *memoria* of *virtus* which is being recorded (*Agr*. 1, 2).

There is, thus, no inconsistency in Tacitus writing that the treatise is *honori Agricolae destinatus* (c. 3, 3). Still less should we be misled into seeing the *Agricola* as a special kind of 'biographical encomium'[6] or as a literary variant

[1] Nepos, *Pelop*. 1, 1. *vereor si res explicare incipiam ne non vitam eius enarrare sed historiam videar scribere*.

[2] See F. Leo, *Griechische-römische Biographie* (1901).

[3] W. Steidle, *Sueton u. d. antike Biographie*² (1963), pp. 3 ff.

[4] Cf., e.g., Cicero, *ad Fam*. 5. 12, 3 ff. See D. R. Stuart, *Epochs of Greek and Roman Biography* (Sather Classical Lectures, 1928), pp. 235 ff.

[5] 10. 21, 8.

[6] So Gudeman in his German and English editions of the *Agricola*,

of the funeral laudation.[1] It is true that shortly before writing the *Agricola* Tacitus, as consul, had pronounced the funeral laudation of Verginius Rufus;[2] it is also true that the *Agricola* contains several commonplaces which recur in epitaphs (see c. 4, 2 n.; 46, 1 n.). But it lacks some of the fundamental elements which Quintilian and the Greek rhetoricians demand of a *laudatio*. It is fanciful to see in the digression on Britain (cc. 10–12) a covert comparison with Caesar;[3] yet such comparison (σύγκρισις) was indispensable. Nor are there any oracular forewarnings of future greatness. Laudations at actual funerals had to be short since they were usually recited by the children of the deceased, and it is clear from Cicero (*de Orat.* 2. 341) that there did not exist at Rome a tradition of literary, as distinct from occasional, laudations. The *Agricola* may well owe something to the spirit of the funeral laudation but is certainly not a specimen of one.

There was also at Rome another literary form which may have contributed something to the *Agricola*, but with which it should not be confused. During the last years of Tiberius it became fashionable to write short accounts of the deaths of great and good men who had perished undeservedly as victims of the régime,[4] and we hear later of the *exitus illustrium virorum* composed by Titinius Capito.[5] The genre had a vogue under Nero and was evidently used by Tacitus as a source in Books 15 and 16 of the *Annals*.

who attempts to relate it to the formal rules of later Greek rhetoricians (e.g. Theon, Menander) and J. Cousin (*R.É.L.* 12 (1934), 326 ff.), who compares the scheme for *laudationes* outlined by Quintilian (*Inst. Or.* 3. 7).

[1] E. de St.-Denis, *L.É.C.* 9 (1941), 14–30; refuted by H. Bardon, *L.É.C.* 11 (1943), 3–7 and 127–8; ibid. 12 (1944), 273 ff.

[2] Pliny, *Epist.* 2. 1, 6.

[3] So P. Couissin, *Rev. Phil.* 6 (1932), 97–117.

[4] F. A. Marx, *Philologus* 92 (1937), 83 ff.; Syme, *Tacitus*, 1. 19 ff.; Paratore, *Pensiero politico e oratoria nell'Agricola di Tacito* (1961–2).

[5] Pliny, *Epist.* 8. 12, 5 *funebribus laudationibus seris quidem sed veris*. Suet. *Nero* 47 ff. appears to be based on such a work.

It is a probable assumption that the works written by Arulenus Rusticus about Thrasea Paetus and by Herennius Senecio about Helvidius Priscus (c. 2. 1 n.) belonged to this category rather than to biography proper. The only analogous specimens which survive are the fragmentary Acts of the Pagan Martyrs (*Acta Alexandrinorum*), although these deal primarily with the trials of the martyrs, but it is possible to reconstruct their chief features. They focused attention on the death, not on the life, of the man, emphasizing the manner of his end and celebrating his last words. They contrasted his personality with the cruelty and wickedness of the emperor. We may detect their influence in Tacitus' account of Agricola's death-bed and in the denigration of Domitian. But the *Agricola* remains first and foremost an account of Agricola's life and achievements.

Yet the contents and the tone of the treatise have raised difficulties in the minds of scholars. In literary form or in purpose, or in both, it has been considered to be not a genuine biography.

(1) It was long ago pointed out that the main portion of the book (cc. 10–38) is more historical than biographical. The ethnological and geographical description of Britain (cc. 10–12), the survey of the earlier history of the conquest and government of the island (cc. 13–17), the annalistic account of Agricola's campaigns (cc. 18–38), the episode of the Usipian cohort (c. 28), the speeches of Calgacus and Agricola before the Battle of Mons Graupius (cc. 30–34) and the detailed description of the battle itself (cc. 35–38) read at first sight as history rather than biography. In style,[1] too, this central section of the *Agricola* reflects the narrative influence of Livy and above all of Sallust, whereas the opening and the conclusion of the work are in an oratorical vein with frequent Ciceronian echoes. On closer examination, however, this heterogeneity is not so

[1] See below, p. 24.

pronounced as scholars have claimed.[1] The archaic *elogia* of the Scipios[2] show how the tendency prevailed to commemorate men by a record of their deeds and, particularly, to single out their greatest achievements. A contemporary example in monumental form is the Arch of Titus which is the expression in stone of the phrase which Tacitus applies to Agricola: *narratus et traditus superstes erit* (c. 46, 4). It is the function of history and biography alike to keep alive the record of great men, to preserve the *memoria* of *virtus* (*A*. 3. 65, 1 *ne virtutes sileantur*). Agricola's claim to fame was that he was the conqueror of Britain and, therefore, the life of Agricola was to a large extent the history of the conquest of Britain. The digression on ethnography and geography and the survey of earlier governors are necessary preliminaries in which the magnitude and value of the undertaking (c. 11, 4; 12, 2; 13, 2; cf. c. 12, 6 n. *pretium victoriae*) and Agricola's own part (c. 10, 3; 10, 4) are continually stressed. Those same two themes—Britain as the extremity of the world and Agricola as responsible for its subjugation—dominate the main narrative. The Romans were fascinated by the vision of pushing their frontiers to the edge of the earth,[3] and Tacitus exploits this interest: it is the refrain of Calgacus' and Agricola's speeches (c. 30, 3; 31, 3; 33, 3; 33, 5) and it accounts for the prominence of the episode of the Usipi (see n. on c. 28). Agricola was the man who by his personal exertions brought the Romans success. Hence, only Agricola's actions are noted, the doings of others are passed by in silence. It is Agricola who is almost always the subject of the sentences (c. 24, 1 n.). As Anderson says, 'everything, or nearly everything, serves in one

[1] It should only be necessary to refer in passing to Andresen's theory that the central section was originally part of a preliminary study for the historical work on Domitian's reign which Tacitus mentions as already in contemplation (*Agr*. 3, 3), and which ultimately appeared as the *Histories*.

[2] Ernout, *Receuil de textes latins archaiques* (1947), nos. 13 ff.

[3] Cf. Seneca, *Medea* 364 ff.

way or another to set in relief the hero's character and achievements' (p. xxviii). A true history would not have been so selective.[1] Nor are the variations of style surprising: there is no need to take refuge in Tacitus' conventional profession of immaturity (c. 3, 3 n.). Latin literature, because it is a literature in which genre determines style, allows of much flexibility and mixture. Such a range of styles is evident, for instance, in Horace's *Satires* or within the history of Livy himself. It is, therefore, natural that Tacitus should juxtapose the styles appropriate for consolation and narrative, for ethnography and oratory as his material required.

(2) It has also been held that Tacitus' purpose was to write more than just a biography of his father-in-law. Agricola—and Tacitus himself—had held high office under Domitian. Now Domitian was dead and those who had suffered under his tyranny lusted for vengeance. Furneaux developed the view: 'Agricola had been tribune at the time of the trial of Thrasea (c. 6, 3 n.) and had shown no such generous impulse as Arulenus Rusticus. He had served Domitian in Britain when he had a great army and might have set up the standard of revolt like Antonius Saturninus. After his return to Rome, his eight years of non-resistance, or (as they would put it) of servile acquiescence in the senate, his conduct in relation to his pro-consulate, his nomination of the emperor as coheir in his will (a degradation which, it would be pointed out, bolder spirits had spurned), would all be matter of invective. Nor would the son-in-law be without his share of censure. He had owed to Domitian a praetorship, a priesthood[2] . . . he had been at Rome as a senator during the last and worst years of the reign of terror and had been no bolder than those around him.' Other scholars have gone farther and claimed that Agricola

[1] So also G. L. Hendrickson, *University of Chicago Decennial Publications* 6 (1904).

[2] And, perhaps, nomination to the consulship: see p. 9.

was actually a personal friend of Domitian's[1] or that he was under obligations to Domitian which Tacitus has been at pains to distort and obscure.[2] But most[3] are agreed that the biography had a political motive, that it is an apologia for the moral and political ideals of the new class of men, mainly provincial aristocrats like Agricola or Trajan or Julius Frontinus, who sought to serve the State whatever the régime, instead of either standing aside altogether from public life (cf. *Hist.* 4. 5, 2) or adopting a doctrinaire and fatal policy of resistance (*Agr.* 42, 4). The principate did not prevent good men from making their own contributions to the government of the Roman empire but it required of them *quies* and *obsequium*. The *Agricola*, it is argued, is a political pamphlet designed to justify this attitude and to clear Tacitus himself.

There is, of course, much that is topical and opportune in the *Agricola*: we may note the use of current political slogans—*felicitas temporum* and *securitas publica* (c. 3, 1 nn.). But it is very doubtful whether after the death of Domitian there was any serious criticism of those who had held office under him. The sources speak only of attacks on the professional *delatores*[4] and, apart from a handful of extremists, most of Tacitus' contemporaries had been 'members of a senate who had all been slaves together'.[5] There was in all probability no external need for Tacitus to defend his own and his father-in-law's careers.

[1] T. A. Dorey, *Greece & Rome* 7 (1960), 66 ff. There is no evidence for this.

[2] E. Paratore, *Tacito*, pp. 71 ff.

[3] See especially Syme, *Tacitus*, 1. 19 ff., 2. 585 ff.; Forni, *Introduzione*, 24 ff. Somewhat differently E.-R. Schwinge, *Rh. Mus.* 106 (1963), 362 ff., sees Agricola's career as personifying the tensions between the capital and the provinces under Domitian. W. Liebeschuetz (*C.Q.* 16 (1966), 126 ff.) argues that the work is a study on the theme of *libertas*, showing that the loss of independence, both personal (for citizens of Rome under the autocracy of the Emperors) and national (for the Britons), includes moral degeneration.

[4] Pliny, *Epist.* 9. 13, 2.

[5] *Hist.* 4. 8, 5.

The case for supposing the *Agricola* to have an ulterior purpose rests largely on the supposition that it can be shown to contain several tendentious distortions.[1] Such distortions, it is claimed, only have point if Tacitus is trying to make a defence. On detailed investigation none of these alleged distortions of fact can be substantiated.[2] The achievements of Agricola's governorship are solid and documented. The Battle of Mons Graupius, so far from being a fictitious 'literary' battle, is confirmed by external evidence (see c. 35, 1 n.). There is no reason to doubt the story that the possibility of the governorship of Syria was held out to him (c. 40, 1 n.), the claim that he was forced to withdraw as a candidate for Africa or Asia (c. 42, 1 n.), or the rumours that attended his death (c. 43, 2 n.). Nor is it reasonable to prefer the garbled and abbreviated version which Dio (39. 50) gives of Agricola's career. Wherever we can test the facts given by Tacitus, they are reliable. They are not affected by the pessimistic innuendoes with which he is apt to gloss them (see n. on cc. 1–3).

Distortion of a different and more subtle kind has recently been suggested by H. Nesselhauf,[3] who sees a conflict between a Flavian and an anti-Flavian view of the conquest of Britain. Flavian propaganda claimed that the decisive victories had been won before Agricola reached the island as governor; that Vespasian, as legate under the emperor Claudius,[4] had broken the chief resistance; that the Forth–

[1] E. Paratore, *Tacito*, pp. 71 ff.; F. Grosso, *In Memoriam A. Beltrami*, 97–145; T. A. Dorey, loc. cit.; H. W. Traub, *Class. Phil.* 49 (1954), 255 ff. See the salutary remarks of B. Zanco, *Aevum* 33 (1959), 252–63. The fairest assessment remains that by A. G. Woodhead, *Phoenix* 2 (1947–8), 45 ff.

[2] See also c. 6, 5 n.; 7, 3 n.; 18, 6 n.

[3] *Hermes* 80 (1952), 222–45. He makes use of some arguments advanced by E. Birley, *Durham Univ. Journal* (1946), 79–84 = *Roman Britain and the Roman Army*[2] (1961), pp. 10 ff. The relevant passages are Pliny, *N.H.* 4. 102; Silius Ital. 3. 597 ff.; cf. Val. Flaccus, *Argon.* 1. 7 ff.

[4] Momigliano, *J.R.S.* 40 (1950), 41–42.

Clyde line had been reached by Cerialis; that the main pacification had been achieved by Vespasian's governors. Tacitus, on the other hand, it is alleged, aims to minimize these successes and to extol the personal contribution made by Agricola in spite of imperial discouragement so that no credit should seem to accrue to Domitian and his house; the events of A.D. 44–77 are cursorily dismissed and the seven campaigns of Agricola given an importance which they did not historically have. In fact, however, archaeology has shown that the account given by Tacitus is substantially accurate. Knowledge of Scotland will have come to the ears of some of Agricola's predecessors (and that is all that Pliny's *notitia* implies) but the first permanent penetration into Scotland was the work of Agricola (see c. 22 n.). Within the scope of a biography Tacitus does allow both for their achievements (see nn. on cc. 13–17) and for Agricola's own role.

The *Agricola* is, then, neither a political pamphlet nor a personal apologia. It may have been pregnant with lessons for Tacitus' contemporaries: that is incidental. It is certainly permeated by Tacitus' views on life and on history (cf., e.g., note on cc. 1–3; 13, 2 n.). But neither political nor personal expediency accounts for such distortions and exaggerations as can be proved: for they are of a quite different kind. The ancients, not least Tacitus, believed that character was fixed, that a man is at any one point in his life what he always was and always will be. It was, therefore, natural that in biography, as in history, some interpretation and some exaggeration of the events should be allowed in order that the final character of the actors should appear homogeneous and plausible. It is well known that in his most mature work, the *Annals*, Tacitus was obsessed with a view of Tiberius which was at variance with the historical evidence. Tacitus saw him as a tyrant and accordingly believed that he must always have had the attitude, ambition, and characteristic behaviour of a tyrant.[1] So in the *Agricola* a coherent picture of a devoted public servant emerges which is contrasted with the jealous despotism of Domitian. Qualities are attributed to Agricola which are as much conventional hall-marks of the good soldier and the good administrator as particular characteristics of the man himself. He is depicted as combining a stern public demeanour with an affable private manner (c. 9, 3), as making full use of his time (c. 9, 3), as leading his troops in person (c. 18, 2), as promptly following up initial successes (c. 18, 2), as framing modest communiqués (c. 18, 6), as conscientiously bestowing praise and punishment (c. 21, 1), as reconnoitring the route (c. 20, 2), and as personally supervising the selection of sites (c. 22, 2). Agricola, like Aeneas, *may* have done all this and more but, whether he did or not, it could safely be claimed that he had done so because that is what good generals are supposed to do. Tacitus was later to find in Germanicus another model of loyalty and efficiency and we find that he ascribes to Agricola many of the traits which are also ascribed to Germanicus (c. 9, 3 n.). Similarly Domitian was a tyrant. Tacitus appropriates sayings and incidents from Caligula (c. 39, 2 n.; 44, 5 n.) and Nero (c. 45, 1 n.) and uses them of Domitian because that is how tyrants behave.

There is, therefore, a tendency in the *Agricola*, as in his other writings, for the events to be accommodated to the characters as Tacitus saw them, and this tendency is liable to entail distortion. But Tacitus knew and loved Agricola and the biography which he has left us is an intimate and penetrating record of a man.

[1] Cf., above all, the discrepancy between fact and interpretation in the *maiestas* cases. See B. Walker, *The 'Annals' of Tacitus*, pp. 188 ff.

4 · LANGUAGE AND STYLE

In Roman literature style was largely determined by subject-matter. Different genres had their own appropriate styles and vocabularies. Thus, although the idiosyncrasies of the author will shine through, it is necessary also to keep in mind the demands of the genre in which he is writing. The Horace of the *Satires* is very different from the Horace of the *Odes*; yet they are recognizably the same poet. So too the *Dialogus* and the *Annals* come from the same hand but their styles are widely divergent. To appreciate the *Agricola*, therefore, it is prudent to begin by seeing what literary character was expected of a biography rather than by analysing the similarities and dissimilarities to Tacitus' other works.[1]

Biography had a strong tradition at Rome (see Section 3), but of Tacitus' predecessors only the works of Cornelius Nepos survive. These do not provide a true analogy since they are written without literary pretensions. Nor can Sallust be invoked as a comparison since, as their titles and contents show, his two monographs on Catiline and Jugurtha were not biographies. We cannot, therefore, be sure how earlier biographers had interpreted their task.[2] Greek biographies, such as Xenophon's *Agesilaos*, were often

[1] For a survey of Tacitus' style with references to earlier discussions see Syme, *Tacitus*, 1. 125 ff., 198 ff.; App. 42–60. There is an important assessment by E. Löfstedt, *Roman Literary Portraits* (O.U.P., 1958), pp. 157 ff. F. Kuntz, *Die Sprache des Tacitus* (Diss. Heidelberg, 1962, phototype) is a useful dissertation. There are some points of interest in A. Salvatore, *Stilo e Ritmo in Tacito* (Naples, 1950) and a handy collection of parallels is given by G. B. A. Fletcher, *Annotations on Tacitus* (Collection Latomus, vol. 71), pp. 99 ff. For details see A. Draeger, *Über Syntax und Stil des Tacitus*[3] (Leipzig, 1882).

[2] Pliny's judgement on the works of C. Fannius that they lay *inter sermonem historiamque medios* (*Ep.* 5. 5, 3) should not be invoked in this context since the works were not strictly biographies (see above, p. 13) and *sermonem* means 'ordinary speech' not 'oratory'.

a blend of oratory and history. That is natural. A biography is likely to contain both a picture of the man himself, which can be more effectively sketched in oratorical tones (cf. Cicero's descriptions of Verres and M. Caelius Rufus), and a narrative of events and actions, which is more suited to an historical manner. And it is precisely these two strands which predominate in the *Agricola*. The opening and the close of the work are oratorical and show close affinities with the periods and diction of Cicero: the central narrative of Agricola's doings in Britain recalls the literary technique of Sallust and Livy. These two categories are not kept rigidly distinct. It would be inelegant to have made a crude juxtaposition of different styles and Tacitus discreetly blends them so that the reader feels a change of emphasis and notices that one part is more historical and another more oratorical. But although they are not truly separable and indeed are united by qualities which are distinctive of Tacitus himself, it may be helpful to consider them separately.

Tacitus had been trained as an orator and had won distinction as an orator.[1] As the *Dialogus* shows, he was steeped in the works of Cicero, and in the same year that he composed the *Agricola* he had delivered the funeral oration of Verginius Rufus. The introduction to the *Agricola*, a justification for writing biography, recalls the literary discussions in Cicero's *Brutus* and *Orator* or in Quintilian's *Institutio*. It is a rhetorical disquisition. The epilogue is a Consolation in the traditional manner, with reminiscences of the Consolation on the death of L. Crassus in the third book of the *de Oratore*. The account of Agricola's early years (cc. 4–9) is a résumé of his career rather than a full historical narrative. In all these sections the influence of oratory, and of Cicero in particular, is most strong. It can be seen in the richness of expression and in the accumulation of virtual synonyms. Thus we have *vicit ac supergressa est* (c. 1, 1), *comitio ac foro* (c. 2, 1), *fiduciam ac robur* (c. 3, 1), *sinu indulgentiaque* (c. 4, 2), *incensum ac flagrantem* (c. 4, 3), *sublime et erectum, pulchritudinem ac speciem, magnae excelsaeque* (c. 4, 3), *decus ac robur* (c. 6, 1), *subsidium ac solacium* (c. 6, 2), *quiete et otio* (c. 6, 3, etc.), *vim ardoremque* (c. 8, 1), *vulgus et populus* (c. 43, 1), *intervalla ac spiramenta* (c. 44, 5), *formam ac figuram* (c. 46, 3). It can be seen, too, in the use of rhetorical devices such as alliteration (e.g. c. 46, 4 *inglorios et ignobiles oblivio obruet*), apostrophe (e.g. c. 45, 3 *tu vero felix, Agricola*), anaphora and the *tricolon auctum* (e.g. c. 2, 2 *vocem populi Romani et libertatem senatus et conscientiam humani generis*; 45, 4 *adsidere valetudini, fovere deficientem, satiari vultu complexuque*; 46, 2 n.). Above all it is evident in the preference for the clausulae favoured by Cicero, particularly the double cretic (cf. c. 1, 3 *obtrect-ationi fuit*; 2, 3 *audiend-ique commercio*; 46, 2 n.; 46, 4 n.) and the double trochee (cf. c. 2, 1 *urerentur*; 2, 3 *quam tacere*; 45, 4 *figeremus*). Hence it is not surprising that they also contain echoes of Ciceronian phraseology in addition to the remarkable imitation of the *de Oratore* already mentioned. Such are *tam saeva et infesta virtutibus tempora* (c. 1, 4), *habuerunt virtutes spatium exemplorum* (c. 8, 2), *incolumi dignitate* (c. 44, 4), *inglorios et ignobiles* (c. 46, 4). Balanced clauses and sonorous periods suit the mood of these sections.

But an historical narrative, such as the account of Agricola's governorship, demanded different qualities. As Sallust was the first Roman to see, an individual style had to be forged for history. 'To tell a story properly calls for speed and variety' which the periodic structure of oratory cannot sustain. History is an affair of rapidity and surprise, of sudden changes of interest and fortune. Sallust devised a style to match it, combining brevity and incisiveness with a striking vocabulary.[1] Livy, in his turn, was deeply influenced by Sallust and, although his temperament

[1] See the survey by Syme, *Sallust* (Sather Classical Lectures,

was more open so that he avoided many of the crabbed and obscure mannerisms of Sallust and developed a more flowing technique to fit the larger canvas of his history, he was in essentials a disciple of Sallust. He used the same methods of rapid syntax, he shared the same propensity for an effectively coloured language.[1] Tacitus followed their lead. Even though the fragmentary preservation of both Sallust and Livy enjoins extreme caution in assigning particular debts (phrases which are only paralleled in Livy may well have occurred in the missing portions of Sallust's *Histories* and many phrases are common to both authors), it is possible to detect some of the influences that moulded Tacitus' own style.[2]

In looking back to Sallust Tacitus was to some extent going against the fashion of his age. Livy was enshrined as the *eloquentissimus veterum* (c. 10, 3) and the qualities of his style were praised by contemporary critics for imitation by historians, whereas Sallust was often censured (Quint. 4. 2, 45) and rarely imitated.[3] But Tacitus' dark nature was more in tune with Sallust than with Livy, and Sallust foreshadowed some of the stylistic quirks which are thought of as peculiarly Tacitean.

Sallust had made great use of the historic infinitive as a means of rapid narrative. It is true that it is found also in Cicero and Livy, but it is distinctive of Sallust and had

1964), pp. 240 ff., 305 ff. The stylistic differences between history and oratory are similarly interpreted by Pliny, *Ep.* 5. 8, 9–11.

[1] See the brief summary by Ogilvie, *A Commentary on Livy 1–5* (O.U.P., 1965), pp. 20 ff. with the references given there.

[2] For Sallustian influences see, in addition to Syme, *Tacitus*, App. 53; Wölfflin, *Philologus* 26 (1867), 122 ff.; Urlichs, *De Vita et Honoribus Taciti* (Progr. Würzburg, 1879); W. Heraeus, *Archiv f. lat. Lex.* 14 (1906), 273 ff.; Bardon, *Mélanges de la faculté . . . Poitiers* (1946), 195 ff.

[3] But the fact that Hadrian preferred Caelius to Sallust was thought odd (*S.H.A. Hadr.* 16, 6) and Plutarch and Zenobius among the Greeks of the time certainly read him.

passed largely out of use by the first century A.D. (Probus in Σ Virg. *G.* 4. 134). Tacitus revived it and employed it to great effect (cf. especially c. 38, 1).[1] Sallust was apparently the first Roman historian to break up a long account into sections by the insertion of digressions which serve to separate off the main episodes of the action. In this he was followed by Livy—and by Tacitus, whose digressions on the ethnography of Britain (cc. 10–12) and on the mutiny of the Usipi (c. 28) mark off the climaxes of Agricola's career. Sallust used epigrams to pass universal comments on events. In the same way Tacitus extracts the general implications of a particular action by arresting *sententiae*. But it is in his adaptation of specific passages of Sallust that Tacitus most clearly acknowledges his master. The description of Romans put in the mouth of Calgacus recalls the letter of Mithridates (c. 30, 4; cf. also c. 31, 1), while some of the sentiments of Agricola on the same occasion are taken from the *Catiline* and the *Jugurtha* (c. 33, 4; 33, 5; 33, 2). Parts of the account of the Battle of Mons Graupius are modelled on the battle against Jugurtha (c. 37, 2; cf. c. 36, 3). But throughout the whole narrative portion of the *Agricola* echoes of Sallustian language recur, e.g. *pro salute de gloria certare* (c. 26, 2), *multus in agmine, nihil quietum pati* (c. 20, 2), *quibus volentibus erat* (c. 18, 2), *id sibi maxime formidulosum* (c. 39, 2).

Sallust's works were few and brief. Livy's history was a monumental achievement which had early established its position as a classic. He had consecrated a vocabulary which was to become the stock-in-trade of subsequent historians. He left his mark on Curtius Rufus and Florus and Justin. An incipient historian could hardly overlook him, and in the *Agricola*, more even than in the *Histories* and the *Annals*, Tacitus exploits the rich resources of Livy's

[1] Perret, *R.É.A.* 56 (1954), 90 ff.; Perrochat, *R.É.L.* 13 (1935), 261–5.

language.[1] We note such military phrases as *hostiliter populatur* (c. 7, 1), *erexit aciem* (c. 18, 2), *terga occasioni patefecit* (c. 14, 3), *firmatis praesidiis* (c. 14, 3), *ignavis et imbellibus* (c. 15, 3), *sustinuit molem* (c. 17, 2), *debellari* (c. 24, 3), *infesta itinera* (c. 25, 1), *castella adorti* (c. 25, 3), *tulisse opem . . . eguisse auxilio* (c. 26, 2), *exuere iugum* (c. 31, 4), *furto noctis* (c. 34, 1), *vastum silentium* (c. 38, 2), *silvae fugientes texissent* (c. 26, 2), *novissimae res et torpor defixere aciem* (c. 34, 3), *ad arma discursum* (c. 35, 1), *gnari circumveniebant* (c. 37, 4), *dies faciem victoriae aperuit* (c. 38, 2). But Livy was not just a military historian. Indeed his understanding of war was much less sure than that of 'the most unmilitary of historians', Tacitus. He brought all aspects of Roman life into the framework of his history and his crisply formulated phrases can be heard throughout the *Agricola*, e.g. *subit dulcedo* (c. 3, 1), *grande mortalis aevi spatium* (c. 3, 2), *monstratus fatis* (c. 13, 3), *his vocibus instincti* (c. 16, 1), *conscientia et timor* (c. 16, 2), *otio lasciviret* (c. 16, 3), *inritatis animis* (c. 27, 2), *nata servituti* (c. 31, 2), *insigniretur* (c. 41, 3), *praeceps in iram* (c. 42, 3), *nihil adfirmare ausim* (c. 43, 2). Tacitus emulated not only the elegance of Livy's language and the variety of his narrative tempo (the full periods in which the setting is outlined leading up to the clipped, staccato sentences which convey the dramatic highlights of the action, as in c. 37, 1–5). He also adopted his practice of speeches. Speeches had been a part of historical writing ever since Herodotus. Sallust had deployed them to great effect in his *Histories* as well as in his two monographs. But Livy lavished unusual stylistic care on them. He constructs them on exact rhetorical principles and adds touches of verisimilitude by introducing reminiscences of familiar masterpieces.[2] It was not for nothing that he too had studied Cicero and had written a work on Style (Quint.

[1] See Syme, *Tacitus*, App. 54; G. Andresen, *Woch. klass. Phil.* (1916), 402 ff.

[2] Ogilvie, op. cit., p. 19.

10. 1, 39). It was Livy's example that inspired Tacitus to elaborate the two contrasting speeches of Calgacus and Agricola. On examination these prove to be based on the speeches which Livy gave to Scipio and Hannibal before Ticinum (see notes on cc. 30–32; 33–34). Furthermore there is also an unmistakable echo of the opening of Cicero's *pro Marcello* (c. 30, 1) and other characteristically rhetorical commonplaces (c. 30, 4; 33, 2, etc.).

There remain a number of usages which are anticipated by Sallust and Livy, such as *memoratu* (c. 1, 2), *egregius cetera* (c. 16, 2), *ancipiti malo* (c. 26, 2), *memorabile facinus* (c. 28, 1), *defensantium* (c. 28, 2). Compared with the debt to Cicero, Sallust, and Livy, the influence of other authors, such as Nepos,[1] Curtius,[2] and Seneca,[3] although presumed by scholars, is slight and disputable. There is, however, one further stylistic ingredient which can be isolated. As the *Germania* and the digressions in Sallust and Livy illustrate, popular ethnography had its own terms and its own conventions. These are to be recognized in the section on the ethnography of Britain. We may note the distinctive syntax of c. 12, 1 and typical phrases such as *rerum fide* (c. 10, 1), *obtenditur* (c. 10, 2), *qui mortales initio coluerint* (c. 11, 1), *parum compertum* (c. 11, 1), *in universum aestimanti* (c. 11, 3).

It would, however, be quite wrong to think of the style of the *Agricola* as a mere pastiche of elements taken from earlier authors and from other genres. Such elements can be detected but they have been fused into an organic original by the genius of Tacitus himself. In the process he reveals some tendencies which are common to his age. During the century that had elapsed since the death of

[1] C. W. Mendell, *T.A.P.A.* 52 (1921), 53 ff.

[2] See notes on c. 30, 4; 33, 4.

[3] Wölfflin, *Archiv. f. lat. Lex.* 12 (1902), 119 ff.; P. Keseling, *Phil. Woch.* (1932), 1466. Cf. c. 30, 4 *si locuples hostis est, avari, si pauper, ambitiosi*, etc.

Cicero many new words and forms had come into literary circulation. In the *Agricola* we may observe noteworthy abstract plurals as *fulgores* (c. 33, 1), *pallores* (c. 45, 2); comparative forms, as *porrectior* (c. 35, 4), *inrevocabilior* (c. 42, 3); extended meanings, as *anxius* (c. 5, 1), *dissociabilis* (c. 3, 1); recent usages, as *derisui* (c. 39, 1), *audentissimi* (c. 33, 1), *silere* with acc. of person (c. 41, 2), *intrepidus* (c. 22, 3), *aspicit* (= *spectat*, c. 24, 1), *attollerent* (= *extollerent*, c. 25, 1), *decentior* (c. 44, 1), *spiramenta* (c. 44, 5), *inlacessitus* (c. 20, 3); choicer forms, as *portione* (c. 45, 3), *adstruere* (c. 44, 3), *iactantia* (c. 39, 1), *adfectu* (c. 32, 1), *condicione* (c. 45, 5). All these are symptoms of the development of the Latin language.

Another aspect of the age was the liking for poetical phrases which added colour and piquancy to the language and conveyed a depth of association. This taste was prevalent even in oratory as Tacitus himself states (*Dial.* 20, 5): *exigitur enim ab oratore poeticus decor, non Accii aut Pacuvii veterno inquinatus sed ex Horatii et Vergilii et Lucani sacrario prolatus. horum igitur auribus et iudiciis obtemperans nostrorum oratorum aetas pulchrior et ornatior exstitit.* It is dangerous to try to pin down specific borrowings since so much of the poetry available to Tacitus is now lost, but some phrases can be certainly identified while others can be shown to have had earlier instances in poetry but not, so far as our knowledge goes, in prose. Virgil,[1] for example, was surely the source for *cruda ac viridis senectus* (c. 29, 4), *aliquando etiam victis ira virtusque* (c. 37, 3) and perhaps for *vires ministrantem* (c. 14, 3), *famam fatumque* (c. 42, 3), *silvas saltusque* (c. 34, 2). Lucan may have inspired *incerta fugae vestigia* (c. 38, 2), *spargi bellum* (c. 38, 2) and the picture in c. 25, 1. Other phrases, such as *avia petiere* (c. 37, 5), *secreti* (c. 38, 2), *nihil virtuti invium* (c. 27, 1), *solacium tulit* (c. 44, 5), *sanguine perfudit* (c. 45, 1),

[1] The Virgilian reminiscences have been re-examined by N. P. Miller, *Proc. Virg. Soc.* 1 (1961), 25 ff.

can only be classed more cautiously as 'poetical'. It is significant that the majority of these examples come from passages of high emotion such as the aftermath of Mons Graupius or the last hours of Agricola. In this Tacitus followed the precedent of Livy.[1] Such scenes fired his imagination and he strove to capture them in the most expressive colours.

Tacitus differs chiefly from Livy in his urge to say the unexpected in an unexpected way. This habitually leads him to seek an imbalance of construction.[2] A sentence such as *virtute in obsequendo, verecundia in praedicando extra invidiam nec extra gloriam erat* (c. 8, 3) is a comparative rarity. More often Tacitus will deliberately vary the two parts of an antithesis. Thus we find prepositional phrases contrasted with plain ablatives, as *ostentanda virtute aut per artem* (c. 9, 4), *non studiis privatis nec ex commendatione* (c. 19, 2), *temeritate aut per ignaviam* (c. 41, 2), *non per alienam materiam et artem sed tuis ipse moribus* (c. 46, 3), and with adverbs, as *neque ambitiose neque per lamenta* (c. 29, 1). Or he will vary his prepositions, *nihil appetere in iactationem, nihil ob formidinem recusare* (c. 5, 1), *per abrupta sed in nullum usum* (c. 42, 4). The same mannerism is evident in his variation of persons, as *vinctos di vobis tradiderunt . . . inveniemus nostras manus* (c. 32, 2–3), *tu vero felix . . . amissus est* (c. 45, 3–5), of tenses, as *ruere . . . pellebantur* (c. 34, 2),[3] of nomenclatures, as *Livius . . . Fabius Rusticus* (c. 10, 3), *Carus Mettius . . . Messalini* (c. 45, 1) and of antithetical clauses, as *ingens victoriae decus citra Romanum sanguinem bellandi et auxilium si pellerentur* (c. 35, 2), *transisse aestuaria pulchrum ac decorum in frontem, ita fugientibus periculosissima quae hodie prosperrima sunt* (c. 33, 5: see note ad loc.).

[1] Ogilvie, op. cit., p. 20.

[2] This phenomenon has been exhaustively treated by G. Sörbom, *Variatio Sermonis Tacitei* (Upsala, 1935).

[3] But *impellitur* (c. 25, 1) is probably corrupt.

Variety is pursued at all costs and the result is memorable and exciting. The quest for 'point',[1] for pithy sharpness, enjoyed, of course, a contemporary vogue. It looms large in the writings of Lucan, Seneca, and Juvenal, and, as Tacitus and Quintilian make clear, fashion required orators to be terse and epigrammatical. In the *Dialogus* Aper says that an audience expects to carry home *aliquid illustre et dignum memoria . . . sive sensus aliquis arguta et brevi sententia effulsit* (c. 20, 4). Tacitus had been brought up in this school and had excelled in it. The traces of it are omnipresent in the *Agricola*. Sometimes he will construct verbal antitheses which on inspection are seen to embody little antithesis of thought, as *Britanniam potest videri ostendisse posteris, non tradidisse* (c. 13, 1), or he will repeat in starker terms an idea that has already been expressed, as *singuli pugnant, universi vincuntur* (c. 12, 1), *nec poena semper sed saepius paenitentia contentus* (c. 19, 3), or will introduce gratuitous antitheses that have little relevance in the context. The outstanding example of this habit is the frequency of epigrams (*sententiae*) with which the work is studded. Tacitus was here again following the precedent of Sallust and was conforming to the taste of his age which relished such things (cf. Quint. 8. 5, 31 *non multas plerique sententias dicunt sed omnia tamquam sententias*; *Dial.* 20, 4 quoted above) and which excused plagiarism if it resulted in a neater epigram (Sen. *Contr.* 9. 1, 13). And his epigrams do serve to universalize the action and to point a moral from the biography. Hence he often concludes a section of the work with a maxim of this kind (c. 1, 4 n.). Many of his *sententiae* became and have remained household sayings, e.g. *et in luctu bellum inter remedia erat* (c. 29, 1), *ubi solitudinem faciunt, pacem appellant* (c. 30, 4), *omne ignotum pro magnifico* (c. 30, 3), *periculosius esse deprehendi quam audere* (c. 15, 5).

[1] B.-R. Voss, *Der pointierte Stil des Tacitus* (Orbis Antiquus, Heft 19, 1963).

But this ingrained quest for memorable brevity is not always so smoothly handled in the *Agricola* as it is in the more practised writing of the *Histories* and the *Annals*. It leads to harsh ellipses, as *ut nulla ante nova pars inlacessita transierit* (c. 20, 3: see note ad loc.), or awkward zeugmas, as *nec spem modo ac votum securitas publica sed ipsius voti fiduciam ac robur assumpserit* (c. 3, 1), *summa rerum et gloria in ducem cessit* (c. 5, 3), *naturam margaritis deesse quam nobis avaritiam* (c. 12, 6). It sometimes contributed to historical inaccuracies, as in c. 5, 2 and c. 32, 4 (see notes ad locc.). But chiefly it is responsible for the obscurities of thought and ambiguity of expression which make the *Agricola* such a difficult work to translate. It is enough to call attention to phrases such as *medio rationis atque abundantiae* (c. 6, 4), *tristitiam . . . exuerat* (c. 9, 3), *emere ultro frumenta ac luere pretio* (c. 19, 4), *recessus ipse ac sinus famae* (c. 30, 3), *virtute et auspiciis imperii Romani, fide atque opera nostra* (c. 33, 2).

The *Agricola* was Tacitus' first work. His writing was to mature and to be refined. But the style of the *Agricola* should be judged more on its own terms and less by comparison with the developments that can be traced in his later works.[1] It is the only biography that Tacitus was to write and, as such, it is an assured and homogeneous masterpiece. He has taken elements from different fields and impressed them with a distinctive and single character of his own which, as Gibbon wrote, 'will delight and instruct the most distant posterity'.

5 · TACITUS AND THE GEOGRAPHY OF BRITAIN

Geographical information in the *Agricola* is largely incidental to the biographical and rhetorical nature of the work.

[1] Syme, App. 44–52, following Wölfflin, and Löfstedt, *Syntactica*, 2, pp. 276 ff.

Some passages are, however, rewarding. The sea-lochs of the West Highlands (c. 10, 6) and the isthmus between Clyde and Forth (c. 23) are admirably described for an audience accustomed to discourse and lacking maps. On the other hand, the sea-voyage of the Usipi (c. 28), which confirmed that Britain was in fact an island, or the official expedition (c. 10, 4; 38, 3) of A.D. 84, which included the viewing of Thule (c. 10, 4), are mentioned in terms so general as to be positively imprecise.

Some measure of the activity in exploration which was at this time in progress, but which Tacitus does not record, is afforded by Plutarch.[1] In the early summer of 83, Demetrius, a γραμματικός, or grammar-school teacher, journeying homewards from Britain to Tarsus and discussing at Delphi the decay of oracles, recounted some of his experiences during 'an Imperial expedition of inquiry and survey'.[2] He observed that 'many of the islands round Britain were deserted and scattered, while some were named after spirits and heroes'. He then described a visit to 'one of the nearest, whose few inhabitants were considered holy and inviolate by the Britons'. The landing was followed by a severe storm of thunder and lightning, squalls and waterspouts, and the islanders considered this to mark the extinction of one of the Greater Spirits.[3] 'As a burning light', they said, 'carries no terror, while its quenching vexes many, so Great Spirits burn brightly for good and

[1] *De Defectu Oraculorum* 410 A, 434 C. The dramatic date of the dialogue has been much discussed; for a recent reconsideration see *Phoenix* 21 (1967).

[2] The employment of Greeks on the staff of Roman provincial governors was normal from the time of the Republic (Bowersock, *Augustus and the Greek World*, pp. 1 ff., 30 ff.).

[3] This is a common West Highland and Irish belief (T. P. Cross, *Motif-Index of Early Irish Literature*). Cf. also Pepys's *Diary* 19 October 1663: 'Waked with a very high wind and said to my wife, "I pray God I hear not of the death of any great person, this wind is so high"'.

not for harm; but their quenching and extinction, as now, often breeds storms and squalls or poisons the air with disease.' Demetrius also told that in one of the islands 'Cronus lay fast asleep in prison guarded by Briareus. Sleep enchained him, and many spirits accompanied him as attendants and servants'. This picturesque but manifestly authentic account identifies Demetrius, as is generally agreed, with that *Scrib(onius) Demetrius* who dedicated at York two bronze tablets once hung upon offerings and inscribed in Greek 'to Oceanus and Tethys' and 'to the Gods of the Governor's Residency' respectively. The offerings have perished, but the implications of the dedications remain clear and unequivocal.[1] It cannot be a coincidence that the marine deities are precisely those honoured by Alexander the Great on reaching the islands which were the furthest point of his voyage in the Indian Ocean, representing for him the limit of the world:[2] so a reference to the corresponding limit of the western world becomes apparent. The gods of the Governor's Residency, on the other hand, are linked with the objective of the professional activities of Demetrius in Britain which was the realization of Agricola's educational programme for sons of the British aristocracy (c. 21, 2). The boys, half students and half hostages, would reside in the governor's household for this purpose, and no doubt the voyage itself had been for the schoolmaster a summer trip.

The statement of Demetrius, that some islands were named after spirits and heroes, wins corroboration from the twenty-seven western islands named in the Ravenna Cosmography, of which four carry names or epithets of Celtic deities; and there should also be added Dumna, now Lewis and Harris—another divine name. That others were deserted is confirmed by the statement on the Orkneys preserved by Solinus:[3] 'vacant homine: non habent silvas,

[1] *RIB* 662–3.
[2] Diod. 17, 104.
[3] *Collect. Rer. Mem.* 22, 16.

tantum iunceis herbis inhorrescunt: cetera earum nudae harenae et rupes tenent'. These islands are mentioned by Pliny, who specifies forty, 'modicis inter se discretae spatiis', Pomponius Mela (3. 6, 54) recording thirty and Solinus mentioning three, as if *triginta* had dropped out. The true number is difficult to estimate. The 1961 *Census* gives 'about 70' of which 25 are inhabited: the *Census* of 1861 gave 49, of which 22 were deserted. In the ancient sources they are not specifically named, unless Bergi and Berrice (Pliny, *N.H.* 4. 104) were among them. A named group, however, is the five *Ebudae*, between Britain and Ireland (Ptolemy 2. 2, 11), which comprised a pair, each named *Ebuda*, Ricina, or Eggaricenna, Maleus, and Epidium. This might suggest Tiree and Coll, Rum, or Eigg (Egga) and Rum (Ricenna), Mull, and Islay (the last perhaps combined with Jura). A description of their inhabitants, preserved by Solinus,[1] is such a report as Demetrius himself might have made: 'Nesciunt fruges, piscibus tantum et lacte vivunt. Rex unus est universis, nam quotquot sunt omnes angusta interluvie dividuntur. Rex nihil suum habet, omnia universorum; ad aequitatem certis legibus stringitur, ac, ne avaritia devertat a vero, discit paupertate iustitiam, utpote cui nihil sit rei familiaris, verum alitur e publico. Nulla illi femina datur propria sed per vicissitudines in quamcunque commotus sit usurariam sumit; unde ei nec votum nec spes conceditur liberorum.' It can be set against another such report, again preserved by Solinus, on Lundy Island:[2] 'Siluram quoque insulam, ab ora quam gens Brittana Dumnonii tenent, turbidum fretum distinguit. Cuius homines etiam nunc custodiunt morem vetustum. Nummum refutant: dant res et accipiunt: mutationibus necessaria potius quam pretiis parant: deos percolunt: scientiam futurorum pariter viri ac feminae ostentant.' It is in the light of such passages as these that the activities of Agricolan exploration must be understood.

[1] *Collect. Rer. Mem.* 22, 12–15. [2] Ibid. 22, 7.

The voyage of Demetrius probably came before the summer of 83, if he was already at Delphi then, and thus in 82 at latest. It cannot, therefore, be linked either with the exploration ordered by Agricola in 84, just as it has nothing to do with the famous voyage of the mutinous cohort of Usipi, dated by Tacitus to 83 and mentioned by Martial (6. 61, 1–4) as common knowledge in 90. It must, however, be noted that Dio, in describing the adventure of the Usipi, mentions a plurality of expeditions mounted thereafter by Agricola. There were, in short, at least four northern voyages, that by Demetrius, the accidental venture of the Usipi, and two more ordered by Agricola, one of the last being dated by Tacitus to 84.

Compared even with the incidental information retailed by Plutarch, the account of Britain and its people by Tacitus is disappointing. The *Agricola* employs geographical terms only when the narrative demands them, and so names no more than five tribes,[1] three rivers,[2] two islands,[3] one hill,[4] and one harbour.[5] It is ironical that three should still remain unlocated. Yet there are important additions to contemporary knowledge of the general physical geography of Britain; namely, the general shape of Scotland north of the Forth–Clyde isthmus, the isthmus itself, the deeply indented western coast, the landing on the Orkneys and the sighting of Thule, undoubtedly the Shetlands. The long days and short nights of the North are, however, noted without precision, and the solar parallax, though not unhappily described, is ill explained in terms of relationship of sun to earth. Climate and agriculture are well defined, probably after inquiry from Agricola. Mineral resources,

[1] c. 11, 1 *Caledoniam habitantes*; 11, 2 *Silures*; 18, 1 *Ordovices*; 31, 4 *Brigantes*; 38, 2 *Boresti*.

[2] c. 23 *Clota, Bodotria*; 22, 1 *Taus*.

[3] c. 18, 3 *Mona*; 10, 4 *Thule*; cf. also c. 10, 4 *Orcades*.

[4] c. 29, 2 *Mons Graupius*.

[5] c. 38, 4 *Trucculensis*; see note ad loc.

on the other hand, which were not the governor's concern but the procurator's, receive the barest mention; while fiscal burdens and conscription (c. 13, 1) are related not to yield but to the provincial reaction, which is appraised as shrewdly as their social quality. Salient anthropological characteristics are also sensibly used as indicators of the Gallic, Germanic, and Iberian types to be recognized among certain British tribes.[1]

This treatment, very skilfully and deftly combined with the main purpose and subject of the work, may be contrasted with the slightly earlier information on Britain in Pliny's *Natural History*, which is related solely to general physical geography, without any account of political geography such as is given for Gaul. Other details are entirely incidental to the multifarious subjects discussed in the work, and make a very disconnected list. Pliny comments upon the long days, mentioning 17 hours as the maximum (2, 186): he also notes the eighty-cubit (120-foot) tides of Pytheas (2, 217); British magic (30, 13), and painting of the body with woad for certain funeral rites (22, 2) or the ring worn on the middle finger (33, 24); dressing of fields with marl (17, 42) and chalk (17, 45); the wild goose (10, 56); oysters (9, 169; 32, 62); coracles (7, 206); pearls (9, 116); introduction of the cherry (15, 102); amber (37, 35); tin (4, 104); and controlled production of lead (34, 164). These items are interesting in themselves, but their presentation has no relation whatever to a systematic account of Britain. This neither Tacitus nor Pliny intended to give, though Pliny by his own standards might in fact have been expected to give more than he does.

A more interesting product of the Agricolan occupation of the north is the information retailed by the geographer Ptolemy. Although his work was in process of compilation between A.D. 127 and 141, his sources for Britain are undoubtedly earlier. It is agreed that they are derived from

[1] c. 11, 1–4.

the authority acknowledged by Ptolemy as the most recent available to him, namely, Marinus of Tyre. His work, compiled about A.D. 100, is praised by Ptolemy for its care and thoroughness (1. 6, 1), while criticized on a number of points, as, for example, its treatment of the British Noviomagus (1. 15, 7). The information itself, however, is susceptible of a considerably closer dating.

In southern Britain the only position mentioned among the Silures is *Βούλλαιον*—*Burrium*, the pre-Flavian site at Usk, while among the Durotriges the sole site recorded is *Δούνιον*, the native stronghold of Maiden Castle, where life faded out in Flavian times. An interesting omission is Gloucester, which was presumably in the source as a legionary fortress, but cancelled in the light of an up-to-date list, since it had not yet been re-founded as a *colonia* under Nerva. The section might indeed seem to include *πόλεις* which became towns in A.D. 79 or a little before; but the term *πόλις* is frequently used by Ptolemy for forts, and it was exactly from forts that these towns sprang. If, then, for example, *πόλις Κορίνιον* of the Dobunni means the fort evacuated about A.D. 65, the southern section could well go back into Nero's reign, and would be entirely consistent within itself and presumably derived by Marinus from a source of that date. In northern Britain, on the other hand, *Πτερωτὸν στρατόπεδον* (2. 3, 13) is shown by the Latin form *Pinnatis* to be a Greek translation of *Pinnata Castra*, comparable in Ptolemy's Britain with *Μέγας λιμήν* for *Portus Magnus* (2. 3, 4), or *Ὕδατα θερμά* for *Aquae Calidae* (2. 3, 28). This site is identifiable as Inchtuthil, the legionary fortress constructed after A.D. 83 and abandoned while a-building soon after A.D. 87, together with all positions beyond the Earn: and it may further be noted that the date of these sites is not earlier than A.D. 83, since *Οὐικτωρία* (2. 3, 9), north of the Forth, cannot refer to any victory before that year. Accordingly, it becomes clear that the northern section represents the new geographical and

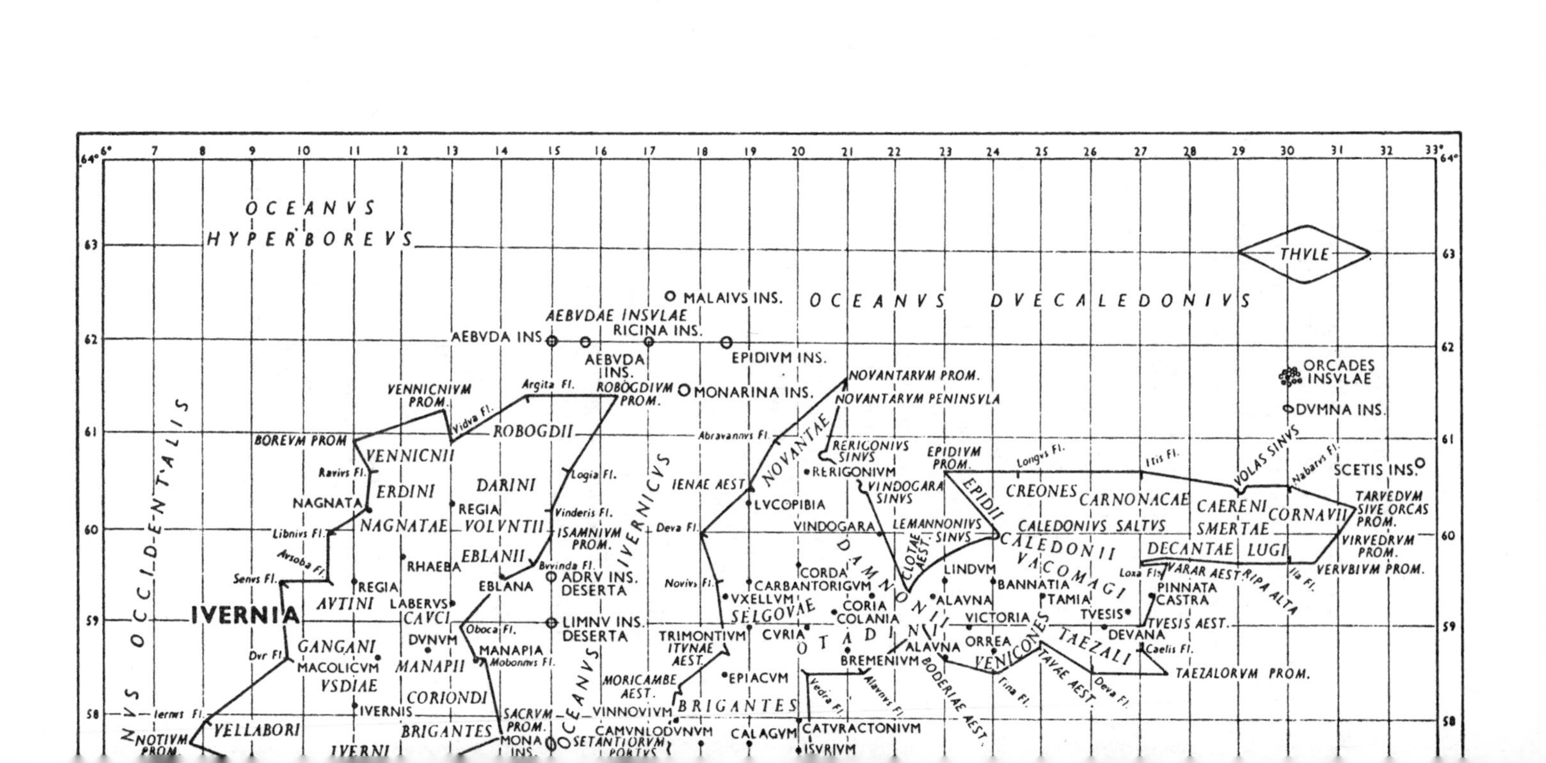

OCEANVS HYPERBOREVS
THVLE
OCEANVS DVECALEDONIVS
MALAIVS INS.
AEBVDAE INSVLAE
AEBVDA INS.
RICINA INS.
AEBVDA INS.
EPIDIVM INS.
ORCADES INSVLAE
DVMNA INS.
SCETIS INS.
MONARINA INS.
NOVANTARVM PROM.
NOVANTARVM PENINSVLA
VENNICNIVM PROM.
ROBOGDIVM PROM.
BOREVM PROM
ROBOGDII
VENNICNII
ERDINI
DARINI
NAGNATA
REGIA
NAGNATAE
VOLVNTII
ISAMNIVM PROM.
EBLANII
RHAEBA
EBLANA
ADRV INS. DESERTA
LIMNV INS. DESERTA
REGIA
AVTINI
LABERVS
CAVCI
IVERNIA
GANGANI
MACOLICVM
VSDIAE
DVNVM
MANAPIA
MANAPII
CORIONDI
IVERNIS
BRIGANTES
IVERNI
VELLABORI
NOTIVM PROM.
SACRVM PROM.
MONA INS.
OCEANVS OCCIDENTALIS
OCEANVS IVERNICVS
IENAE AEST.
NOVANTAE
RERIGONIVS SINVS
RERIGONIVM
VINDOGARA SINVS
LVCOPIBIA
VINDOGARA
LEMANNONIVS SINVS
CLOTAE AEST.
EPIDIVM PROM.
EPIDII
CREONES
CARNONACAE
CAERENI
CORNAVII
TARVEDVM SIVE ORCAS PROM.
VIRVEDRVM PROM.
VERVBIVM PROM.
CALEDONIVS SALTVS
SMERTAE
CALEDONII
DECANTAE
LVGI
VACOMAGI
VOLAS SINVS
VARAR AEST.
RIPA ALTA
LINDVM
BANNATIA
PINNATA CASTRA
TAMIA
TVESIS
TVESIS AEST.
DEVANA
TAEZALI
TAEZALORVM PROM.
TAVAE AEST.
VENICONES
ORREA
ALAVNA
VICTORIA
BODERIAE AEST.
DAMNONII
CORDA
CARBANTORIGVM
VXELLVM
SELGOVAE
CORIA
COLANIA
OTADINI
CVRIA
TRIMONTIVM
ITVNAE AEST.
BREMENIVM
EPIACVM
MORICAMBE AEST.
BRIGANTES
VINNOVIVM
CAMVNLODVNVM
SETANTIORVM PORTVS
CALAGVM
CATVRACTONIVM
ISVRIVM
Argita Fl.
Vidva Fl.
Logia Fl.
Vinderis Fl.
Bvvinda Fl.
Oboca Fl.
Mobonnvs Fl.
Ravivs Fl.
Libnivs Fl.
Avsoba Fl.
Senvs Fl.
Dvr Fl.
Iernvs Fl.
Abravannvs Fl.
Deva Fl.
Novivs Fl.
Longvs Fl.
Itis Fl.
Nabarvs Fl.
Loxa Fl.
Ila Fl.
Caelis Fl.
Deva Fl.
Tina Fl.
Alavnvs Fl.
Vedra Fl.

I. Ptolemy's map

topographical knowledge acquired as a consequence of Agricola's conquests and their immediate development. For the *Agricola* it is this section which is of concern.

The North British picture is divisible into physical and political features.[1]

1. *The physical features* are greatly distorted in plotting. Ptolemy was an astronomer who intended his map to be a basis for astronomical observation, but it is clear that he must have had to depend on the work of his predecessors. The earliest map of Britain, compiled by Eratosthenes from material gathered by Pytheas *c.* 325 B.C., had portrayed the island as stretching very far away to the north-east with what is actually the west coast facing north. This was the result of faulty measurements which gave an obtuse angle of 120° at the corner of Kent. The more accurate observations and measurements made by Caesar had altered this picture as far as England was concerned, so that it was approximately truly orientated, but he had no information about Scotland, and the older view prevailed in subsequent discussions (Pliny, *N.H.* 4. 102–3; Strabo 4. 5, 4, p. 201; Mela 3. 6, 53; the Map of Agrippa). Marinus, however, combined the correct views of Caesar as to the orientation of England with the incorrect view, going back to Eratosthenes, for the orientation of Scotland which appears in consequence in Ptolemy as running west–east from the north of England.[2] Neither Marinus nor Ptolemy had the opportunity for checking this by observation (μετεωροσκοπικόν) with astrolabes and gnomons, although they were able to supplement what was known about the country with information from itineraries and coastal surveys and from the accounts of travellers and soldiers.

[1] I. A. Richmond, *Roman and Native in North Britain* (Studies in History and Archaeology), pp. 131 ff.

[2] This is the explanation of J. J. Tierney, *J.H.S.* 79 (1959), 132 ff. For a different explanation see I. A. Richmond, *Proc. Soc. Ant. Scot.* 56 (1922), 288 ff.

For the *Agricola*, however, the essential matter is to appreciate the knowledge conveyed. It is manifestly derived from coastal surveys, fairly detailed on the east, north and south-west coasts, but sketchy between Clyde and Dunnet Head.

On the east coast are mentioned the Wear (*Οὐέδρα*), the Aln (*Ἄλαυνος*), the Forth estuary (*Βοδερία εἴσχυσις*), the Tyne (*Τίνα*), which is out of position, the Tay estuary (*Ταούα εἴσχυσις*), the Dee (*Δηούα*) and Buchan Ness (*Ταιξάλων ἄκρον*), connected with the tribal name *Ταίξαλοι*.[1] It is not clear upon what basis the selection was made by Ptolemy, and it is curious that the Tweed at Berwick and Esk at Montrose should be omitted. Then follow the Deveron (*Κελνίος*), Spey (*Τούαισις*), the Beauly Firth (*Οὐάραρ εἴσχυσις*), the Lossie (*Λόξα*), Tarbat Ness (*ὄχθη ὑψηλή*), the Ilidh or Helmsdale River (*Ἴλα*), Noss Head (*Οὐερουβίουμ ἄκρον*), Duncansby Head (*Οὐιρουέδρουμ ἄκρον*), and Dunnet Head (*Ταρουεδου(νου)μ* or *Ὀρκὰς ἄκρον*). The transposition of the Beauly Firth and the Lossie and the absence of the Findhorn will be noted. The information is now scantier, but comprises the rivers Nabhar or Naver (*Νάβαρος*), Loch Broom (*Οὔολσας κόλπος*), Loch Etive (in position *Ἶτις*, but not in etymology) and Loch Fyne (*Λόγγος*), the Mull of Kintyre (*Ἐπίδιον ἄκρον*), Loch Long (*Λεμαννόνιος κόλπος*), and the Firth of Clyde (*Κλώτα εἴσχυσις*). In this section also there is confusion or transposition, or both, as between *Longus*, *Epidium*, and *Lemannonius sinus*. From the Clyde southwards information is more complete, comprising Irvine Bay (*Οὐινδόγαρα κόλπος*), Loch Ryan (*Ῥεριγόνιος κόλπος*), the Rhinns of Galloway (*Νοουαντῶν χερσόνησος*), the Mull of Galloway (*Νοουαντῶν ἄκρον*), Luce Bay (*Ἀβραουάννος*) and Wigtown Bay (*Ἰηνᾶ εἴσχυσις*), the river Dee (*Δηούα*),

[1] The MSS. of Ptolemy have *ταιζ-* of the place and *ταιξ-* of the tribe. A decision does not seem possible since either spelling is philologically unexplained (K. Jackson, *The Problem of the Picts*, p. 136 (ed. F. T. Wainwright)).

the Nith (*Νοουίος*), and the Solway Firth or Eden (*'Ιτούνα εἴσχυσις*). Transposition may perhaps have taken place between *Abravannus* and *Iena aestuarium*.

The list is impressive and curious points emerge. First, while in comparison with other coasts that between Dunnet Head and Clyde was much less well known, it happens to cover the very area in which the *Agricola* so tellingly describes the exceptionally complex relationship between land and sea. On the other hand, there is fairly detailed information about the western islands, as if surveying might have concentrated upon them and kept away from the sea-lochs. This would raise the question whether Tacitus based his description of the West Highlands upon report of a land reconnaissance, since in fact a few high viewpoints in the west would furnish exactly the landsman's vision conveyed by the wording: *influere penitus atque ambire et iugis etiam ac montibus inseri velut in suo* (c. 10, 6). The restricted coastal information would also suggest that the offensive descents and exploration by Agricola's fleet had been confined to the east coast and that, on reaching the Pentland Firth, the exploratory cruise had headed for Scapa Flow, the Orkneys, and the northern sea. *Dispecta est Thule* (c. 10, 4) has long been recognized to mean the Shetlands: it is not, however, usually observed that when the outline of the Shetlands is considered in relation to Ptolemy's Thule it fits them well enough. Their position, like that of Skye, Long Island, or the Orkneys, seems haphazard, but only because none has been adjusted to the distorted mainland.

2. *Places and tribes* go together. The Novantae cover the land between the Nith and the Rhinns of Galloway, containing Lucopibia and Rerigonium. Rerigonium, on Loch Ryan (Rerigonius Sinus), is doubtless the missing fort at the west end of the road through south-west Scotland. Lucopibia is presumably the important intermediate fort needed between the west and Glenlochar. The Selgovae

embraced Dumfriesshire Eskdale but chiefly the upper Tweed and Teviot basins. Among their four *πόλεις*, Trimontium is certainly Newstead, named from the triple-peaked Eildons; Uxellum, Corda, and Carbantorigum (for Carbantoritum = Chariot-ford) are unknown. With the Selgovae marched the Votadini, extending from Lothian to the Tyne. Ptolemy, however, lists them to east (his south) of the Damnonii. Votadinian *πόλεις* are Bremenium, the fort of High Rochester; Curia, an unidentified tribal hosting-place, whose name was probably transferred to a Roman site; and Alauna, which is misplaced north (Ptolemy's east) of the Forth, probably by a misreading of latitude, since it should go with the river Alauna, now the Aln.

The Damnonii included Ayrshire, the Forth–Clyde Isthmus, if not the territory northwards to the Earn. *Οὐανδούαρα*,[1] going with Irvine Bay (*Οὐινδόγαρα κόλπος*), is the terminal fort of the road from Castledykes to the coast by Loudoun Hill. Colania is on the Isthmus: Coria, Alauna, Lindum, and Victoria represent forts on or near the northward road. Intervals are no guide, owing to the distortion: only the tribal territories and relative positions remain important. On this criterion, Alauna may be equated with Ardoch on the Allan, and Victoria with Strageath: Coria would lie further south, perhaps at the Forth crossing, and Lindum at Camelon, overlooking the marshy mouth of the Carron. The Vacomagi of Strathmore are placed by Ptolemy both there and in Banffshire, clean against topographical likelihood; and this is again due to the placing of Pinnata Castra in the far north. Tuessis, which is certainly on the Spey (*Τούαισις*), just as *Δηούανα*[2] is on the Dee (*Δηούα*), is accordingly put in Vacomagian territory. But if Pinnata Castra is assigned to Strathmore and identified as Inchtuthil, *Βαννατία* and *Τάμεια* must

[1] The MSS. vary between *Οὐανδόγαρα* and *Οὐανδούαρα*.

[2] K. Jackson, *Language and History in Early Britain*, p. 34.

come further south with it and Tuessis is left on the Spey, either among the Taezali or Tacitus' Boresti. Bannatia is connected with a prominent westward (Ptol. northward) peak and might well be Dealgin Ross, named from the adjacent peak of Dundurn. The second has a river-name which is not the Tay. South of the Vacomagi (Ptolemy's west) lay the Vennicones, and their πόλις, Ὄρρεα, the stores-base (*horrea*). This implies a port and suggests Carpow, or an analogous Flavian site. It does not lie on the Tay, as Ptolemy plots it, but the Tay lies too far north of the Forth and this may be the result of the revised half-degree corrections, inconsistently applied throughout the British sheet and based on the double readings for Κατουρακτόνιον and Λονδίνιον and divorcing the latter from the Thames.

Tribes associated with place-names do not exhaust the list. The Caledonii, who marched with the Vacomagi, were masters of the Highland plateau. Surviving place-names link them with the upper Tay basin from Dunkeld to Schichallion, but they extended to the Beauly Firth and also gave their name to the Western (Ptol. northern) ocean (ὠκεανὸς Δουηκαληδόνιος). The west (Ptol. north) coast tribes were the Epidii of Kintyre, whose name also connected them with Jura and Islay, the Cerones (with a ghost-doublet the Creones), the Carnonacae of Wester Ross, the Caereni of north-west Sutherland, and the Cornavii of Caithness. North of the Caledonii came the Decantae of Easter Ross, the Lugi of Sutherland and the Smertae, whose name survives in Carn Smeart in Loch Oykell. The Caledonian confederacy mentioned in the *Agricola* is thus specifiable: even if the remote Epidii stood aside, it would still comprise ten tribes, since the Taezali are presumably to be included. From these must have been drawn the *omnium civitatium vires* assembled to meet Agricola at Mons Graupius.

A striking contrast is thus apparent between the further regions, where only tribal names were recorded, and the hither territory replete with place-names. Nor can it escape note that place-names do not extend beyond the sphere of Agricola's campaigning. It is certain on archaeological grounds (see p. 62) that his armies reached the Aberdeenshire Dee, and a highly reasonable inference that they reached the Spey. Tuessis and Devana might, then, be considered as points where the land and sea forces met, since every recognizable place-name in the north can be regarded as denoting a Roman military position, as opposed to a native site. Indeed, the Roman origin of the entire body of information is unmistakably indicated by the failure to transpose into Greek the Latin terminal endings of the three northern capes, and by the specifically Latin place-names of *Pinnata Castra*, *Victoria*, *Horrea*, and *Trimontium*. Two other names for natural features also reveal a Latin origin; ὄχθη ὑψηλή must reproduce *Ripa alta*, the long high bluff of Tarbat Ness, while Νοουαντῶν χερσόνησος is the equivalent of *Novantarum peninsula*. But most of the names are Latinized forms of native names; and it is to be observed that there is everywhere a high proportion of Celtic names, though the tribal name Taezali, in the very heart of Pictland, is one of those which Celtic philologists find hard to accept. Incidentally Cape Orcas, of which the name goes back to Pytheas through Diodorus,[1] receives the new name of *Tarvedum*, preserved more correctly by Marcian in the form *Tarvedunum*;[2] the native fort (*dunum*) attested by this second part of the word adds a local touch, which Pytheas may not have had opportunity to experience.

The map of Ireland by Ptolemy confirms in striking fashion the statement of Tacitus, *aditus portusque per commercia et negotiatores cogniti*.[3] The broad outlines of the whole island, with fifteen rivers and six capes, are evident and its over-all size is not incorrect. The sixteen tribes are well distributed, and there are ten πόλεις, of which three

[1] 5. 21, 3.

[2] *Periplus Maris Exteri* 2, 45.

[3] c. 24, 2.

lie on the coast while seven are inland (μεσόγειοι). The πόλεις are not related specifically to tribal territories, though it must be accepted that *Eblana* was connected with the *Blanii*, *Manapia* with the *Manapii*, and *Nagnata* with the *Nagnatae*. Δοῦνον (Dunum) implies at least one Iron-Age fortress of importance. But the choice of the term *regia* for two royal seats attests a Latin source, from which the nouns in Ῥηγία and ἑτέρα Ῥηγία have been transliterated, but not translated, into Greek. The information, however transmitted, thus derives from Latin-speaking merchants. The account of Ireland can accordingly be divided into three sources; a coastal survey, an account of its people, and merchants' accounts of inland trading and places visited for the purpose. The sources are too detailed to date earlier than the Roman occupation of Britain, and must be due to the early interest in Ireland which culminated in Agricola's consideration of conquest. The whole character of the information exactly fits the definition by Tacitus.

6 · THE CONQUEST OF BRITAIN

(i) *The First Phase*

In the *Agricola* no reason is offered by Tacitus for the conquest of Britain, whatever he may, or may not, have chosen to write later in the lost books of the *Annals*. He treats the matter as if, after Julius Caesar's venture, the conquest was a natural development, as, indeed, with varying results, it has seemed to ambitious Continental powers ever since; and the abstention of Augustus and Tiberius or the instability or failure in generalship of Gaius are listed as mere delaying factors in the march of events. But there are significant remarks strewn through the work: the essential similarity of British tribes to one or other Continental group—three are specified[1]—and their chronic

[1] c. 11, 2.

disunity,[1] or the observation on Ireland, attributed to Agricola himself,[2] that it was a good thing to establish Roman power everywhere and to remove freedom from view—all these might imply points to be made in justification of conquest, if required. Economic advantages, such as they were, are firmly regarded as the reward of victory,[3] not as its objective, despite the caustic remarks which follow on British pearls and Roman greed.

Nor is much said about the earlier stages of the conquest.[4] The first advance under Plautius and Scapula, which is more specifically related to the great rivers in the *Annals*,[5] is not unfairly summarized in the words *redactaque paulatim in formam provinciae proxima pars Britanniae* (c. 14, 1). The work of Didius Gallus or his legates[6] in Wales is observed though merely classed as *paucis admodum castellis in ulteriora promotis* (c. 14, 2) in the context of expansion of territory. Here Paulinus is given some credit, yet to be specifically linked with actual sites,[7] though most of the narrative concerned with him is naturally devoted to the disaster of the Boudiccan revolt and the firm recovery from it, while the opportunity is not missed for a fine *suasoria* both on the evils of subjection and the unique opportunity for rebellion. Cerialis and Frontinus receive their due,[8] with appropriate emphasis upon the formidable character of the enemy in each case, and the stage is then set for a more detailed account of Agricola's governorship and prowess.

In terms of physical geography the initial stages had been easy. No severe obstacle existed to impede any army south-east of the Severn–Trent line, and the task was to organize the newly acquired territory in terms of legionary

[1] c. 12, 2. [2] c. 24, 3. [3] c. 12, 6.

[4] D. R. Dudley and G. Webster, *The Roman Conquest of Britain*, 1965.

[5] 12. 31, 2. [6] *A*. 12. 40.

[7] I. A. Richmond in I. Ll. Foster and G. Daniel, *Prehistoric and Early Wales*, 1965, pp. 151 ff.

[8] c. 17.

fortresses, auxiliary forts, and roads to link them. The burden of road-making under ruthless military supervision fell hardly upon the natives and the grievance was not missed.[1] The earliest forts are well exemplified by Hod Hill[2] or Waddon Hill[3] in Dorset, the area first conquered by Vespasian. Where the legions lay at this moment is not fully demonstrated archaeologically. The Twentieth was certainly at Colchester (Camulodunum) and moved in 49 to Gloucester (Glevum). The Second Augusta lay in the south-west and the Ninth in the north-east, the Fourteenth somewhere in the Midlands. But it is in the middle or later years of Nero that the fortresses of the two last can be recognized for certain at Lincoln (Lindum) and at Wroxeter (Viroconium)[4] respectively. The great sea-port stores-base for the island was at Richborough (Rutupiae) where from the first the *classis Britannica* must have played a most important part in commissariat.[5] But there were other harbour bases also, as remains at Fishbourne (Sussex)[6] and Fingringhoe (Essex)[7] indicate or suggest; while the early road-system points to a similar use for Hamworthy at Poole Harbour (Dorset), where Claudian coins and pottery have been found;[8] Topsham near Exeter is likely to have been another base.[9]

To a discernible extent, though this must not be exaggerated, the flag had followed trade, while trade had been linked with pre-conquest political relations. If Caesar had

[1] c. 31, 1.

[2] Sir Ian Richmond, *Hod Hill*, vol. 2 (1967).

[3] G. Webster, *Dorset N. H. and Arch. Soc.* 82 (1960), pp. 88 ff., and 86 (1964), pp. 135 ff.

[4] See p. 77.

[5] *Richborough IV* (Soc. Ant. Res. Rep. 16, 1949), p. 5 and pp. 18 ff.; *Richborough V* (Soc. Ant. Res. Rep.), forthcoming.

[6] B. Cunliffe, *Ant.* 39 (1965), 178.

[7] *V.C.H. Essex*, vol. iii (1963), p. 4 and p. 131.

[8] G. Webster, *Arch. Journ.* 115 (1958), 57 (note 7).

[9] C. A. R. Radford, *Proc. Devon Arch. Exploration Soc.* 3 (1937–47), 10.

dedicated British pearls in his *forum*,[1] the Rome of Augustus had seen official visits of British princes in search or in consequence of formal recognition as *reges*.[2] There is little doubt that the title *rex* used on the coins of Eppillus and Verica[3] was intended to signify official recognition of just that kind. Caesar's tribute, again, fell into desuetude, but it was followed under Augustus and Tiberius by no small income from Gallic customs on British imports and exports.[4] This is fully borne out by the mass of imported pottery at Colchester (Camulodunum),[5] the capital of Cunobellinus, and by the grave-goods of native chieftains,[6] which also indicate very substantial imports into his realm. While more allowance for survival may have to be made on imported Arretine wares than has sometimes been made in the past, imported pottery did have a fairly wide dispersion and travelled to other tribal areas. This is attested for the

[1] Pliny, *N.H.* 9. 116.

[2] Strabo, 4, p. 200. The history of Britain between the invasions of Caesar and Claudius is discussed by C. E. Stevens in *Aspects of Archaeology* (Essays presented to O. G. S. Crawford, 1951; ed. W. F. Grimes), pp. 332 ff. Stevens argues that Augustus decided not to follow up Caesar's invasions by the annexation of the island but rather to assert the Roman claim to it by extending recognition to dynasties which were favourably disposed to Rome.

[3] D. F. Allen, *Arch.* 90 (1944), 8 ff. Cunobellinus also minted one issue with the legend *Rex* (Allen, op. cit., p. 21). Stevens doubts whether Cunobellinus, to whom he attributes consistently nationalistic ambitions, could ever have been entitled to the rank, but we do not know enough about the shifting politics of Britain in this period to reject the prima-facie implication of the coins and he is similarly called *Britannorum rex* by Suetonius (*Calig.* 44, 2). His recognition as a client-king may have been genuine enough but short-lived.

[4] Strabo, loc. cit.

[5] C. F. C. Hawkes and M. R. Hull, *Camulodunum* (Soc. Ant. Res. Rep. 14, 1947), pp. 27 ff.

[6] e.g. Welwyn, *Arch.* 63 (1912), 1 ff.; Mount Bures, *V.C.H. Essex*, iii (1963), p. 60; Lexden, *Arch.* 76 (1927), 241 ff.; Snailwell, *P. Camb. A. S.* 47 (1953), 25 ff.; Prior's Wood, *T. East Herts. A. S.* 14 (1957), 1 ff.

Atrebates at Silchester,[1] the Dobunni at Bagendon (the British forerunner of Cirencester),[2] the Coritani at Leicester,[3] and the Parisi at Ferriby[4] or even Dragonby.[5] At Colchester, indeed, not only do the written names of traders survive on pottery but coin-designs suggest that the later moneyers active in the royal mints of Colchester, and perhaps also of Verulamium, had learned their craft in Roman workshops.

It could, then, have been argued, as it was by Strabo,[6] that Britain was well left alone. But four factors had altered the situation by A.D. 43. The expanding ambitions of Cunobellinus and his sons, notably Togodumnus and Caratacus, were threatening or disrupting the political arrangements of the south of the island, explicitly recognized and implicitly guaranteed by Rome: while Cunobellinus had expelled a son, Adminius, with pro-Roman followers, and the refugee prince had made submission to Gaius,[7] another philo-Roman king, Verica of the Sussex area, had been expelled and had fled to Claudius.[8] Secondly, the Britons had demanded extradition of refugees and on refusal were creating trouble (*Britanniam tumultuantem*).[9] Thirdly, the apparent caprice of Gaius in abandoning the invasion of Britain had lowered Roman prestige.[10] Finally, Claudius wished to rival the glory of his famous ancestors by succeeding where Julius Caesar himself had fallen short.[11] It was these factors, all tending in a single direction and the first three abetting the last, itself paramount, which

[1] G. C. Boon, *Roman Silchester*, 1957, p. 58 and Fig. 5.

[2] E. M. Clifford, *Bagendon: a Belgic Oppidum*.

[3] K. M. Kenyon, *The Jewry Wall Site, Leicester* (Soc. Ant. Res. Rep. 15, 1945), p. 9; but see also G. Webster in *The Civitas Capitals of Roman Britain* (ed. J. S. Wacher), p. 42.

[4] *Antiq. Journ.* 18 (1938), 270.

[5] J. May, *Dragonby 1964* and *Dragonby 1965*, Board of Extra-Mural Studies, Nottingham.

[6] Loc. cit.

[7] Suet. *Calig.* 44.

[8] Dio, 60, 19.

[9] Suet. *Claud.* 17.

[10] c. 13, 2 and note.

[11] Suet. *Claud.* 17.

determined and precipitated the invasion. Commercial interests may have looked for opportunities of exploitation by investment or development, but they must be viewed as following opportunity rather than creating it.

The *Agricola* gives no account of the actual invasion and the relevant text of the *Annals* is lost. Cassius Dio is epitomized, with inevitable loss of links in the narrative.[1] Thus, while the invasion force started in three divisions, the landing is not described, though Vespasian and Hosidius Geta are present at the first important battle which appears to include the whole invading force. Specific details of the kings who surrender to Claudius and presumably figured in his triumph are missing and his own claim of eleven[2] includes neither personal nor territorial names. But it is not difficult to imagine the possibilities. Even excluding the new client-kings, Cogidubnus of the *regnum* in west Sussex,[3] Prasutagus of the Norfolk Iceni, and (if so soon) Cartimandua of the Brigantes,[4] there were to be reckoned the Cantiaci, Atrebates, Catuvellauni, Trinovantes, Coritani, Parisi, Durotriges, Belgae, Dumnonii, Dobunni, and Cornovii.[5] The Britons were indeed caught by surprise when the invasion, for a while postponed, actually occurred: but on the Roman side the initial campaigns seem to have been carried out with precision.

From the bases of Colchester, where Legio XX was at first quartered, and London the further conquest of the island was carried on, and the road-system might suggest that at least three prongs were employed. The left wing (Legio II Augusta, under Vespasian) overran the south, as far, probably, as Exeter and the fringe of south Wales.[6]

[1] 60, 20.

[2] c. 13, 3 and note.

[3] c. 14, 1 and note.

[4] c. 17, 2 and note.

[5] For these people see A. L. F. Rivet, *Town and Country in Roman Britain*[2] (1964), pp. 34 ff.

[6] c. 13; *H.* 3. 44; Suet. *Vesp.* 4. Troops from this Legion were on duty in the Mendip mines under Nero (*CIL.* xiii. 3491).

The centre—Legio XIV—crossed the Midlands.[1] The right wing, Legio IX, moved northwards towards Lincoln. These lines of advance agree with the three main routes which radiate from London, and suggest that they were now organized: that is, (1) a south-west route, running to Silchester, where it branched to Winchester, Dorchester, and Exeter, and to Bath and Gloucester, whither Legio XX was moved in A.D. 49;[2] (2) a north-west route across the Midlands, through St. Albans towards Lichfield and ultimately to Wroxeter, near Shrewsbury; (3) a north-east route, through Colchester, Cambridge, and Castor, near Peterborough, to Lincoln. It is doubtful whether the legions were placed in forward positions immediately—the case of Gloucester is instructive: but Tacitus attests that by A.D. 49 or 50 the Roman arms had reached the basins of the Trent, the Severn, and even the Dee.[3] Here there was a long pause.

(ii) *The Occupation of Wales, North England, and Southern Scotland*

The Roman forces were now faced on the west by the hills of Wales, on the north by those of Derbyshire and northern England. These hills and their stubborn inhabitants delayed progress for nearly a generation, and the Romans may well have felt that the existing limits of the province were satisfactory in themselves for the time being.

[1] See p. 77. [2] See p. 77.

[3] This is evident from the fact that by A.D. 50 Scapula (A.D. 47–52) had to deal (*A.* 12. 31, 2) with the Iceni of Norfolk, the Brigantes north of the Trent, the Decangi of Flintshire, and the Silures of Monmouthshire. It is this state of affairs that so strongly supports the emendation of *A.* 12. 31, 2, where the single MS. reads *cunctaque castris Antonam et Sabrinam fluvios cohibere parat* to *cunctaque cis Trisantonam* etc., the British name of the Trent being *Trisantona.*

Time was also needed to absorb the lowland area, and in particular to take in the client-kingdoms or protectorates which here, as everywhere, were viewed as stepping-stones to annexation. In A.D. 57 the decision was made to advance into Wales, but after three years' campaigning the attempt proved abortive. Lax control by the central government bred misrule and discontent in the rearward areas, till a great rising under Boudicca in A.D. 60–61 almost ended Roman dominion in the island. The subsidence of this upheaval, under two years of gentler rule, was followed by a period of easygoing administration during which the *status quo* was maintained and consolidated till Nero fell, and the great crisis of the Empire, the civil war of A.D. 69, was over. At the end of Nero's reign the Roman boundaries varied little from what they had been at the beginning. The western limit ran close to a line drawn from Newport through Shrewsbury to Chester, and the northern to one through Chester, Derby, and Lincoln.

Under the new dynasty of Vespasian, progress was resumed: *magni duces, egregii exercitus, minutae hostium spes* (c. 17). In seven years (71–77) the Roman arms had penetrated the Derbyshire hills and eastern Pennines, and the Ninth Legion had been moved by Cerialis to York; in the west more than half of Wales was subdued by Frontinus;[1] and over the conquered lands was spread a network of forts followed in due course by roads (see Fig. 2). Then Agricola assumed command. His first work was to complete the conquest of North Wales by the virtual annihilation of the Ordovices and the annexation of Anglesey, areas with which his command of the Twentieth Legion, presumably at Wroxeter, and his service with Paulinus had familiarized him. Now, in preparation for northward moves, he established a legionary fortress on the fort-site of Chester, where lead pipes of 79 bear his name and seem to chime in with a fragment of an Imperial

[1] c. 17.

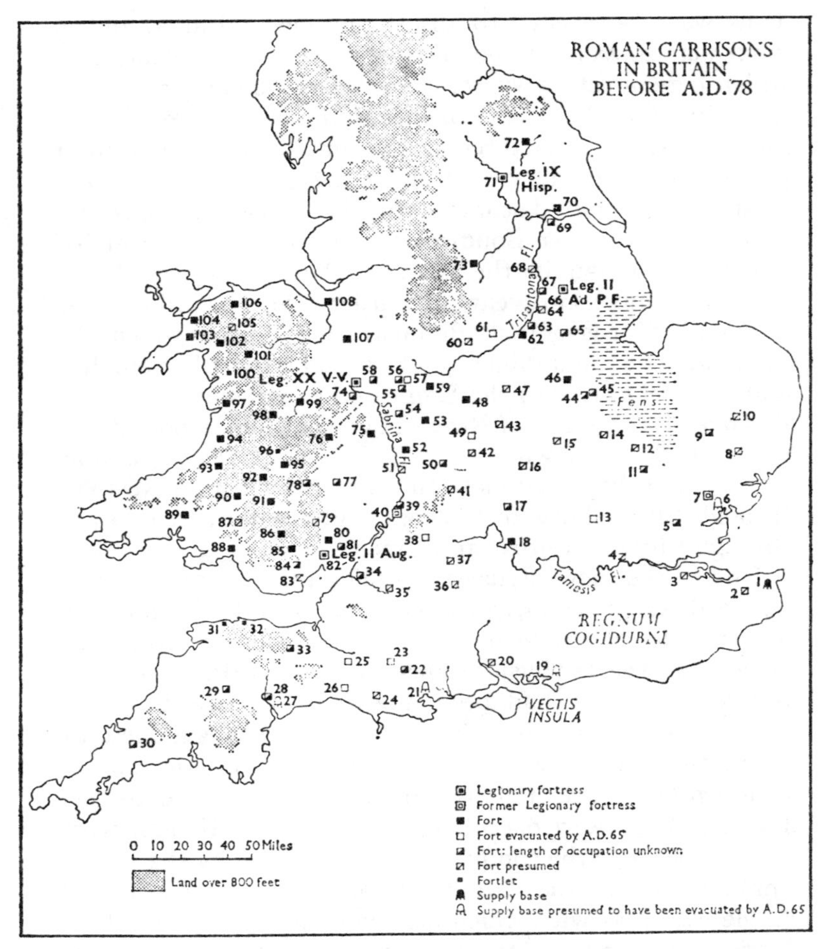

ROMAN GARRISONS
IN BRITAIN
BEFORE A.D. 78
Leg. IX
Hisp.
Leg. II
Ad. P. F.
Leg. XX V.V.
Leg. II Aug.
Trisantona Fl.
Sabrina Fl.
Tamesis Fl.
Fens
REGNUM
COGIDUBNI
VECTIS
INSULA
Legionary fortress
Former Legionary fortress
Fort
Fort evacuated by A.D. 65
Fort: length of occupation unknown
Fort presumed
Fortlet
Supply base
Supply base presumed to have been evacuated by A.D. 65
0 10 20 30 40 50 Miles
Land over 800 feet

Fig. 2.

building-inscription.[1] The new fortress commanded not only access to North Wales and to the Irish Sea, but the western route to the north, essential for the effective occupation of the territory of the Brigantes.

The nature of the conquest of the Brigantes by Cerialis is shadowy, and has not yet been fully determined by archaeology. He is described as 'having embraced a large part' of them 'by victory or warfare'.[2] While the campaigning can be traced to the stronghold of Stanwick,[3] near Catterick, and beyond it across Stainmore[4] to Carlisle, where pottery has been claimed, and denied, to betoken pre-Agricolan[5] occupation, permanent dispositions in Lancashire, West Yorkshire, Westmorland, and Cumberland are recognizable as Agricolan. Pottery and structures of this period have been noted from at least 28 forts ranging from North Yorkshire to the Tyne and from the Mersey to the Eden.[6] Only Brough on Humber, Malton (N. Riding),[7] and York[8] emerge as securely tied to Cerialis, as if his permanent occupation had enveloped or protected the Parisi of East Yorkshire, shielding them by a legionary fortress which was to be the base for subsequent extension of the permanent

[1] *RIB* 463; I. A. Richmond in I. Ll. Foster and G. Daniel, *Prehistoric and Early Wales*, 1965, pp. 155 ff.; see c. 46, 4 n.

[2] c. 17, 2.

[3] R. E. M. Wheeler, *The Stanwick Fortifications* (Soc. Ant. Res. Rep. 17, 1954). For an account of Cerialis' campaigns see J. Clarke in *Roman and Native in North Britain* (ed. I. A. Richmond, 1958), pp. 37 ff.

[4] I. A. Richmond and J. McIntyre, *C.W.*[2] 34 (1934), 57 ff.; *J.R.S.* 41 (1951), 54.

[5] For different views see: J. P. Bushe-Fox, *Arch.* 64 (1912–13), 295 ff.; F. Haverfield and D. Atkinson, *C.W.*[2] 17 (1917), 235 ff.; E. Birley, *T.D.G.A.S.* 29 (1950–1), 46 ff.; R. Hogg, *C.W.*[2] 64 (1964), 14 ff. and especially 58.

[6] See Fig. 3.

[7] P. Corder, *The Defences of the Roman Fort at Malton*, 1930, p. 64; M. Kitson Clark, *A Gazetteer of Roman Remains in East Yorkshire* (Roman Malton and District Report No. 5, 1935), pp. 28 ff.

[8] *Eburacum: Roman York* (R.C.H.M., 1962), p. 6.

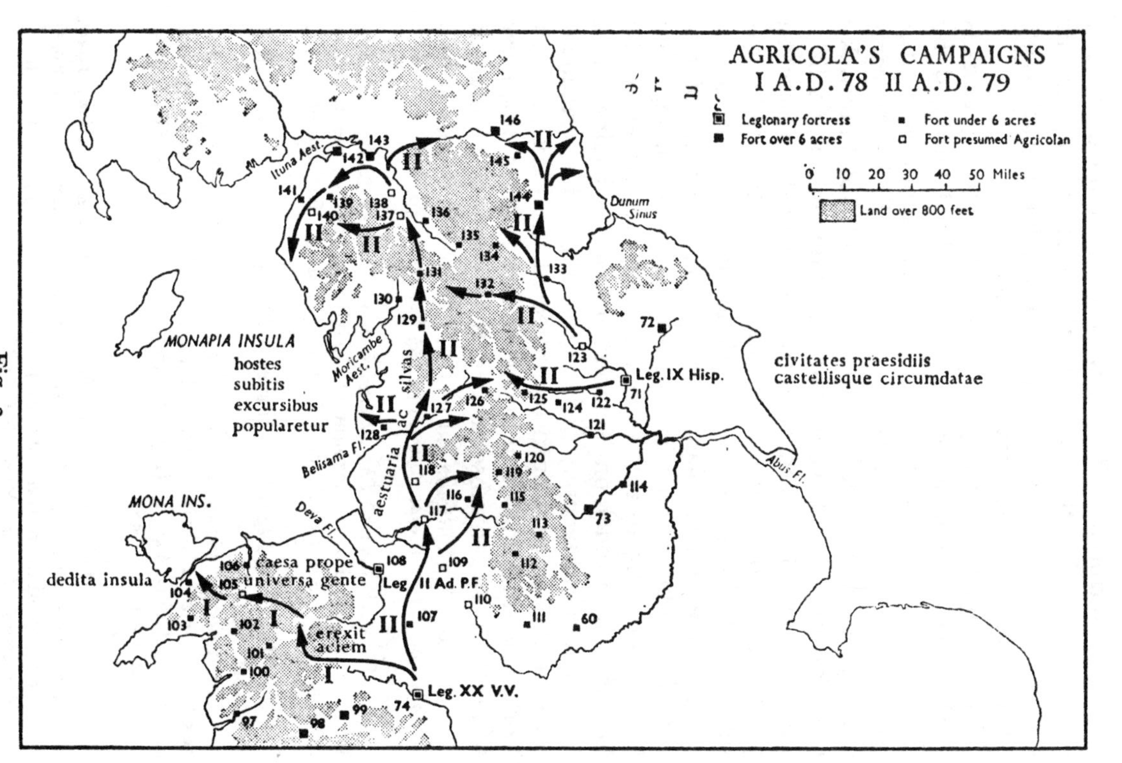

AGRICOLA'S CAMPAIGNS
I A.D. 78 II A.D. 79
Legionary fortress
Fort over 6 acres
Fort under 6 acres
Fort presumed Agricolan
0 10 20 30 40 50 Miles
Land over 800 feet
Ituna Aest.
Dunum Sinus
MONAPIA INSULA
hostes
subitis
excursibus
popularetur
Moricambe Aest.
silvas ac aestuaria
Leg. IX Hisp.
civitates praesidiis
castellisque circumdatae
Belisama Fl.
Abus Fl.
MONA INS.
Deva Fl.
dedita insula
caesa prope
universa gente
Leg. II Ad. P.F.
erexit
aciem
Leg. XX V.V.

Fig. 3.

system. The 28 forts marked on the map are certainly not a complete total, for numerous sites have not yet yielded the requisite evidence, and it might be thought that at least 10 more were involved, making 38 in all. The system was a simple one: the principal lines of communication were seized and patrolled and tribal movements, even in insignificant numbers, were paralysed. That is what is meant by the phrase used in connexion with the consolidation of A.D. 79, *civitates . . . praesidiis castellisque circumdatae*, following Agricola's second campaign.

To the north of the Tyne–Solway isthmus, the first objective was the conquest of the eastern and central Lowlands of Scotland, inhabited by the Votadini and Selgovae respectively.[1] These are not named; but they are the *novae gentes* of the third campaign, in contrast with the *ignotae* of the fifth. The campaign reached to the estuary of the Tavus,[2] now the Tay, and, if the later road-system is any guide,[3] the plan of action was a pincer movement starting from Carlisle and from Corbridge and ending on the Forth at Inveresk, followed by probing action across the Forth from Stirling by way of Strathallan and Strathearn to the Tay. The territory thus covered was scarcely smaller than that of the Brigantes and its consolidation, behind the new Forth–Clyde line (see p. 323), took place the following season, described as the fourth summer.

The account of the third summer's work lays particular emphasis upon the planting of *castella*, following a campaign afflicted by bad weather (see Fig. 4). The eastern route was

[1] J. Clarke, op. cit., pp. 47 ff.

[2] The corrected reading *Taum*, probably for *Tavum*, in *E* is preferable to *Tanaum*; and philologists have no doubt of its equation with Ptolemy's *Ταούα εἴσχυσις* (*Geogr.* 2. 3, 57). W. J. Watson (*Celtic Place Names of Scotland*, p. 50), E. Ekwall (*English River Names*, p. 394), and K. Jackson (*Language and History in Early Britain*, p. 369) all concur.

[3] S. N. Miller (ed.), *The Roman Occupation of S.W. Scotland* (1952), p. 210.

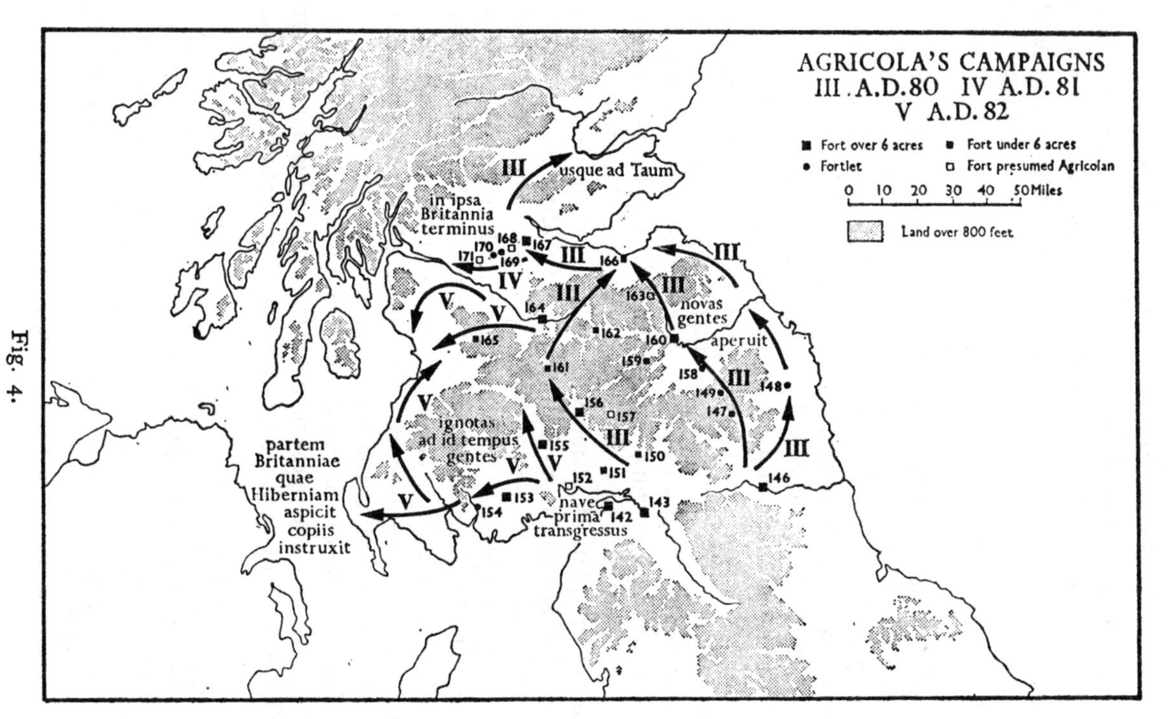

AGRICOLA'S CAMPAIGNS
III A.D.80 IV A.D.81
V A.D.82
Fort over 6 acres
Fort under 6 acres
Fortlet
Fort presumed Agricolan
0 10 20 30 40 50 Miles
Land over 800 feet
usque ad Taum
in ipsa Britannia terminus
novas gentes
aperuit
ignotas ad id tempus gentes
nave prima transgressus
partem Britanniae quae Hiberniam aspicit copiis instruxit

Fig. 4.

secured by the fort at High Rochester,[1] fortlets at Chew Green and Cappuck, a very large fort at Newstead on Tweed, a possible fortlet at Oxton, and a terminal fort at Inveresk: the western by forts at Birrens, Tassiesholm (Milton), Crawford, and Castledykes. They are linked through the Tweed valley, with a fort at Easter Happrew, and with outliers on the Yarrow at Oakwood and on the Dumfriesshire Esk at Broomholm and Raeburnfoot. The southern territory of the Votadini is garrisoned by a fort at Learchild, the northern by no fort as yet known, whether this represents freedom from occupation or a blank in knowledge. Certain it is that the northern area, known as the Merse, was very thickly populated. The most important garrison post among the Selgovae was Newstead,[2] whose name *Trimontium* was taken from the triple peaks of the Eildon Hills, the North Eildon being crowned by a large native *oppidum*, once one of the major centres of the tribe.

The fifth campaign has been more discussed than the others, and its starting-point is obscured by a passage which may be corrupt and which perhaps originated from the misunderstanding of a place-name.[3] However that may be, the troops arrived, among tribes hitherto unknown, on the coast facing Ireland. These points, it must be recognized, rule out anywhere but Galloway and Ayrshire (see Fig. 4), where substantial remains of Agricolan occupation have now been found, in Galloway at the large fort of Dalswinton on the Nith, and at Glenlochar on the Dee, and in Ayrshire at Loudoun Hill. Further west in Kirkcudbrightshire a fortlet has been excavated at Gatehouse-of-Fleet on the way to the Rhinns of Galloway, where the terminal fort of this system must have existed. Here lies the stretch of coast which can most appropriately

[1] For references to these sites see p. 338.

[2] *Trimontium* is placed in the territory of the Selgovae by Ptolemy (*Geogr.* 2. 3, 8).

[3] c. 24, 1 and note.

be described as 'that which looks at Ireland', since the island is in full view across the North Channel. Coastal and inland links with Ayrshire, isolating the wild mass of Carrick, might be expected and are beginning to emerge,[1] while in Ayrshire itself the coastal fort on Irvine Bay, which is the obvious goal of the road from Castledykes to Loudoun Hill and the west, is still to seek. Nevertheless, the essential pattern of the conquest and consolidation of the south-west has begun to emerge.

The sixth and seventh campaigns are concerned with the area beyond the Forth–Clyde isthmus which Tacitus thrice describes as Caledonia[2] and with which contact was made after enveloping the *civitates trans Bodotriam sitas* (c. 25, 1). This fits well the evidence both of Scottish place-names, which associate the people named *Caledonii* with the upper Tay valley from Dunkeld to Schichallion,[3] and of Ptolemy, who places their northern boundary in the Great Glen.[4] It would in fact make them the premier tribe or coalition of tribes in the Highlands as they plainly continued to be in the early third century.[5] By the substantive *Caledonia*, however, Tacitus (and no doubt primarily his father-in-law) clearly meant the entire Highland region and all its people: for in the context of physical geography he employs the term[6] to mean all northern Scotland, while politically, having noted that the confederacy commanded by Calgacus included *omnium civitatium vires* (c. 29, 3), he puts into the leader's mouth the phrase *quos sibi Caledonia viros seposuerit* (c. 31, 4).

The *civitates trans Bodotriam sitas* presumably comprise those south-east of the Highland plateau, that is, the people of Fife, Fortrenn, Angus, and the Mearns (see Fig. 5). In this corridor, the main route of access to the north, and indeed beyond it, lie the remains of Roman temporary

[1] *J.R.S.* 55 (1965), 202.
[2] c. 25, 3; 27, 1; 31, 4.
[3] W. J. Watson, op. cit., p. 21.
[4] *Geogr.* 2. 3, 12.
[5] Dio 66, 12.
[6] c. 10, 3.

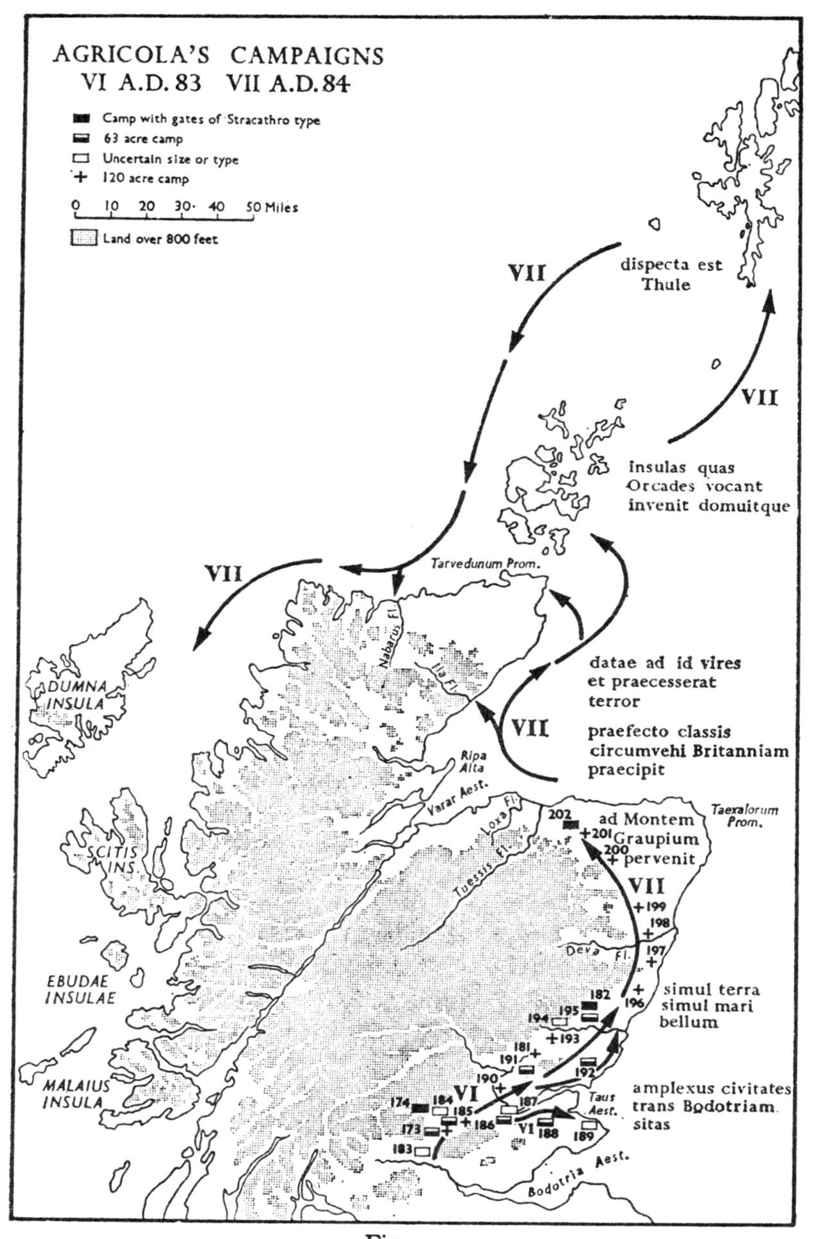
AGRICOLA'S CAMPAIGNS
VI A.D. 83 VII A.D. 84
Camp with gates of Stracathro type
63 acre camp
Uncertain size or type
120 acre camp
0 10 20 30 40 50 Miles
Land over 800 feet
VII
dispecta est Thule
VII
insulas quas Orcades vocant invenit domuitque
VII
Tarvedunum Prom.
Nabarus Fl.
Ila Fl.
DUMNA INSULA
datae ad id vires et praecesserat terror
VII
praefecto classis circumvehi Britanniam praecipit
Ripa Alta
Varar Aest.
Loxa Fl.
Taexalorum Prom.
ad Montem Graupium pervenit
SCITIS INS.
Tuessis Fl.
VII
Deva Fl.
EBUDAE INSULAE
simul terra simul mari bellum
MALAIUS INSULA
VI
amplexus civitates trans Bodotriam sitas
Taus Aest.
Bodotria Aest.
202
201
200
199
198
197
196
195
194
193
192
191
190
189
188
187
186
185
184
183
182
181
174
173

Fig. 5.

camps erected by armies on campaign, first planned and published in the posthumous work of General William Roy, issued in 1793, under the title of *The Military Antiquities of the Romans in North Britain*. The works, excepting a few in northern England, run from the Borders to Aberdeenshire, but they have in this generation been very richly supplemented by the discoveries of Dr. J. K. St Joseph, Director in Aerial Photography in the University of Cambridge. Dr St Joseph observes[1] that the camps north of the Forth, not without counterparts further south, fall into three sizes. The largest cover 120 acres, the medium-sized 63 acres, the smallest 40 acres or less. A criterion of the number of troops involved is furnished by the 26-acre temporary camp of Reycross,[2] which plainly contained a legion and extra troops amounting to about a cohort. Thus the medium-sized camps could contain two legions with auxiliaries, while the largest size could plainly house a force representing the entire provincial army of three legions and a substantial number of auxiliaries or other troops, in other words an Imperial expedition. There is, indeed, some evidence that these largest works post-date the Flavian period, for at Ardoch one of them overrides a signal-station which must be Flavian at earliest and might be Antonine.[3] They can best be associated with Severus.

Here, however, must be taken into account another category of temporary camps in Scotland, of which the common characteristic is not size, except that the scale of their defences is optimistically modest, but a highly unusual gateway-plan, wherein the ditch forms on one side an external *clavicula* and on the other an oblique external spur. These are associated with proved Agricolan sites, at Dalswinton on the Nith and at Stracathro on the North

[1] *J.R.S.* 48 (1958), 93.

[2] See p. 55, n. 4.

[3] *Arch. Journ.* 93 (1937), 313 (Fig. 3) and 314.

Esk[1] (p. 67) respectively, and there can be little doubt that they do indeed record movements of Agricola's troops.

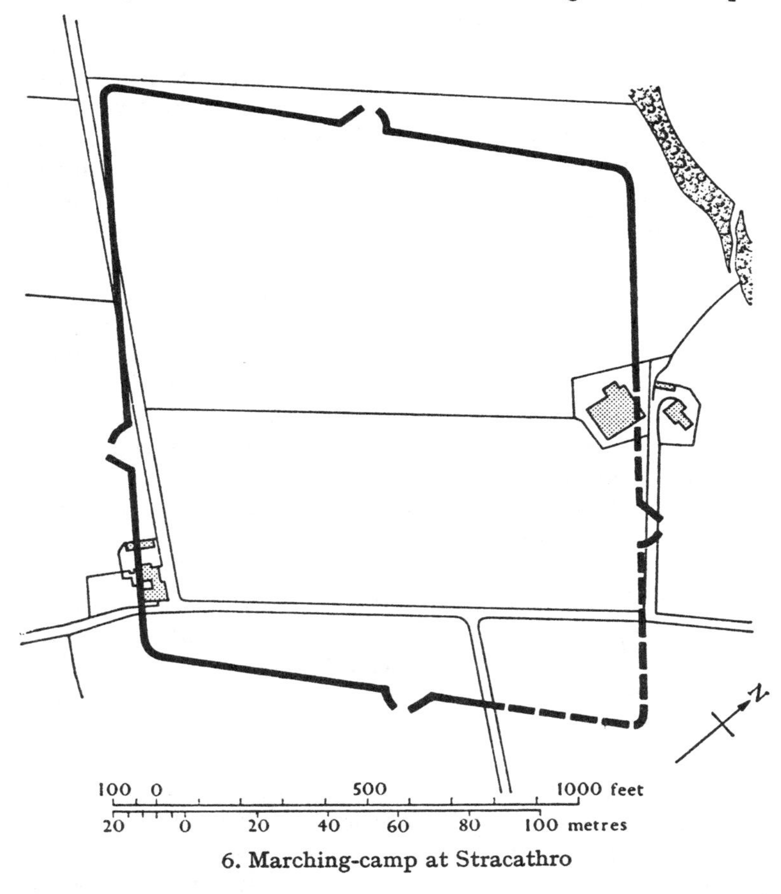

6. Marching-camp at Stracathro

Size thus becomes an interesting point. A 25-acre work of

[1] For Dalswinton see I. A. Richmond and J. K. St Joseph, *D.G.N.H.A.S.T.*[3] 34 (1955–6), 9 ff.; for Stracathro, *J.R.S.* 48 (1958), 92; further examples have been found at the Lake of Monteith, Ythan Wells, and, possibly, Dunning.

this kind appears at Dealgin Ross,[1] at the head of Strathearn. Stracathro is 40 acres in size. A third example, at Auchinhove[2] in the Pass of Grange, which leads from Aberdeenshire to the sea-plains of Banff and Moray, is not yet fully delimited but contains not less than 27 acres. None of these works, however, is large enough to have accommodated a force of the size that fought at Mons Graupius: nor, again, do they occupy the coastal positions described in the sixth campaign and shared by sea-forces and land-forces alike. On the other hand, the 25-acre size certainly recalls the tactics adopted late in the sixth campaign, when the forces were triply divided and the Ninth Legion, in reduced strength, was operating alone.[3] The fort of *Victoria*,[4] in the territory of the Damnonii, which was identified with Dealgin Ross by Roy, may well commemorate the battle in which victory was snatched from defeat. While, however, it can be recognized that these camps took their place in the Agricolan picture, the place itself can hardly be assigned. There are, for example, the camps of fort-builders to take into account. It is also certain that, apart from these camps with peculiar gateways, Agricolan troops at Chew Green and Newstead built works with ordinary traverse gateways.[5] The solid gain is the important fact that Agricola's operations extended not merely to Kincardineshire but to Banffshire and that the culminating battle of Mons Graupius must have been fought much nearer Moray than was once thought possible.

Two points are implicit in the account by Tacitus of the site of that much-sought battlefield. Firstly, the name itself, even if it were more correctly *Craupius* than *Graupius*,[6] though the Romans cared little for the difference between

[1] J.R.S. 48 (1958), 93. [2] Ibid.

[3] c. 25, 4. [4] c. 26, 2; Ptolemy, *Geogr.* 2. 3, 7.

[5] For Chew Green see I. A. Richmond and G. S. Keeney, *Arch. Ael.*[4] 14 (1937), 129 ff.; for Newstead, R.C.H.M., *Roxburghshire* ii (1956), pp. 312 ff.

[6] See note on c. 29, 2.

these Celtic consonants, would not imply a spectacularly lofty hill. Secondly, the fact that Calgacus had chosen the position in advance would indicate that it lay at an obvious and to that extent unavoidable point of encounter on the main line of advance to the north-east. In British strategy the pre-selected position was something of a habit. The rebellious Iceni chose a battleground in 47; in 51 Caratacus did the same thing among the Ordovices, and some twenty years later the Brigantian leader Venutius followed very much that choice of action at Stanwick at the dividing of the ways north of York.[1] Calgacus was, in fact, following an established pattern of native strategy. In relation to the terrain of north-east Scotland this implies that he was making a stand at some point before the coast-lands of Moray were reached, and this would fix the battlefield somewhere in the pass between Auchinhove and Keith.[2] The reason why Calgacus felt compelled to stop the invaders at this point is clearly a determination to protect the coastal plain, not only because this was the last major tract of good land in Scotland, but also because it was the last area in which the combined strength of the Highland tribes could be congregated. When Tacitus puts into his mouth the words *nunc terminus Britanniae patet* (c. 30, 3), he was presumably repeating yet another truth about British topography learnt from his father-in-law, whose achievement in bringing about a pitched battle, by means of his irritating sea-borne raids, was the essential condition of success. Agricola deserves the fullest credit as a strategist: *veniunt e latebris extrusi* (c. 33, 4) is another authentic statement, the prelude to a victory which was as final for its generation as Culloden.

[1] *A.* 12. 31 (Iceni), and 33 (Ordovices); and see above, p. 55, n. 3 (Brigantes).

[2] So also Sir David Henderson-Stewart, *Trans. Anc. Mon. Soc.* 8 (1960), pp. 75 ff. O. G. S. Crawford, *Topography of Roman Scotland*, 1949, pp. 130 ff. tentatively located it at Raedykes.

The planning of the battle has its own special interest. It is the first recorded instance of a tactical use of auxiliaries which kept the legionaries entirely in reserve and won a victory *citra Romanum sanguinem* (c. 35, 2).[1] Exactly the same principle is shown in more than one battle figured on Trajan's Column.[2] These were the tactics of the developing Imperial army, and are a preparation for campaigning entirely with auxiliaries, as described by Arrian in his *Order of Battle against the Alani*. But the *Agricola* is the sole literary source in which the principle is clearly and explicitly stated. The dispositions following the battle are more obscure. The *Boresti*, into whose territory the army was now led, might well have occupied the Moray coast, since orders to the fleet are coupled with the reception of hostages.[3]

While the sixth and seventh campaigns thus present a strategy designed to meet the particular problems of Highland geography, little is said by Tacitus of permanent arrangements. The pattern of these had no doubt begun to take shape under Agricola: at least one *castellum* is mentioned in the territory north of the Forth,[4] but the *hiberna* mentioned after the campaign[5] probably mean the ordinary permanent quarters of the troops concerned, and do not necessarily imply any structure north of the isthmus. Beyond the Forth, as has long been recognized, the essential feature is the arterial route along the natural corridor from Stirling to the north-east. At Stirling the fort guarding the crossing of the Forth remains unidentified, though temporary camps at Dunblane[6] indicate that the route itself followed the traditional line by Strathallan to Ardoch; and so to Strageath, at the Earn crossing, and Bertha[7], at the

[1] See also E. Birley, *Roman Britain and the Roman Army*[2] (1961), p. 17, note, commenting on Cerialis' use of auxiliaries against Civilis (Tac. *H*. 5. 16).

[2] c. 35, 2 and note.

[3] c. 38, 3–4.

[4] c. 25, 3.

[5] c. 38, 3.

[6] *J.R.S.* 41 (1951), 62.

[7] The signal stations guarding the Gask ridge road date from the Flavian period (*Disc. and Exc.*, 1966, p. 37).

Tay crossing above Perth. The line now heads for the Mearns, through Strathmore, with a fort on the Dean at Cardean, near Meigle; crossing the South Esk at Finavon, the site of a temporary camp, and reaching the North Esk and the approach to the Cairn-o'-Mount pass at Stracathro, where a six-acre Flavian fort has been identified. Cardean could be considered as watching Glen Isla and the pass to Braemar, but elsewhere the arterial route stands clear of the Highland passes. To block these effectively a series of forward positions was established. Inchtuthil lies at the mouth of the Dunkeld gorge, Fendoch closes the Sma' Glen, Dealgin Ross guards the head of Strathearn, Bochastle lies near the important pass of Leny.[1] The policy is self-evident; to close all the principal exits from the Highlands by a *cordon militaire*. This granted, it is clear that some elements still await discovery. The head of the Forth and the Balloch gap demand similar blocking forts, as also does the north-east end of the Mearns, at or near Stonehaven. Again, coastal positions and harbours have yet to make, as in the south-west, their contribution to the picture. It should be added that a Flavian date for most of the positions is well attested. Ardoch has yielded not only Flavian pottery but an early tombstone, Stracathro one fragment of Flavian pottery. Among the forward stations Inchtuthil is associated with Flavian pottery and coins, Fendoch with Flavian pottery, Dealgin Ross with Flavian coins, and Bochastle with Flavian pottery.[3] Three of the sites, moreover, have produced substantial remains of internal buildings: indeed, at Ardoch these are the sole structures of the Flavian fort, its defences having been totally obscured by those of Antonine date. But at Fendoch the remains of an

[1] See Fig. 7.

[2] For Ardoch, *P.S.A.S.* 32 (1898), 436 ff. and *RIB* 2213; Stracathro, see above, p. 63 n. 1; Inchtuthil and Fendoch, see below, pp. 69 ff.; Dealgin Ross, *P.S.A.S.* 58 (1923–4), 326; Bochastle, W. A. Anderson, *T. Glasgow A. S.*, N.S., 14 (1956), 35.

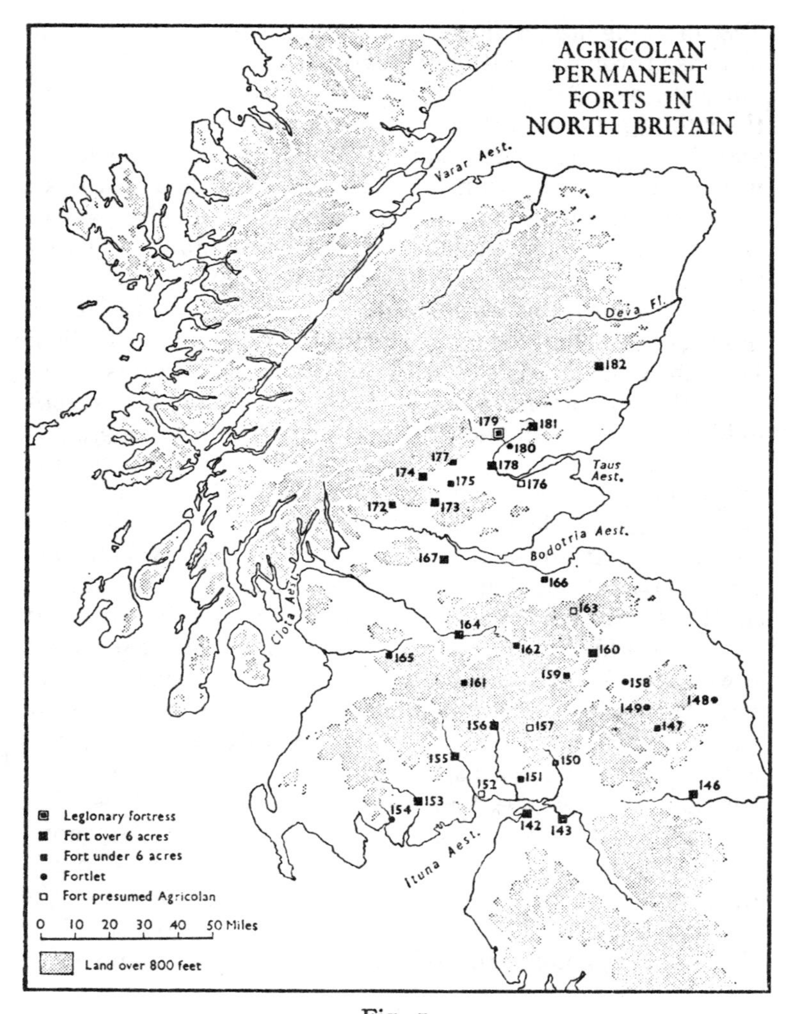

AGRICOLAN PERMANENT FORTS IN NORTH BRITAIN
Varar Aest.
Deva Fl.
Taus Aest.
Bodotria Aest.
Clota Aest.
Ituna Aest.
182
179
181
180
177
178
174
175
176
172
173
167
166
163
164
162
160
165
159
161
158
149
148
156
157
147
155
150
151
152
146
153
154
142
143
Legionary fortress
Fort over 6 acres
Fort under 6 acres
Fortlet
Fort presumed Agricolan
0 10 20 30 40 50 Miles
Land over 800 feet

Fig. 7.

auxiliary fort have been recovered, while at Inchtuthil remarkable details of an unfinished legionary fortress have come to light, illustrating the measures taken by Agricola and his successor in the furthest north and the fate which befell their arrangements.

Fendoch,[1] at the mouth of the Sma' Glen (see Plate VIII), through which the river Almond emerges from the Highlands, guards an important route to the upper Tay valley, later taken by the Military Road designed by General Wade. The fort, squeezed on to a flat and isolated hummock, is not quite regular and measures 598 by 320 feet over its rampart, 17 feet wide, and is further defended by a double or single ditch. The internal buildings, all timber-framed, comprise ten barracks for the ten *centuriae* of a milliary cohort, two kit-stores, a headquarters building, commandant's house, a pair of granaries, a hospital, and three minor sheds. Ovens lie at the back of the rampart. There is an annexe of fair size, for convoys and settlers. A signal-tower[2] lies to north, with a commanding view of the narrow glen and a wide view eastwards towards the west end of Strathmore. No road of permanent construction had yet reached the fort before it was systematically dismantled, after a short occupation.

At Inchtuthil[3] the story is more complicated and the site much larger. The Roman works occupy an isolated plateau on the north bank of the Tay, 11 miles north of Perth, at the first point outside the Dunkeld gorge where there is a good defensible site raised above the flood-plain of the river and large enough to contain not only a 50-acre legionary fortress, 1,500 feet square, but the camp of its

[1] I. A. Richmond and J. McIntyre, *P.S.A.S.* 73 (1938–9), 110 ff.

[2] This was first recognized by Mr. C. H. Millar.

[3] An early report in *P.S.A.S.* 36 (1901–2), 182. It has been systematically excavated by I. A. Richmond and J. K. St Joseph since 1952. See the annual summaries in *J.R.S.* 43 (1953) and subsequent years. The final report will be published by Dr. St Joseph. For the plan see Fig. 9.

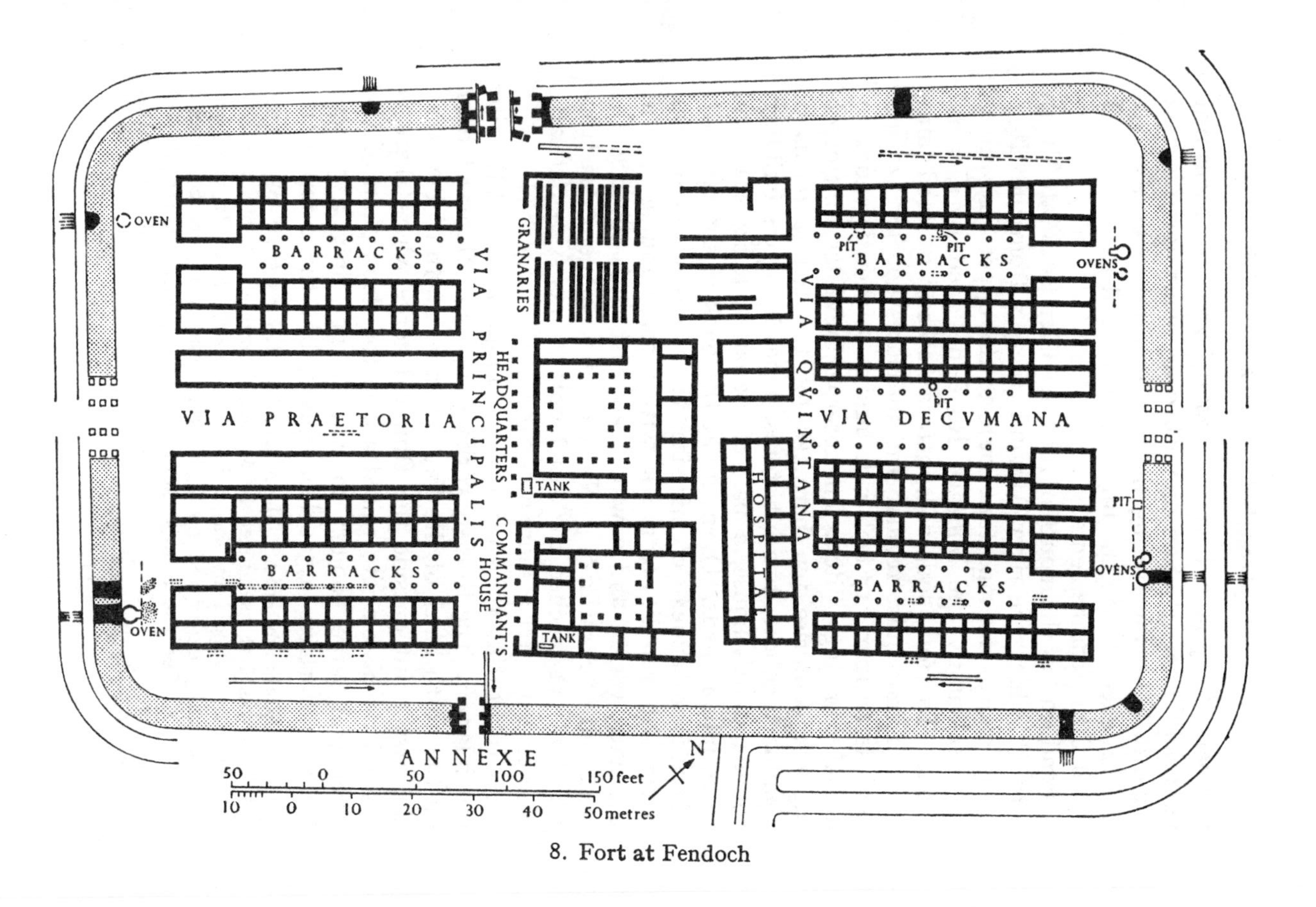

8. Fort at Fendoch

builders, the stores compound, and the residential compound for senior officers supervising the work. The sure Roman military reconnaissance which picked the sites of Vetera or Vindonissa is once more visible; here it vindicates, as at so many other sites, the testimony to Agricola's eye for *opportunitates locorum*. Praise that may have been a commonplace here measures up to the facts.

The fortress was unfinished and its demolition-layers are associated with fresh bronze coins of A.D. 86 or 87. It was equipped almost entirely with timber-framed buildings, erected with seasoned standard timbers brought to the spot. These comprise 64 large barracks, for the 54 *centuriae* of nine quingenary and ten *centuriae* of one milliary cohort, four tribune's houses, six large granaries, a great hospital (*valetudinarium*), a construction shop (*fabrica*), a drill-hall (*basilica exercitatoria*), and headquarters building (*principia*): also 180 store-rooms fronted by colonnades on the main streets. All the accommodation for the legionaries and their daily affairs had thus been erected. The defences, comprising originally a massive turf rampart, single 20-foot ditch, and elaborate timber gateways, had, however, been remodelled by substituting a stone defensive wall throughout the 2,000-yard perimeter, while outside the fortress the builders' camp, now abandoned, had been cut by a cross-dyke more rigidly defining the extra-mural area. Work was in full progress when demolition began. Ovens were built but not all yet fired, the site for the legionary legate's residence was being prepared, though the building itself was not yet erected, a bath-house in the residential compound awaited the installation of its hot-water system. Two more tribune's houses and probably two more granaries still had to be erected, while the area behind and around the *principia* remained undeveloped. Evidently, headquarters staff had not yet moved out of the fortress from which the legion had been drawn to occupy Inchtuthil. The identity of the legion itself is as yet unattested by any

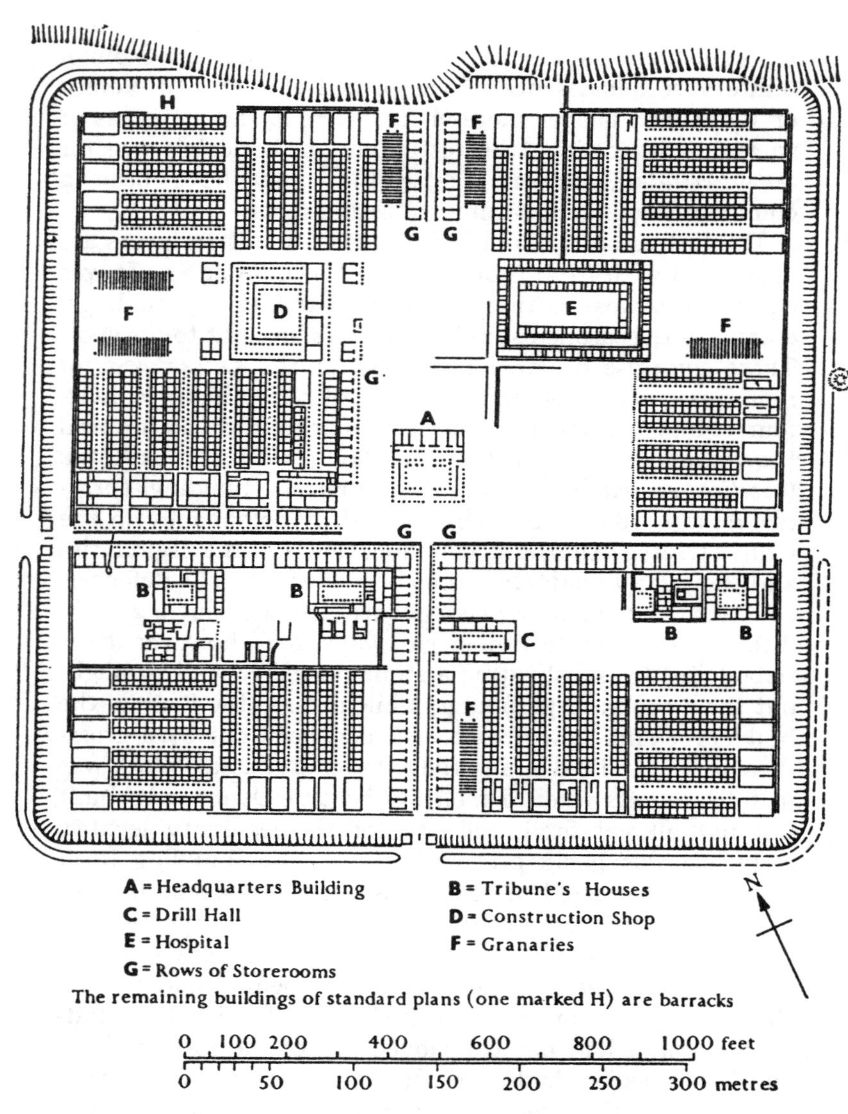

9. Legionary fortress at Inchtuthil

relic at Inchtuthil itself, but it may be regarded as virtually certain that the garrison was in fact the Twentieth Legion.

The demolition is well attested everywhere within the fortress. The post-holes of the main uprights of the timber-framing, whether in the gateways or the internal buildings, show abundant signs of removal, while the foundation trenches in which they are set are accompanied by a litter of nails bent by withdrawal with a claw or claw-hammer. At the south gate the planking dividing tower and rampart was allowed to collapse after the tower-posts had been withdrawn, and was then burnt. In the *principia* the post-holes at first left empty by withdrawal of the posts were found to be filled with burnt wattle, which had been washed into them from adjacent bonfires of that material. The stone main drain from the hospital was found to have been filled tightly with gravel after stripping of its top, while the stone fortress-wall had been systematically taken down with many stones laid on the berm. On the *via principalis* the pottery and glass-ware from stores had been dumped in the gutter of the colonnade and deliberately pounded into minute fragments. In the residential compound stone walls had been demolished and their empty foundation-trenches choked with broken flue-tiles and large fragments of concrete floor. The cold-room of the bath-house had been stripped of its flagging, leaving only the impress on the concrete bedding, while the walls had been demolished and stripped of their unused flue-tiles. Finally, a surplus of about one million nails of all sizes and types had been consigned to a pit in the construction-shop and buried, in order to prevent them from falling into the hands of the enemy. The evident intention was to leave not an empty fortress but a *tabula rasa*. Seldom, if ever, has clearer evidence for systematic and final evacuation been obtained. On the other hand, the fortress as planned conveys an impression of an intended permanent occupation no less striking than that offered by the planning of Fendoch.

The difference between Inchtuthil and Fendoch is manifestly that while at Fendoch the fort was completely occupied and used for a short period as a going concern, at Inchtuthil the main body of troops was still in the act of moving in, while preparations to receive the headquarters staff had hardly begun. It is indeed self-evident that *castella* could in fact be erected and equipped far more rapidly than a legionary fortress, which covered well over ten times their average area and contained far more complicated buildings. But in terms of time, if the evacuation of Inchtuthil was taking place very soon after A.D. 86, it must be supposed that the decision to plant a fortress here was Agricola's, even if the bulk of its construction was achieved under his successor, just as the decision to create the fortress of Chester was probably made by Frontinus, Agricola's predecessor. Two seasons' work is undoubtedly represented by the changes wrought in the defences and the residential compound, while continuous occupation of the plateau must be envisaged in order to protect the works under construction. It is therefore possible to recognize in the permanent works, completed and moving towards completion, Agricola's plan for the occupation of Scotland.

The logicality of that plan is as striking as its simplicity. The Highlands, whose western seaboard is so graphically and incisively described[1] and whose central plateau must have seemed so uninviting to reconnaissance parties on such heights as Schichallion, were to be left unoccupied and isolated by blocking-forts established in all the principal adits. The essential corollary was to deny to their war-bands any position outside the Highland plateau which could be used as a spring-board for raids. It is instructive to observe how, over a century later, the strategy of Severus in the same area was directed to exactly the same end, namely, to destroy any possibility of a coalition between the Highland Caledonii and the Maeatae of Angus

[1] c. 10, 6–7.

and Fortrenn. The essential requirement was to maintain on the spot a reserve striking force manifestly large enough to deal with any major invasion of Strathmore, which was in fact the spring-board for any invasion of the province from the north. This reserve was to be provided by the legion housed in the new fortress at Inchtuthil.

The success of the plan depended upon keeping it in working order: that was not to be. In 86–87 came two crushing defeats on the Danube, the second during a Roman counter-attack which miscarried. The next counterblow was not mounted until 88 and was rewarded by the victory of Tapae. It was either then or in preparation for the Suebo–Sarmatic war of 92–93 that the transfer of *Legio II Adiutrix* from Britain took place. There is indeed no doubt that the legion took part in the war of 92;[1] but a centurion from it was decorated also in the war concluded in 89.[2] How does this fit evidence from Inchtuthil? The coins of A.D. 86–87 were indubitably in mint condition when dropped and their connexion is with the demolition of the fortress. This all takes place so suddenly that it must be connected with an emergency change of plan. If the Twentieth Legion was now precipitately called back with no thought of return, as the very thorough demolition implies, the reason can only be that *Legio II Adiutrix* had already left Chester and was to be replaced at once. On the whole this would seem to fit 87 as the moment of transference better than a date as late as 90. For the demolition at Inchtuthil can only be understood as the sequel and not the preliminary to the departure of the legion.

With the evacuation of Inchtuthil goes the demolition of the fort at Fendoch, less graphically but no less certainly attested; and this carries with it automatically all else to the north-east. What happened was the abandonment of Strathmore, that is, the jettisoning of the strategical plan upon which the real security of the northern frontier

[1] *ILS* 2719. [2] Ibid. 9193.

depended. Not all Scotland was now evacuated. There is good reason to think that the system of auxiliary forts continued to be held as far north as Strathearn until after the turn of the century. It is the primary abandonment, however, which fits the bitter phrase *perdomita Britannia et statim missa* which helps to make up the catalogue of disasters in the west with which Tacitus opens the *Histories*. In the chronology of that list it is placed fairly and squarely between the disasters of 69–70 and the Suebo-Sarmatic invasion of 92. This gives a proper force and point to the whole phrase: for to Agricola, whose achievement is summed up in the proud *perdomita Britannia*, it must have meant that the very keystone of his frontier structure had been removed, and there was time before his death in 93 for him to have communicated his feelings to Tacitus. The context shows quite certainly that the phrase can have nothing to do with the evacuation of all Scotland about 105, quite apart from the difficulty which *statim* would then produce. *Statim*, in the context of a legionary fortress and its entire command abandoned before it was even finished, wins a hitherto unappreciated immediacy.

7 · THE ARMY OF AGRICOLA

(i) *The Legions and Auxiliaries*

The Claudian army of occupation in Britain, following the invasion of A.D. 43, comprised four legions, *II Augusta*, *XIV Gemina*, *XX Valeria*, from the Rhineland, and *IX Hispana* from Pannonia on the Danube frontier. Each legion is to be reckoned as some 5,000 to 6,000 strong. With them went the auxiliaries, that is, infantry *cohortes* 500 or 1,000 strong,[1] sometimes part-mounted, and cavalry *alae* normally 500 strong. These must be calculated at not less

[1] But see E. Birley, *Corolla Memoriae E. Swoboda*, 1966, pp. 54 ff.

than the same strength, representing at least some 40 units and in due course certainly more.

Until the time of Agricola only two major factors of change had operated, apart from the steady increase in the size of the occupied area: firstly, the Boudiccan revolt, which must have been followed by a considerable reorganization of the provincial garrison; even Tacitus notes 2,000 replacements in *Legio IX* and among auxiliaries a supplement of eight cohorts and two *alae* (*Ann.* 14. 38). Secondly came the Civil Wars, which called for both troop movements and subsequent reorganization. Eight cohorts of Batavi (*Hist.* 1. 59) and *Legio XIV* went from Britain to Italy in A.D. 68 (*Hist.* 2. 11), the legion returning to Britain for a few months in 69 (*Hist.* 2. 66) and then withdrawn for good in A.D. 70 (*Hist.* 4. 68). In A.D. 71 *Legio II Adiutrix p*(*ia*) *f*(*idelis*) had taken its place.[1]

The permanent fortresses of the legions had been established by Agricola's time as follows. *Legio II Augusta* lay at Caerleon (*Isca*), founded under Frontinus about A.D. 75. *Legio II Adiutrix*, having been quartered at Lincoln from 71 until at least 76 (*RIB* 258), was transferred to Chester, where it was building its fortress under Agricola in 79.[2] Tombstones of this legion have been found both at Lincoln (*RIB* 253, 258) but more especially at Chester, where thirteen are certain (*RIB* 475–87). *Legio IX Hispana* lay at York, as inscriptions attest (*RIB* 659, 665, 673, 680). *Legio XX*, having been quartered first at Colchester (*RIB* 200) and in A.D. 49 at Gloucester (Tac. *Ann.* 12. 32; *RIB* 122), has left a tombstone at Wroxeter (*RIB* 293), where it presumably succeeded *Legio XIV* after 70, at the time

[1] The title *Adiutrix* was probably chosen because it had been formed from crews of the fleet who were not citizens, and was therefore not constituted in the usual manner, but in the fashion of auxiliaries (to 'aid' the regular troops). The titles *pia fidelis* rewarded the prompt adhesion of the marines to Vespasian.

[2] A Bath tombstone presumably indicates a visitor for health or pleasure (*RIB* 157).

when Agricola was its legate. When *Legio II Adiutrix* left Chester, *Legio XX* succeeded it, having begun meanwhile to build a new northern fortress at Inchtuthil (see above).

Of the auxiliaries it is impossible to compile a nominal list. But certain units deserve note. Tacitus names Tungrians and Batavians (c. 36, 1) in the battle of Mons Graupius. According to the best MS. four Batavian cohorts were engaged. This large number recalls the eight cohorts of Batavians, *quartae decimae legionis auxilia* (*Hist.* 1. 59, 64), which after Nero's fall joined the revolt of Civilis (*Hist.* 4. 19), and seem to have been cashiered by Vespasian. Thus, the Batavians serving under Agricola were probably new creations.[1] A *cohors I Batavorum quingenaria* was in Britain in A.D. 124; but it seems more probable that these were *miliariae* and are to be recognized in those numbered I, II, and III (all *miliariae*), which by the end of the first century or later were in other provinces. Tacitus further mentions two Tungrian cohorts, raised from the district of Tongres (*Aduatuca Tungrorum*) in Belgian Limburg. These were *miliariae* and are doubtless the two Tungrian cohorts (*Hist.* 2. 14), active on the Roman side throughout the Civil War and identical with the *cohortes I* and *II Tungrorum miliaria equitata*, which formed part of the army of Britain until the fourth and third centuries respectively. The *cohors Usiporum* (c. 28, 1; 32, 3), which mutinied during its initial training, can hardly be counted on the strength. Finally, three passages mention British auxiliaries (c. 29, 2; 32, 1, 3), though without specifying actual formations. When it is considered what a strain Agricola had already put upon his auxiliary quota (see below), it becomes more than likely that these were newly recruited British units. As Professor Birley has observed, one might be the *cohors I Brittonum miliaria*,[2] to which

[1] Probably already in the attack upon Mona, employing their *patrius nandi usus* (c. 18, 4).

[2] *Roman Britain and the Roman Army*[2] (1961), p. 21.

the *cohors II Brittonum miliaria* could properly be added. This would make up the eight thousand infantry present at the battle. Nothing specific is stated of cavalry, except that at least ten *alae quingenariae* were employed. So far as Britons were concerned, however, it is unlikely that cavalry were present, since this was not an arm in which British recruiting was strong.

(ii) *The Garrison Posts*

If, however, no contemporary list of Agricola's army has come down to us, archaeological information has a good deal to tell of Agricolan garrison posts, the *praesidia* and *castella* by which the native communities were enmeshed or cordoned. If a list of these is compiled, it becomes clear that while at least 40 were required to garrison northern Britain up to and including the Tyne–Solway isthmus, about 30 more were needed to reach and hold the Forth–Clyde line and 12 more to block the Highland gates, apart from holding the coastal positions necessary for supplies. These are minimal figures, and something like a 10 per cent. increase would not be impossible, but the minimum total is 82. If it is asked whence these troops came, it is not irrelevant to compare a map of what is known, or may fairly be inferred, of the garrison of Britain in 78, before the northern advances had taken place. It will then be clear that an approximately equivalent strength had been required for the garrisoning of the Midlands and the South, though it seems certain that here the minimum figure of 76 is at present liable to a greater increase than in the north, where knowledge of fort-sites is better. It thus becomes evident that Agricola's northern occupation was attained only at the cost of stripping the South, and the significance of his emphasis upon civilization becomes the more telling: the policy *ut homines . . . in bella faciles quieti et otio per voluptates adsuescerent* has just as much meaning for the military as for the civil programme (see Fig. 2 on p. 54).

The other side of the picture is that of engineers' supplies. It is to be considered that the construction of an auxiliary fort for a *cohors miliaria*, as at Fendoch, demanded for the defences 42 gateway timbers, each not less than 30 feet long, 1,650 feet of corduroy for the rampart-walk and the same length of hurdling for the parapet; while for the internal buildings the requirement in round figures was 11,000 feet, or just over two miles, of timber-framed wattle and daub walling. At Inchtuthil the earthwork defences, as first designed, required 82 gateway timbers and a mile and a quarter of corduroy rampart-walk and wattle parapet, while the *barracks alone* demanded 71,680 feet or over 13½ miles of timber-framed walling. When, again, it is considered that between A.D. 79 and 84 over eighty forts had to be built and that while many were smaller than Fendoch not a few were substantially larger, the scale of requirements in provision, transport, and erection of supplies of seasoned timber begins to emerge. These questions of supply never emerge in the biography, for it was not the Roman convention to reveal them. But they constitute a quantum of organization such as no previous governor in Britain had needed to exercise. This is the special tribute which archaeology can pay to the genius of a famous commander-in-chief.

8 · THE MANUSCRIPTS OF THE *AGRICOLA*

(i) *History*

The history of the rediscovery of the minor works of Tacitus in the Middle Ages is still obscure. Tacitus, by comparison with Cicero or Livy, was little read, but there existed a manuscript of the *Germania* at the great library of Fulda, in Hessen, in the ninth century because Ruodolfus, a monk of Fulda (d. 865), quotes extensively from *Germ.* 4 in his

Translatio S. Alexandri (*Mon. Germ. Hist.*, Scriptores, ii. 675). It is probable that there was also a manuscript of the *Agricola* at Fulda in this period since Adam of Bremen (*c.* 1075) appears to adapt *Agr.* 11 in the opening chapters of his history of the church of Hamburg (*Patrol. Lat.*, 146, 459 ff.). There is, however, clear evidence of a manuscript of the *Agricola* at the Abbey of Monte Cassino. In the prologue to *Vita S. Severi episcopi et confessoris* (*c.* 1135), Peter the Deacon, the Librarian of Monte Cassino, quotes freely from the first two chapters of the *Agricola*.[1] Monte Cassino had close links with Fulda and it is a reasonable hypothesis (but no more than an hypothesis) that the Monte Cassino *Agricola* was ultimately derived from Fulda.

The *Agricola* was unknown to humanist scholars at the beginning of the fifteenth century, but from about 1422 onwards rumours circulated that there were unpublished Roman historical manuscripts in Germany. In particular Poggio received information about such manuscripts from a monk from Hersfeld (near Fulda) and wrote to Niccolò Niccoli on 3 November 1425: 'inter ea volumina est Iulius Frontinus et aliqua opera Cornelii Taciti nobis ignota'. These manuscripts seem to be the same as those that Panormita in a letter to Guarino in April 1426 mentions as having recently been located; he lists the *Germania*, the *Agricola, quidam dyalogus de oratore . . . ut coniectamus, Cor. Taciti, Sex Iulii Frontonis* [sic] *de aquaeductibus, item eiusdem Frontonis liber alter*, and *liber Suetonii Tranquilli de grammaticis et rhetoribus*. In the following years there are references in Poggio's correspondence to his attempts to secure these manuscripts from the Hersfeld monk, but his efforts were unavailing. Niccolò, however, persevered and in 1431 drew up with Poggio's aid a catalogue (*commentarium*)[2] of manuscripts which were thought to be in

[1] H. Bloch, *Class. Phil.* 36 (1941), 185.

[2] See Sabbadini, *Storia e Critica*, 1 ff.; R. P. Robinson, *Class. Phil.* 16 (1921), 251.

five German and Danish monasteries, including Hersfeld, and gave a copy of it to Cardinal Giuliano Cesarini, who was about to visit Germany to organize attacks on the Hussites. It was hoped that, as an ecclesiastical dignitary, the Cardinal might get access to the monastic libraries and their treasures. The catalogue lists with detailed specifications the following works at Hersfeld (*monasterio hispildensi*): *Julii Frontini De aquae ductis . . . liber .j., item eiusdem frontini liber*, a codex containing the *Germania* (xij folia), the *Agricola* (xiiij folia), the *Dialogus de Oratoribus* (xviij folia), Suetonius *De grammaticis et rhetoribus*, and Ammianus Marcellinus. No more is heard of this codex or of the *Agricola* until 1455 when Pier Candido Decembrio[1] noted that he had seen at Rome in that year a codex comprising the *Germania* ('foliorum xii in columnellis'), the *Agricola* ('opus foliorum decem et quattuor in columnellis'), the *Dialogus* ('opus foliorum xiiii in columnellis, post hec deficiunt sex folia . . . post hec sequuntur folia duo cum dimidio'), and Suetonii Tranquilli *de grammaticis et rhetoribus liber*. This specification of the number of leaves and the order of the works agrees with the description of the codex in Niccolò's catalogue, except that Niccolò gives 18 leaves to the *Dialogus*; Decembrio gives 16½ but he has evidently disregarded blank leaves which represented the lacuna in the archetype.[2] There is thus no doubt that the Hersfeld MS. did eventually reach Rome and that it was this manuscript that Decembrio saw.

New evidence has come to light which explains how the manuscript came to Rome and what happened to it subsequently.[3] In 1475 or 1476 Poggio's son, Jacopo, made

[1] Ambros. R 88 Sup. fol. 112.

[2] A note in the Codex Vindobonensis of the *Dialogus*, written in 1466, confirms that the lacuna was 1½ leaves: *hic est defectus unius folii cum dimidio*. Robinson, *Germania*, 12–13.

[3] N. Rubinstein, *Italia Medioevale e Umanistica* 1 (1958), 383 ff.; B. L. Ullman, ibid. 2 (1959), 309 ff.; G. Brugnoli, *Riv. Cult. Class. e Med.* 3 (1961), 68 ff.

a revision of Niccolò's catalogue of 1431, listing manuscripts which were still unsecured and noting those that had been discovered. The list was addressed to Filippo Sagromoro, Milanese orator in Florence, but was evidently meant for the attention of the Duke of Milan. In it Jacopo gives the same description of the Hersfeld manuscripts as given in the 1431 catalogue but adds: 'verum nostris temporibus Pii Pontificis opera in Italiam venit Suetonius hic: et Cornelius de situ et origine germanorum; et de oratoribus. Sed corruptus et laceratus'. That is to say, the Hersfeld codex, in poor condition and *lacking the Agricola* came to light again after 1458 (Aeneas Sylvius was elected Pope, as Pius II, in August 1458). This is precisely confirmed by a note (f. 1^{v}) written by Jovianus Pontanus in March 1460 in the manuscript which he had copied of the *Dialogus*, the *Germania*, and Suetonius, *de Grammaticis* (= Leidensis xviii Periz. Q. 21): 'hos libellos Iovianus Pontanus excripsit nuper adinventos et in lucem relatos ab Enoc Asculano, quamquam satis mendosos'. Elsewhere (f. 47^{v}) he notes that the Suetonius reappeared only shortly after the death of Bartolomeo Fazio, which occurred in the autumn of 1457. It is thus clear that Enoch of Ascoli, who was commissioned by Pope Nicolas V (1447–55) to collect manuscripts,[1] brought the codex from Hersfeld. He returned with manuscripts from Germany in 1455 and it was in that year that Decembrio saw it in Rome. Thereafter the Hersfeld codex disappears from view, presumably because Calixtus III, who succeeded to the Papacy in 1455, was antipathetic to Humanism so that Enoch was forced to adopt other measures to find a purchaser. Enoch himself died in December 1457, and when the Hersfeld codex turned

[1] On Enoch see, most recently, J. Perret, *Recherches sur le texte de la 'Germanie'*, 135 ff. The new evidence lessens the doubts of Robinson (*Germania*, 351 ff.), Till, Mendell (*Tacitus*, 248 ff.), and others who thought it unlikely that Enoch was responsible for bringing the MS. from Hersfeld to Italy.

up again in the following year it had lost the leaves that contained the *Agricola*.[1] Presumably Enoch had detached the *Agricola* between January 1456 and December 1457 for separate sale; and this would be easy to do if the codex was a double codex and the *Agricola* formed the last section of the first half (see below).

It is not known what became of the rest of the manuscript but copies[2] were made of the *Agricola* before it passed into the hands of Stefano Guarnieri, chancellor of Perugia from 1466, who with his brother Francesco founded the library at Jesi. The exact date at which Guarnieri acquired the manuscript is uncertain, but is likely to have been between 1462, when Guarnieri is first heard of as interested in manuscripts, and 1474, when *T* was copied: it is probably significant that Guarnieri was in Rome in 1472. The condition of the manuscript had, however, deteriorated further by then, for Guarnieri was obliged to supply a new beginning and end. The *editio princeps* of Tacitus (Bononiensis, 1472)[3] did not include the *Agricola*, which was first printed by Puteolanus = Francisco dal Pozzo (Milan, 1475–80?).

(ii) *The Manuscripts*

There are four extant manuscripts of the *Agricola*.

1. *E* and *e* = *Codex Aesinas* (*Aesinas Latinus* 8, library of Count Balleani in Jesi). The manuscript is made up as follows: ff. 1–51 Dictys Cretensis, *Bellum Troianum* (of these ff. 1, 5–8, 11–40 are in Carolingian minuscule, 2 was in Carolingian minuscule but has been erased, 2^r being left blank and 2^v containing a second copy of the prologue of

[1] This disposes of Perret's theory (op. cit. 142 ff.) that the Hersfeld codex was not dismembered but that Pius II being uninterested in the *Agricola* arranged only for the copying of the other three works.

[2] The sources of *AB* and of *e*; see below, p. 86, where it is argued that the *codex Aesinas* must be identical with the *Agricola* section of the Hersfeld MS.

[3] Perret, op. cit. 17 ff.

the *Bellum Troianum* in a fifteenth-century hand, ff. 3–4, 9–10, and 51 are in a fifteenth-century hand); ff. 52–65 Tacitus, *Agricola* (of these ff. 56–63 are in Carolingian minuscule (c. 13, 1 *munia*—c. 40, 2 *missum*) = *E*, ff. 52–55, 64–65 are in a fifteenth-century hand = *e*); ff. 66–75 Tacitus, *Germania* (of these f. 69 is a palimpsest, which contains the erased text of *Agricola* 40, 2–43, 4 *testamento* following after f. 63ᵛ in Carolingian minuscule, and a secondary text of the *Germania* from 14. 1 *tueri* in a fifteenth-century hand; ff. 66–68 and 70–75 are in a fifteenth-century hand), f. 76 blank (Carolingian minuscule erased. The text was a continuation of the primary script of the *Agricola* after 69ᵛ to the end of the work).[1] The Carolingian hands of Dictys and of the *Agricola* are not necessarily the same but are probably from the same scriptorium. They show affinity with the style of Fulda,[2] and are to be dated *c*. 830–50. The fifteenth-century hand is the same throughout and has been identified with that of Stefano Guarnieri. The *Bellum Troianum* was more or less complete and required only the recopying of the opening leaves, which had become defaced. The surviving Carolingian part of the *Agricola*, however, consists only of a single quaternion and a final *unio* immediately following the quaternion and Guarnieri had to make a new text of the first few leaves and also erased the final *unio* and recopied the text on two new

[1] For a detailed account of the MS. see Robinson, *Germania*, 14 ff.; Mendell, *Tacitus*, 258–9. It was discovered by Sr. Cesare Annibaldi in 1902 and published by him in 1907. A photographic reproduction was published by R. Till in 1943 (*Handschriftliche Untersuchungen zu Tacitus Agricola und Germania*).

[2] Fulda would be appropriate if the exemplar of *E* was in Insular script (see 24, 2 n.) since Fulda was founded by St. Boniface, and his native script is said to have been practised there until the middle of the ninth century. Perret (op. cit. 89 ff.) has argued, on the strength of the variant *Arbonae* in *Germ.* 1, 2, that the exemplar was copied at St. Gall but the evidence is insufficient; Norden (*Sitzb. d. preuss. Akad. d. Wiss.*, phil.-hist. Klasse, 1927, p. 19) postulated a Lombardic origin for it.

leaves. It can be calculated that the complete Carolingian text of the *Agricola* filled 14 leaves but the first few leaves were probably missing by the time that Guarnieri obtained it, so that he had to supply the missing portion from one of the copies, like the exemplar of *AB*, made soon after 1458. The assumption that the Carolingian text was entire and that Guarnieri merely recopied the first few leaves because they were damaged or faded is difficult to square with the following facts. The Carolingian colophon in *E* (f. 76^{v}) reads *de vita Iulii Agricolae*, whereas Guarnieri concluded his text with *de vita et moribus Iulii Agricolae*, a humanist title which is found also in *A* and *B*. Again, Guarnieri wrote down to and including the word *munia* in f. 55^{v} although the Carolingian text begins with *munia* in f. 56.[1] The final *unio* indicates that the Carolingian *codex* either ended with the *Agricola* or was a double *codex*, the first half of which ended with the *Agricola*.

It is thus virtually certain that the *Agricola* section of the Aesinas represents the surviving portion of the Hersfeld MS. brought to Rome by Enoch. The number of leaves (14) agrees with the description of Niccolò and Decembrio and the script points to the vicinity of Fulda. Furthermore, it is clear that the *Agricola* was detached from the Hersfeld MS. after reaching Italy.[2]

So far as the text itself is concerned, it must be observed that *E* was copied from an exemplar which was either in *scriptura continua* or did not have the division of words satisfactorily resolved (cf., e.g., c. 15, 1 *inter prae tando*) and which was already heavily corrupt. It is probable that that

[1] There is also an inexplicable erasure at the end of f. 55^{v}.

[2] Mendell (*Tacitus*, 281 ff.), noting that the *Bellum Troianum* and the *Agricola* were written in the same *scriptorium* at the same period, has attempted to identify the Aesinas with the other manuscript of the *Agricola* known to have been at Monte Cassino on the basis of an entry in the Monte Cassino catalogue (*c.* 1080): *Historiam Cornelii* (i.e. *Taciti*) *cum Omero* (=Dictys, *Bellum Troianum*). '*Omero*' undoubtedly refers to Dares. See now H. Bloch, *A.J.P.* 69 (1948), 74 ff.

exemplar was itself derived from a late classical manuscript in uncials which contained 13 letters to the line see c. 22, 2 n., 30, 1 n., 32, 3 n.).[1] *E* was further corrected by four hands, the scribe himself (= E^1) who made small changes of word-division and deletion or addition of letters, a contemporary corrector (= E^2) who made numerous corrections in the text and orthography and who added some 54 marginal variants, and by Stefano Guarnieri and another fifteenth-century scholar whose contributions are insignificant and have not been recorded. The great majority of the marginal variants cannot be independent conjectures by E^2 himself, first because they are too good for ninth-century conjecture (e.g. c. 19, 4 *auctionem E*: *exactionem* E^{2m}), but, above all, because they are themselves corrupt and must therefore have been copied from corrupted earlier transmission (e.g. c. 36, 2 *foedare et tratis E* : *foede recti trates vel traces* E^{2m}). Nor are they likely to be alternative interpretations of a script which was difficult to read. It is possible, then, that E^2 had access to a second manuscript of the *Agricola*, distinct from the exemplar of *E* and from which he derived the variants.[2] Alternatively, the exemplar of *E* may itself have been contaminated and E^2 has copied out variants from it.[3] But it is important to realize that in either case the readings of E^2 may be regarded as having the same authority as *E* itself.[4]

[2. *T* = Toletanus 49, 2, in the Capitular Library of Toledo. The manuscript is made up as follows: ff. 1–15 *Germania*, 16–36 *Agricola*, 37–63 Io. Antonii Campani Oratio, 64–66 fragment of an oration, 66–221 Letters of

[1] Perret, op. cit. 76 ff. It is tempting to think of the copies of the historian's works ordered to be made for public libraries by his namesake, the Emperor Tacitus. But that may be a fiction.

[2] So Frahm, *Hermes* 69 (1934), 426 ff.

[3] But see Perret, op. cit. 103 ff.

[4] For a full discussion of the sources of error and of the variants in *E* see Robinson, *Germania*, 30 ff.; Till, op. cit. 26 ff.; Perret, op. cit. 98 ff.

Pliny, 222–3 fragment of an oration. There is a subscription after the oration of Antonius: 'scripta per me M. Angelum Crullum Tudertem fulginii pu. scribam Non. Decembr. MCCCCLXXIIII'. The manuscript was discovered in 1897 and published by Leuze, *Philologus*, Supp. Band 8 (1899–1901), 515–56, and by F. F. Abbott, *Decennial Publications, University of Chicago* 6 (1903–4).

Although used by all previous editors *T* is a direct copy of *Ee* and therefore has no primary authority. It has been disregarded in this edition.[1]]

3. *A* = Vaticanus Latinus 3429. A paper manuscript of the *Agricola* only, bound in with a copy of the editio *Veneta Spirensis* of Tacitus (which lacked the *Agricola*). It has 14 leaves. On the first blank page it has the following note: 'Cornelio Tacito della Vita d'Agricola, scritto di mano di Pomponio Laeto, ligato dietro al Tacito stampato. Ful. Urs.' Julius Pomponius Laetus (1425–97) was a distinguished humanist but it is not known from what source he acquired a copy of the text of the *Agricola* nor when he wrote *A* except that it must have been between 1473, the latest date for the *Spirensis*,[2] and 1497. It is almost certainly the manuscript which formed the basis of the notes on the *Agricola* which Fulvio Orsini = Ursinus (1529–1600) published in 1595 (*Fragmenta Historicorum . . . alios*, Antwerp, pp. 460–2) and for which he cited the evidence of a *vetus codex*. It was collated by C. L. Urlichs for his edition (1875) and by Andresen, *Woch. f. klass. Phil.*, 1900, 1299 ff.

4. *B* = Vaticanus Latinus 4498. A fine parchment manuscript of the second half of the fifteenth century which is made up as follows: ff. 1–20 Frontinus, *De aquaeductibus*, 20–35 Rufus, *De provinciis*, 36–45 Suetonius, *De grammaticis*, 45^{v}–63 'C. Plinii Secundi *de viris illustribus*', 63^{v}–77 *Agricola*, 78–97 *Dialogus*, 97^{v}–109 *Germania*, 109^{v}–110 M. Junius Nypsus, *De mensuris*, 111–12 *De ponderibus*,

[1] See Mendell, *Tacitus*, 259–60 for a description of *T*.

[2] See Perret, op. cit. 117 n. 1.

112v–18 Seneca, *Apocolocyntosis*, 119–45 Censorinus, *De die natali*. Collations of *B* have been published by Urlichs and Andresen (vid. sup.).

The precise relationship between *A* and *B* and their relationship with *Ee* is still a matter of dispute. *A* and *B* have certain errors and omissions in common which indicate that they are closely related: e.g. c. 3, 3 *senectutis*; 9, 5 *suis*; 15, 4 om. *felicibus*; 27, 2 om. *se victos*. *B* has a number of omissions and individual errors (e.g. c. 4, 3 om. *in*; 9, 2 *temporis et curarum*; 14, 1 om. *et*; 23 om. *gloria*) which prove that it cannot have been the exemplar of *A*. Perret (R.É.L. 28 (1950), 374) argues that *B* is in fact an emended copy of *A*. Such an hypothesis is difficult to establish, because it is impossible to control what degree of emendation is to be allowed as conceivable. *A* does have certain individual errors not shared by *B* (e.g. c. 10, 3 om. *in*; 16, 1 *Voadicca*; 18, 1 *uterentur*; 33, 4 *aius*; 36, 3 *convinnarii*; 45, 1 *charus*) but these are errors which might be rectified by intelligent conjecture. They are not demonstrably separative. On balance, however, we incline to the opinion that *A* and *B* are derived from a common hyparchetype. What is certain is that both *A* and *B* are ultimately derived from *E*.[1] They share a large number of common errors with *E* (e.g. c. 36, 3 (*a*)*egra diu aut stante*) and, more significantly, choose between the readings of *E* and the marginal variants of E^{2m} (e.g. c. 13, 2 *praecipue EA*: *praeceptum* E^{2m} $A^{m}B$; c. 16, 4 *facta exercitus licentia ducis salute* EA^{m}: *pacti exercitus licentiam dux salutem* E^{2m} *AB*). But if *AB* are indirectly derived from *E*, they cannot be accorded the status of primary manuscripts and they have, therefore, been omitted from the *apparatus* for that section in which *E* is extant, except where they contain conjectural readings which deserve consideration. To print the readings of *A* and *B* along with *E* would enable the reader to form an estimate of the relationship between the ninth-century

[1] At c. 30, 1 *A* and *B* omit *coistis* where *E* is almost illegible.

original and the fifteenth-century copies and so would assist him in evaluating the readings of *ABe* where *E* is missing. This information is, however, readily available in the *apparatus criticus* of other editors (e.g. Lenchantin, Koestermann) and is apt to be more misleading than enlightening. Where only *e* is extant, we have also cited the readings of *A* and *B*. In these sections, however, we have not regularly cited *B*'s readings where *Ae* agree against it. This is contrary to the practice of recent editors but it is clear from the stemma that, whether *B* is an emended copy of *A* or whether *A* and *B* are independently derived from a common hyparchetype, *B*'s *variae lectiones* cannot be authoritative: they are either errors or conjectures, and the conjectures are only cited, as are other conjectures, on their merits.

Note on Orthography

E does not usually assimilate compounds in *ad-*, *in-*, *con-*, *sub-*, etc.; that is, it reads, e.g., *adsequor*, not *assequor*. It is not possible to recover with certainty what Tacitus' own practice was but we have printed the unassimilated forms throughout for which there is manuscript evidence. *E* also prefers the form *-es* for both the nominative and accusative plural of the third declension; *-is* is found in the following places: nom. c. 37, 1 *expertis*; acc. c. 14, 1 *regis*, c. 15, 2 *centurionis*, c. 18, 4 *navis*, c. 25, 3 *hostis*. In the following places E^2 corrects *E*'s original *-is* to *-es*: nom. c. 14, 1 *civitates* (c. 19, 4), c. 18, 2 *Ordovices*, c. 22, 3 *eruptiones*, c. 25, 3 *incolentes*, c. 26, 2 *hostes*, c. 27, 1 *omnes*, c. 13, 3 *gentes*; acc. c. 15, 4 *coniuges*, c. 35, 4 *legiones*. There is some evidence for the nom. in *-is* under the Republic and the acc. in *-is* is well attested from the second half of the first century A.D. but to avoid confusion we have followed *E*'s preference. See G. Tingdal, *Ändelsen -is ackus. plur. hos de efteraugusteiska författarne* (Göteborg, 1916); G. Andresen, *Woch. f. klass. Phil.*, 1916, 1203 ff. Cf. also c. 39, 1 *crines*.

AGRICOLA

SIGLA

E = codicis Aesinatis (Latin. 8) vetus pars (sc. c. 13, 1 *munia*—c. 40, 2 *missum*), saec. IX ineuntis; vide p. 85.
E^2 = corrector eiusdem aetatis.
E^{2m} = variae lectiones a correctore E^2 in margine adscriptae.
E^2 *s.l.* = variae lectiones a correctore E^2 super lineam additae.

e = codicis Aesinatis reliqua pars a St. Guarnierio rescripta, *circ.* a. 1462–74; vide p. 85.
e^m = variae lectiones in margine a St. Guarnierio adscriptae.
e^c = corrector nescioquis eiusdem aetatis.

A = codex Vaticanus, Romae (Vaticanus Latinus 3429), saec. XV exeuntis.
A^m = variae lectiones a prima manu in margine adscriptae.
A^c = corrector nescioquis.

B = codex Vaticanus, Romae (Vaticanus Latinus 4498), saec. XV exeuntis.

codd. = consensus codicum *eAB*.

T = codex Toletanus (Bibl. Capitular. 49. 2), anno 1474. Testimonia codicis *T*, qui apographon est codicis Aesinatis, perraro sunt citata.

Put. = Franciscus Puteolanus in editione sua (Mediolanii, *circ.* a. 1475–80).

Rhen. = Beatus Rhenanus in ed. 1533 sive in ed. 1544 lectionem instituens.

Lips. = Iustus Lipsius, qui septem editiones inter a. 1574 et a. 1600 prodidit, lectionem instituens.

† indicat locum corruptum

[ ] includunt additamenta quae reicienda esse censemus.

⟨ ⟩ includunt supplementa quae credimus necessaria.

CORNELII TACITI
DE VITA
IVLII AGRICOLAE

LIBER

CLARORVM virorum facta moresque posteris tradere, anti- **1**
quitus usitatum, ne nostris quidem temporibus quamquam
incuriosa suorum aetas omisit, quotiens magna aliqua ac
nobilis virtus vicit ac supergressa est vitium parvis mag-
5 nisque civitatibus commune, ignorantiam recti et invidiam.
sed apud priores ut agere digna memoratu pronum magisque 2
in aperto erat, ita celeberrimus quisque ingenio ad proden-
dam virtutis memoriam sine gratia aut ambitione bonae
tantum conscientiae pretio ducebatur. ac plerique suam ipsi 3
10 vitam narrare fiduciam potius morum quam adrogantiam
arbitrati sunt, nec id Rutilio et Scauro citra fidem aut
obtrectationi fuit: adeo virtutes iisdem temporibus optime
aestimantur, quibus facillime gignuntur. at nunc narraturo 4
mihi vitam defuncti hominis venia opus fuit, quam non petis-
15 sem incusaturus: tam saeva et infesta virtutibus tempora.

Legimus, cum Aruleno Rustico Paetus Thrasea, Herennio **2**
Senecioni Priscus Helvidius laudati essent, capitale fuisse,
neque in ipsos modo auctores, sed in libros quoque eorum
saevitum, delegato triumviris ministerio ut monumenta
20 clarissimorum ingeniorum in comitio ac foro urerentur.

Cornelii Taciti de vita Iulii Agricolae liber incipit *e*: Cornelii Taciti de vita et moribus Iulii Agricolae *A*: Cai Cornelii Taciti de vita et moribus Iulii Agricolae prohemium *B* *Vide subscriptionem*

15 *post* incusaturus *distinxit Wex*: *sine distinctione codd.* 17 Senecioni *Rhen.*: Senetioni *codd.*

2 scilicet illo igne vocem populi Romani et libertatem senatus et conscientiam generis humani aboleri arbitrabantur, expulsis insuper sapientiae professoribus atque omni bona arte in exilium acta, ne quid usquam honestum occurreret.
3 dedimus profecto grande patientiae documentum; et sicut 5 vetus aetas vidit quid ultimum in libertate esset, ita nos quid in servitute, adempto per inquisitiones etiam loquendi audiendique commercio. memoriam quoque ipsam cum voce perdidissemus, si tam in nostra potestate esset oblivisci quam tacere. 10

3 Nunc demum redit animus; et quamquam primo statim beatissimi saeculi ortu Nerva Caesar res olim dissociabiles miscuerit, principatum ac libertatem, augeatque cotidie felicitatem temporum Nerva Traianus, nec spem modo ac votum securitas publica, sed ipsius voti fiduciam ac robur 15 adsumpserit, natura tamen infirmitatis humanae tardiora sunt remedia quam mala; et ut corpora nostra lente augescunt, cito extinguuntur, sic ingenia studiaque oppresseris facilius quam revocaveris: subit quippe etiam ipsius inertiae dulcedo, et invisa primo desidia postremo amatur. 20
2 quid, si per quindecim annos, grande mortalis aevi spatium, multi fortuitis casibus, promptissimus quisque saevitia principis interciderunt, pauci et, ut ⟨sic⟩ dixerim, non modo aliorum sed etiam nostri superstites sumus, exemptis e media vita tot annis, quibus iuvenes ad senectutem, senes 25 prope ad ipsos exactae aetatis terminos per silentium
3 venimus? non tamen pigebit vel incondita ac rudi voce memoriam prioris servitutis ac testimonium praesentium bonorum composuisse. hic interim liber honori Agricolae soceri mei destinatus, professione pietatis aut laudatus erit 30 aut excusatus.

11 et *codd.*: sed *Halm* (set *Ed. Bip.*) 12 dissociabiles *A*: dissotiabiles *e*: dissolubiles *B* 22 multi *Lips.*: multis *codd.* 23 ut sic *Wölfflin*: uti *codd.*: ut ita *Urlichs* (*post Rhen. qui in notis etiam* et *delere voluit*) 28 servitutis *e* (*Ursinus*): senectutis *AB*

Gnaeus Iulius Agricola, vetere et inlustri Foroiuliensium **4**
colonia ortus, utrumque avum procuratorem Caesarum
habuit, quae equestris nobilitas est. pater illi Iulius Graecinus senatorii ordinis, studio eloquentiae sapientiaeque
5 notus, iisque ipsis virtutibus iram Gai Caesaris meritus:
namque Marcum Silanum accusare iussus et, quia abnuerat,
interfectus est. mater Iulia Procilla fuit, rarae castitatis. in 2
huius sinu indulgentiaque educatus per omnem honestarum
artium cultum pueritiam adulescentiamque transegit. arce-
10 bat eum ab inlecebris peccantium praeter ipsius bonam
integramque naturam quod statim parvulus sedem ac
magistram studiorum Massiliam habuit, locum Graeca
comitate et provinciali parsimonia mixtum ac bene compositum. memoria teneo solitum ipsum narrare se prima in 3
15 iuventa studium philosophiae acrius, ultra quam concessum
Romano ac senatori, hausisse, ni prudentia matris incensum
ac flagrantem animum coercuisset. scilicet sublime et
erectum ingenium pulchritudinem ac speciem magnae
excelsaeque gloriae vehementius quam caute adpetebat.
20 mox mitigavit ratio et aetas, retinuitque, quod est difficillimum, ex sapientia modum.

Prima castrorum rudimenta in Britannia Suetonio **5**
Paulino, diligenti ac moderato duci, adprobavit, electus
quem contubernio aestimaret. nec Agricola licenter, more
25 iuvenum qui militiam in lasciviam vertunt, neque segniter
ad voluptates et commeatus titulum tribunatus et inscitiam
rettulit: sed noscere provinciam, nosci exercitui, discere
a peritis, sequi optimos, nihil adpetere in iactationem, nihil
ob formidinem recusare, simulque et anxius et intentus
30 agere. non sane alias exercitatior magisque in ambiguo 2

1 CNeus Tulius *e* 3 illi *Wölfflin*: Iuli *eB*: Iulii *A*: *om. Lips.*
6 Silanum *Lips.*: Sillanum *eA*: Sullanum *B* 15 acrius *codd.*:
ac iuris *Pichena* ultra *codd.*: ultraque *Lips.* 16 senatorio *C*,
Heraeus: ac senatori *secl. Gudeman* 19 cautius *Nipperdey*
30 exercitatior *e* (*addito* x, *de quo vide Till, Handschr. Unter-suchungen, 17*), *AB*: excitatior *Buchner*

Britannia fuit: trucidati veterani, incensae coloniae, intersaepti exercitus; tum de salute, mox de victoria certavere.
3 quae cuncta etsi consiliis ductuque alterius agebantur, ac summa rerum et recuperatae provinciae gloria in ducem cessit, artem et usum et stimulos addidere iuveni, intravit- 5 que animum militaris gloriae cupido, ingrata temporibus quibus sinistra erga eminentes interpretatio nec minus periculum ex magna fama quam ex mala.

6 Hinc ad capessendos magistratus in urbem degressus Domitiam Decidianam, splendidis natalibus ortam, sibi 10 iunxit; idque matrimonium ad maiora nitenti decus ac robur fuit. vixeruntque mira concordia, per mutuam caritatem et in vicem se anteponendo, nisi quod in bona uxore
2 tanto maior laus, quanto in mala plus culpae est. sors quaesturae provinciam Asiam, proconsulem Salvium Titia- 15 num dedit, quorum neutro corruptus est, quamquam et provincia dives ac parata peccantibus, et proconsul in omnem aviditatem pronus quantalibet facilitate redempturus esset mutuam dissimulationem mali. auctus est ibi filia, in subsidium simul ac solacium; nam filium ante sublatum brevi 20
3 amisit. mox inter quaesturam ac tribunatum plebis atque ipsum etiam tribunatus annum quiete et otio transiit, gnarus sub Nerone temporum, quibus inertia pro sapientia fuit.
4 idem praeturae tenor et silentium; nec enim iurisdictio obvenerat. ludos et inania honoris medio rationis atque 25 abundantiae duxit, uti longe a luxuria ita famae propior.
5 tum electus a Galba ad dona templorum recognoscenda diligentissima conquisitione fecit ne cuius alterius sacrilegium res publica quam Neronis sensisset.

7 Sequens annus gravi vulnere animum domumque eius 30

1 intersepti *codd.*: intercepti *Put.* 9 degressus *eA*: digressus *A*[c]*B* 19 auctus . . . filia *e*[m]*AB*: nactus . . . filiam *e* 20 ac *e*: et *AB* 22 transiit *B*: transit *eA* 24 tenor *Rhen.*: certior *codd.* 25 medio rationis *eA*: medio luxuriae *B*: modo rationis *Put.*: medio moderationis *Gudeman* 28 fecit *codd.*: effecit *N. Heinsius*

adflixit. nam classis Othoniana licenter vaga dum Intimilios
(Liguriae pars est) hostiliter populatur, matrem Agricolae
in praediis suis interfecit, praediaque ipsa et magnam
patrimonii partem diripuit, quae causa caedis fuerat. igitur 2
5 ad sollemnia pietatis profectus Agricola, nuntio adfectati
a Vespasiano imperii deprehensus ac statim in partes transgressus est. initia principatus ac statum urbis Mucianus re-
gebat, iuvene admodum Domitiano et ex paterna fortuna
tantum licentiam usurpante. is missum ad dilectus agendos 3
10 Agricolam integreque ac strenue versatum vicesimae legioni
tarde ad sacramentum transgressae praeposuit, ubi decessor
seditiose agere narrabatur: quippe legatis quoque consularibus nimia ac formidolosa erat, nec legatus praetorius ad
cohibendum potens, incertum suo an militum ingenio. ita
15 successor simul et ultor electus rarissima moderatione maluit
videri invenisse bonos quam fecisse.

Praeerat tunc Britanniae Vettius Bolanus, placidius quam **8**
feroci provincia dignum est. temperavit Agricola vim suam
ardoremque compescuit, ne incresceret, peritus obsequi eru-
20 ditusque utilia honestis miscere. brevi deinde Britannia 2
consularem Petilium Cerialem accepit. habuerunt virtutes
spatium exemplorum, sed primo Cerialis labores modo et
discrimina, mox et gloriam communicabat: saepe parti exercitus in experimentum, aliquando maioribus copiis ex eventu
25 praefecit. nec Agricola umquam in suam famam gestis 3
exultavit; ad auctorem ac ducem ut minister fortunam
referebat. ita virtute in obsequendo, verecundia in praedicando extra invidiam nec extra gloriam erat.

Revertentem ab legatione legionis divus Vespasianus inter **9**
30 patricios adscivit; ac deinde provinciae Aquitaniae prae-

1 Intimilios *Richmond–Ogilvie*: Intemelios *Lips.*: in templo *codd.*: Intimilium *Mommsen* 5 affectati *AB*: affecti *e*: tati $e^{s.l.}$ 9 dilectus *Lips.*: delectus *codd.* 11 ubi decessor *AB*: ubi c̣ụṃ decessor *e*: ubi . . . narrabatur *susp. Wex* 14 potens *eAB*: potuit e^{m} 17 Bolanus *B*: Volanus *eA* 19 obsequi $e^{c}e^{m}AB$: obsequii *e*, *Ritter* 21 habuit virtutis exemplar e^{m} 27 in exsequendo *Voss.*

posuit, splendidae inprimis dignitatis administratione ac spe
2 consulatus, cui destinarat. credunt plerique militaribus ingeniis subtilitatem deesse, quia castrensis iurisdictio secura et obtusior ac plura manu agens calliditatem fori non exerceat: Agricola naturali prudentia, quamvis inter togatos, 5
3 facile iusteque agebat. iam vero tempora curarum remissionumque divisa: ubi conventus ac iudicia poscerent, gravis intentus, severus et saepius misericors: ubi officio satis factum, nulla ultra potestatis persona; tristitiam et adrogantiam et avaritiam exuerat. nec illi, quod est rarissimum, 10
aut facilitas auctoritatem aut severitas amorem deminuit.
4 integritatem atque abstinentiam in tanto viro referre iniuria virtutum fuerit. ne famam quidem, cui saepe etiam boni indulgent, ostentanda virtute aut per artem quaesivit: procul ab aemulatione adversus collegas, procul a contentione adversus procuratores, et vincere inglorium et atteri sordidum 15
5 arbitrabatur. minus triennium in ea legatione detentus ac statim ad spem consulatus revocatus est, comitante opinione Britanniam ei provinciam dari, nullis in hoc ipsius sermonibus, sed quia par videbatur. haud semper errat fama; 20
6 aliquando et eligit. consul egregiae tum spei filiam iuveni mihi despondit ac post consulatum collocavit, et statim Britanniae praepositus est, adiecto pontificatus sacerdotio.

10 Britanniae situm populosque multis scriptoribus memoratos non in comparationem curae ingeniive referam, sed 25
quia tum primum perdomita est. ita quae priores nondum comperta eloquentia percoluere rerum fide tradentur.
2 Britannia, insularum quas Romana notitia complectitur

9 nulla . . . persona *Rhen.*: nullam . . . personam (m *alteram postea expunxit*) *e*: nullam . . . personam *AB* 10 exuerat *codd.*: effugerat *E. Wolff*: et avaritiam *secl. C. Heraeus*: tristitiam . . . exuerat *secl. Wex* 11 deminuit *e*, *Lips.*: diminuit *AB* 13 fuerit *AB*: fieret fuerit (fieret *postea expunxit*) *e* 14 ostentanda *Rhen.*: ostentandam *codd.* quaesiit *e* 19 ipsius *e*: suis *AB* 21 eligit *e* (*Rhen.*): elegit *AB* consul egregiae *Put.*: consul* (*eras.*) graeciae *e*: consul grate *B*: consul graeciae *A*: gratae *A*[m] 26 ita quae *AB*: itaque *e* 27 fide *eAB*: fides *e*[m]

maxima, spatio ac caelo in orientem Germaniae, in occidentem Hispaniae obtenditur, Gallis in meridiem etiam inspicitur; septentrionalia eius, nullis contra terris, vasto atque aperto mari pulsantur. 3 formam totius Britanniae 5 Livius veterum, Fabius Rusticus recentium eloquentissimi auctores oblongae scapulae vel bipenni adsimulavere. et est eă facies citra Caledoniam, unde et in universum fama: sed transgressis inmensum et enorme spatium procurrentium extremo iam litore terrarum velut in cuneum tenuatur. 10 4 hanc oram novissimi maris tunc primum Romana classis circumvecta insulam esse Britanniam adfirmavit, ac simul incognitas ad id tempus insulas, quas Orcadas vocant, invenit domuitque. dispecta est et Thule, quia hactenus iussum et hiems adpetebat. 5 sed mare pigrum et grave 15 remigantibus perhibent ne ventis quidem perinde adtolli, credo quod rariores terrae montesque, causa ac materia tempestatum, et profunda moles continui maris tardius impellitur. 6 naturam Oceani atque aestus neque quaerere huius operis est, ac multi rettulere: unum addiderim, nus- 20 quam latius dominari mare, multum fluminum huc atque illuc ferre, nec litore tenus adcrescere aut resorberi, sed influere penitus atque ambire, et iugis etiam ac montibus inseri velut in suo.

11 Ceterum Britanniam qui mortales initio coluerint, in- 25 digenae an advecti, ut inter barbaros, parum compertum. habitus corporum varii atque ex eo argumenta. 2 namque rutilae Caledoniam habitantium comae, magni artus Ger-

2 etiam *e* (*Put.*): et *AB* 6 oblongae scapulae *Richmond–Ogilvie*: oblongae scutulae *eA*: oblongae scupulae *B*: oblongo scutulo *Lacey* 7 fama: sed transgressis *Peerlkamp* (*post Doederlein, Feilitzsch*): fama est transgressis unde et universis fama sed *e*, *sed verba* unde . . . sed *per lineam atramento dissimili ductam seclusa* (*de quo vide Till, Handschr. Untersuchungen, 85*), al(ia)s *super* unde *scripto*: fama est transgressis sed *AB*: unde et universis fama *add. A*[m]: fama est transgressa sed *Rhen.* 8 enorme *T*, *Rhen.*: inorme *eB*: īnorme *A* 12 Orcadas *AB*: orchadas *e* 15 perinde *Grotius*: proinde *codd.* 20 dominari (dn̄ari) *eA*[m]: damnari *AB*

manicam originem adseverant; Silurum colorati vultus, torti plerumque crines et posita contra Hispania Hiberos veteres traiecisse easque sedes occupasse fidem faciunt; proximi Gallis et similes sunt, seu durante originis vi, seu procurrentibus in diversa terris positio caeli corporibus 5
3 habitum dedit. in universum tamen aestimanti Gallos vicinam insulam occupasse credibile est. eorum sacra deprehendas ⟨ac⟩ superstitionum persuasionem; sermo haud multum diversus, in deposcendis periculis eadem audacia
4 et, ubi advenere, in detrectandis eadem formido. plus 10 tamen ferociae Britanni praeferunt, ut quos nondum longa pax emollierit. nam Gallos quoque in bellis floruisse accepimus; mox segnitia cum otio intravit, amissa virtute pariter ac libertate. quod Britannorum olim victis evenit: ceteri manent quales Galli fuerunt. 15

12 In pedite robur; quaedam nationes et curru proeliantur. honestior auriga, clientes propugnant. olim regibus parebant, nunc per principes factionibus et studiis ⟨dis⟩tra-
2 huntur. nec aliud adversus validissimas gentes pro nobis utilius quam quod in commune non consulunt. rarus dua- 20 bus tribusve civitatibus ad propulsandum commune periculum conventus: ita singuli pugnant, universi vincun-
3 tur. caelum crebris imbribus ac nebulis foedum; asperitas frigorum abest. dierum spatia ultra nostri orbis mensuram; nox clara et extrema Britanniae parte brevis, ut finem atque 25
4 initium lucis exiguo discrimine internoscas. quod si nubes non officiant, aspici per noctem solis fulgorem, nec occidere et exsurgere, sed transire adfirmant. scilicet extrema et plana terrarum humili umbra non erigunt tenebras, infraque
5 caelum et sidera nox cadit. solum praeter oleam vitemque 30

2 Hispania *Muretus*: hispaniam *eAB*: hiberia e^m Hiberos *eA*: hiberas A^c: iberas *B* 4 vi *Rhen.*: usu *codd.* 7 vicinam *eAB*: vacuam e^m 8 ac *add. Glück, G. F. Schömann* persuasionem *Glück* (*qui et* persuasiones *proposuit*), *Ritter*: persuasione *eAB* 12 in bellis *AB*: bellis *e* 18 distrahuntur *N. Heinsius*: trahuntur *codd.* 21 tribusve *B*: tribusque *eA*

et cetera calidioribus terris oriri sueta patiens frugum pecudumque fecundum: tarde mitescunt, cito proveniunt;
eademque utriusque rei causa, multus umor terrarum caeli-
que. fert Britannia aurum et argentum et alia metalla, 6
5 pretium victoriae. gignit et Oceanus margarita, sed subfusca ac liventia. quidam artem abesse legentibus arbi-
trantur; nam in rubro mari viva ac spirantia saxis avelli,
in Britannia, prout expulsa sint, colligi: ego facilius crediderim naturam margaritis deesse quam nobis avaritiam.

10 Ipsi Britanni dilectum ac tributa et iniuncta imperii **13**
munia impigre obeunt, si iniuriae absint: has aegre tole-
rant, iam domiti ut pareant, nondum ut serviant. igitur
primus omnium Romanorum divus Iulius cum exercitu
Britanniam ingressus, quamquam prospera pugna terruerit
15 incolas ac litore potitus sit, potest videri ostendisse posteris,
non tradidisse. mox bella civilia et in rem publicam versa 2
principum arma, ac longa oblivio Britanniae etiam in pace:
consilium id divus Augustus vocabat, Tiberius praeceptum.
agitasse Gaium Caesarem de intranda Britannia satis con-
20 stat, ni velox ingenio mobili paenitentiae, et ingentes
adversus Germaniam conatus frustra fuissent. divus 3
Claudius auctor tanti operis, transvectis legionibus auxiliis-
que et adsumpto in partem rerum Vespasiano, quod initium
venturae mox fortunae fuit: domitae gentes, capti reges et
25 monstratus fatis Vespasianus.

Consularium primus Aulus Plautius praepositus ac sub- **14**
inde Ostorius Scapula, uterque bello egregius: redactaque
paulatim in formam provinciae proxima pars Britanniae,
addita insuper veteranorum colonia. quaedam civitates
30 Cogidumno regi donatae (is ad nostram usque memoriam

1 pecudumque fecundum *Leuze, Lundström*: pecudumque *e*: fecundum $e^{m}AB$ 5 subfusa e^{m} 10 dilectum e^{c}: delectum *eAB* 11 munia *Ee* (*sed expunxit*): munera *AB* *Hinc incipiunt vetera codicis E folia* 18 praeceptum E^{2m}: praecipue *E* 22 auctor tanti *Bezzenberger*: auctoritate *E*: auctor iterati *Wex*: auctor *Put.* 24 domitae *Put.*: domitiae *E* gentes E^{2}: gentis *E* 30 Togidumno E^{2m}

fidissimus mansit), vetere ac iam pridem recepta populi Romani consuetudine, ut haberet instrumenta servitutis et 2 reges. mox Didius Gallus parta a prioribus continuit, paucis admodum castellis in ulteriora promotis, per quae fama aucti officii quaereretur. Didium Veranius excepit, isque intra 5 3 annum extinctus est. Suetonius hinc Paulinus biennio prosperas res habuit, subactis nationibus firmatisque praesidiis; quorum fiducia Monam insulam ut vires rebellibus ministrantem adgressus terga occasioni patefecit.

15 Namque absentia legati remoto metu Britanni agitare 10 inter se mala servitutis, conferre iniurias et interpretando accendere: nihil profici patientia nisi ut graviora tamquam 2 ex facili tolerantibus imperentur. singulos sibi olim reges fuisse, nunc binos imponi, e quibus legatus in sanguinem, procurator in bona saeviret. aeque discordiam praeposito- 15 rum, aeque concordiam subiectis exitiosam. alterius manus centuriones, alterius servos vim et contumelias miscere. 3 nihil iam cupiditati, nihil libidini exceptum. in proelio fortiorem esse qui spoliet: nunc ab ignavis plerumque et imbellibus eripi domos, abstrahi liberos, iniungi dilectus, 20 tamquam mori tantum pro patria nescientibus. quantulum enim transisse militum, si et se Britanni numerent? sic Germanias excussisse iugum: et flumine, non Oceano de- 4 fendi. sibi patriam coniuges parentes, illis avaritiam et luxuriam causas belli esse. recessuros, ut divus Iulius 25 recessisset, modo virtutem maiorum suorum aemularentur. neve proelii unius aut alterius eventu pavescerent: plus impetus felicibus, maiorem constantiam penes miseros esse. 5 iam Britannorum etiam deos misereri, qui Romanum ducem absentem, qui relegatum in alia insula exercitum detine- 30 rent; iam ipsos, quod difficillimum fuerit, deliberare. porro

1–2 vetere . . . ut haberet *Rhen.*: ut vetere . . . haberet *E* (-sit ut ve- *in ras.*) 3 reges *Rhen.*: regis *E* parta priore E^{2m} 14 e quibus E^2: et quibus *E* 17 centuriones *Rhen.*: centurionis *E* miscere *E*: ciere E^{2m} 20 diripi *N. Heinsius* 22 et se *E*: sese E^2

in eius modi consiliis periculosius esse deprehendi quam audere.

His atque talibus in vicem instincti, Boudicca generis **16** regii femina duce (neque enim sexum in imperiis discernunt)
5 sumpsere universi bellum; ac sparsos per castella milites consectati, expugnatis praesidiis ipsam coloniam invasere ut sedem servitutis, nec ullum in barbaris ingeniis saevitiae genus omisit ira et victoria. quod nisi Paulinus cognito 2 provinciae motu propere subvenisset, amissa Britannia
10 foret; quam unius proelii fortuna veteri patientiae restituit, tenentibus arma plerisque, quos conscientia defectionis et proprius ex legato timor agitabat, ne quamquam egregius cetera adroganter in deditos et ut suae cuiusque iniuriae ultor durius consuleret. missus igitur Petronius Turpilianus 3
15 tamquam exorabilior et delictis hostium novus eoque paenitentiae mitior, compositis prioribus nihil ultra ausus Trebellio Maximo provinciam tradidit. Trebellius segnior et nullis castrorum experimentis, comitate quadam curandi provinciam tenuit. didicere iam barbari quoque ignoscere
20 vitiis blandientibus, et interventus civilium armorum praebuit iustam segnitiae excusationem: sed discordia laboratum, cum adsuetus expeditionibus miles otio lasciviret. Trebellius, fuga ac latebris vitata exercitus ira, indecorus 4 atque humilis, precario mox praefuit, ac velut pacta
25 exercitus licentia, ducis salute, [et] seditio sine sanguine stetit. nec Vettius Bolanus, manentibus adhuc civilibus 5 bellis, agitavit Britanniam disciplina: eadem inertia erga hostes, similis petulantia castrorum, nisi quod innocens

3 Boudicca *Haase*: uo adicca *E*: bouid icta *E*2m 11 tenentibus *E*: tenentibus tamen *Ritter* 12 ne quamquam *Anon. ap. Walch*: nequaquam *E*: nequam *E*2m 13 cuiusque *Wex*: eiusque *E*: quisque *Nipperdey* 14 durius *E*: dubius *E*2m 15 novus eoque (eo *s.l.*) *E*2; novusque *E* 22 lasciviret *E*2: lascivi sed *E* 24 praefuit *E*2: praebuit *E* pacta (*Nipperdey*: facta *E*) exercitus licentia ducis salute *E*: pacti exercitus licentiam dux salutem *E*2m 25 et *secl. John* 26 Bolanus *B*: volanus *E*

Bolanus et nullis delictis invisus caritatem paraverat loco auctoritatis.

17 Sed ubi cum cetero orbe Vespasianus et Britanniam recuperavit, magni duces, egregii exercitus, minuta hostium spes. et terrorem statim intulit Petilius Cerialis, Brigantum 5 civitatem, quae numerosissima provinciae totius perhibetur, adgressus. multa proelia, et aliquando non incruenta; magnamque Brigantum partem aut victoria amplexus est 2 aut bello. et Cerialis quidem alterius successoris curam famamque obruisset: subiit sustinuitque molem Iulius 10 Frontinus, vir magnus quantum licebat, validamque et pugnacem Silurum gentem armis subegit, super virtutem hostium locorum quoque difficultates eluctatus.

18 Hunc Britanniae statum, has bellorum vices media iam aestate transgressus Agricola invenit, cum et milites velut 15 omissa expeditione ad securitatem et hostes ad occasionem verterentur. Ordovicum civitas haud multo ante adventum eius alam in finibus suis agentem prope universam obtri- 2 verat, eoque initio erecta provincia. et quibus bellum volentibus erat, probare exemplum ac recentis legati ani- 20 mum opperiri, cum Agricola, quamquam transvecta aestas, sparsi per provinciam numeri, praesumpta apud militem illius anni quies tarda et contraria bellum incohaturo, et plerisque custodiri suspecta potius videbatur, ire obviam discrimini statuit; contractisque legionum vexillis et modica 25 auxiliorum manu, quia in aequum degredi Ordovices non audebant, ipse ante agmen, quo ceteris par animus simili 3 periculo esset, erexit aciem. caesaque prope universa gente, non ignarus instandum famae ac, prout prima cessissent, terrorem ceteris fore, Monam insulam, cuius possessione 30

4 minuta *E*²: minutae *E* 5 Brigantum *Put.*: Bregantum *E* 8 Brigantum *Put.*: Bregantum *E* 10 subiit *E*² *s.l.*: *om. E* 12 gentem armis *E*²: armis gentem *E* 15 aestate *E*²: atate *E* 17 verterentur *B*: uterentur *E* haut *E*²: aut *E* 27 animus *E*²: animo *E* 30 cuius *E*²: cumius *E*: a cuius *Ed. Bip.*

revocatum Paulinum rebellione totius Britanniae supra memoravi, redigere in potestatem animo intendit. sed, ut in 4 subitis consiliis, naves deerant: ratio et constantia ducis transvexit. depositis omnibus sarcinis lectissimos auxili-
5 arium, quibus nota vada et patrius nandi usus quo simul seque et arma et equos regunt, ita repente inmisit, ut obstupefacti hostes, qui classem, qui naves, qui mare expectabant, nihil arduum aut invictum crediderint sic ad bellum venientibus. ita petita pace ac dedita insula clarus 5
10 ac magnus haberi Agricola, quippe cui ingredienti provinciam, quod tempus alii per ostentationem et officiorum ambitum transigunt, labor et periculum placuisset. nec 6 Agricola prosperitate rerum in vanitatem usus expeditionem aut victoriam vocabat victos continuisse; ne laureatis
15 quidem gesta prosecutus est, sed ipsa dissimulatione famae famam auxit, aestimantibus quanta futuri spe tam magna tacuisset.

Ceterum animorum provinciae prudens, simulque doctus **19** per aliena experimenta parum profici armis, si iniuriae
20 sequerentur, causas bellorum statuit excidere. a se suisque 2 orsus primum domum suam coercuit, quod plerisque haud minus arduum est quam provinciam regere. nihil per libertos servosque publicae rei, non studiis privatis nec ex commendatione aut precibus centurionem militesve
25 adscire, sed optimum quemque fidissimum putare. omnia 3 scire, non omnia exsequi. parvis peccatis veniam, magnis severitatem commodare; nec poena semper, sed saepius paenitentia contentus esse; officiis et administrationibus potius non peccaturos praeponere, quam damnare cum
30 peccassent. frumenti et tributorum exactionem aequalitate 4 munerum mollire, circumcisis quae in quaestum reperta ipso

4 transvexit *E*: tranvexit *E*[2] 13 prosperitate *E*[2]: speritate *E* 14 continuisse ne *E*[2]: continuit sine *E* 21 primum *B*: primam *E* 23 privatis *E*: privatus *E*[2m] 24–25 militesve ascire *Wex*: milites scire *E*: Ne *add.* *E*[2] *s.l.* 30 exactionem *E*[2m]: auctionem *E* aequalitate *E*[2]: mae qualitate *E*

tributo gravius tolerabantur. namque per ludibrium adsidere clausis horreis et emere ultro frumenta ac luere pretio cogebantur. divortia itinerum et longinquitas regionum indicebatur, ut civitates proximis hibernis in remota et avia deferrent, donec quod omnibus in promptu erat paucis lucrosum fieret. 5

20 Haec primo statim anno comprimendo egregiam famam paci circumdedit, quae vel incuria vel intolerantia priorum 2 haud minus quam bellum timebatur. sed ubi aestas advenit, contracto exercitu multus in agmine, laudare modestiam, disiectos coercere; loca castris ipse capere, aestuaria ac silvas ipse praetemptare; et nihil interim apud hostes quietum pati, quo minus subitis excursibus popularetur; atque ubi satis terruerat, parcendo rursus invitamenta pacis 3 ostentare. quibus rebus multae civitates, quae in illum diem ex aequo egerant, datis obsidibus iram posuere et praesidiis castellisque circumdatae, tanta ratione curaque, ut nulla ante Britanniae nova pars inlacessita transierit. 10 15

21 Sequens hiems saluberrimis consiliis absumpta. namque ut homines dispersi ac rudes eoque in bella faciles quieti et otio per voluptates adsuescerent, hortari privatim, adiuvare publice, ut templa fora domos extruerent, laudando promptos, castigando segnes: ita honoris aemulatio pro neces- 2 sitate erat. iam vero principum filios liberalibus artibus erudire, et ingenia Britannorum studiis Gallorum anteferre, ut qui modo linguam Romanam abnuebant, eloquentiam concupiscerent. inde etiam habitus nostri honor et frequens toga; paulatimque discessum ad delenimenta vitiorum, 20 25

1 adsidere *E*²: adsedere *E* 2 luere *E*: ludere *AB*: recludere *Hutter* 8 incuria *E*²: sine curia *E* 14 invitamenta *Acidalius*: irritamenta *E*²: inritamenta *E* 17 circumdatae tanta *E*: circumdatae et tanta *E*²: circumdatae sunt, tanta *Dronke* 18 *ante* inlacessita *add.* pariter *Fröhlich*, perinde *Ritter*, sic *Ernesti* 19 absumpta *Rhen.*: adsumpta *E* 20 bella *Bosius*: bello *E*: bellum *Rhen.* 20–21 et otio *E*²: inotio *E* 23 honoris aemulatio *E*²: honor et aemulatio *E* 28 discessum *E*: descensum *Pichena*

porticus et balineas et conviviorum elegantiam. idque apud imperitos humanitas vocabatur, cum pars servitutis esset.

Tertius expeditionum annus novas gentes aperuit, vastatis usque ad Taum (aestuario nomen est) nationibus. qua 5 formidine territi hostes quamquam conflictatum saevis tempestatibus exercitum lacessere non ausi; ponendisque insuper castellis spatium fuit. adnotabant periti non alium ducem opportunitates locorum sapientius legisse. nullum ab Agricola positum castellum aut vi hostium expugnatum aut 10 pactione ac fuga desertum. crebrae eruptiones; nam adversus moras obsidionis annuis copiis firmabantur. ita intrepida ibi hiems et sibi quisque praesidio, inritis hostibus eoque desperantibus, quia soliti plerumque damna aestatis hibernis eventibus pensare tum aestate atque hieme iuxta pelle- 15 bantur. nec Agricola umquam per alios gesta avidus intercepit: seu centurio seu praefectus incorruptum facti testem habebat. apud quosdam acerbior in conviciis narrabatur; ut erat comis bonis, ita adversus malos iniucundus. ceterum ex iracundia nihil supererat secretum, ut silentium 20 eius non timeres: honestius putabat offendere quam odisse.

(22; 2; 3; 4)

Quarta aestas obtinendis quae percucurrerat insumpta; ac si virtus exercitus et Romani nominis gloria pateretur, inventus in ipsa Britannia terminus. namque Clota et Bodotria diversi maris aestibus per inmensum revectae, 25 angusto terrarum spatio dirimuntur: quod tum praesidiis firmabatur atque omnis propior sinus tenebatur, summotis velut in aliam insulam hostibus. **23**

Quinto expeditionum anno nave prima transgressus igno- **24**

1 balineas *Ritter*: balinea *E*: balnea *E*[2m] 4 Taum *E*[2m]: Tanaum *E* 6 ausi *E*[2]: auxi *E* 7 castellis *E*: telis *E*[2m] 8 ab *E*[2m]: *om. E* 10 ac *E*: aut *E*[2m] crebrae eruptiones *post* ita *transposuit Perret, post* hiems *Halm* 17 conviciis *E*[2m]: convitiis *E* 18 ut erat *E*: et ut erat *Peerlkamp* 19 supererat *E*[2]: erat *E* ut *E*: ac *Wölfflin* 21 percucurrerat *E*: percurrerat *B*: *cf. 37, 1* 22 exercitus *E*: exercituum *E*[2] 26 propior *A*: proprior *E* 28 *aliquid excidisse suspicantur edd.*: Anavam *suppl. Richmond*: Itunam (*pro* nave prima) *P. E. Postgate*

tas ad id tempus gentes crebris simul ac prosperis proeliis domuit; eamque partem Britanniae quae Hiberniam aspicit copiis instruxit, in spem magis quam ob formidinem, si quidem Hibernia medio inter Britanniam atque Hispaniam sita et Gallico quoque mari opportuna valentissimam imperii 5
2 partem magnis in vicem usibus miscuerit. spatium eius, si Britanniae comparetur, angustius nostri maris insulas superat. solum caelumque et ingenia cultusque hominum haud multum a Britannia differunt; [in melius] aditus portus-
3 que per commercia et negotiatores cogniti. Agricola ex- 10 pulsum seditione domestica unum ex regulis gentis exceperat ac specie amicitiae in occasionem retinebat. saepe ex eo audivi legione una et modicis auxiliis debellari obtinerique Hiberniam posse; idque etiam adversus Britanniam profuturum, si Romana ubique arma et velut e conspectu 15 libertas tolleretur.

25 Ceterum aestate, qua sextum officii annum incohabat, amplexus civitates trans Bodotriam sitas, quia motus universarum ultra gentium et infesta hostili exercitu itinera timebantur, portus classe exploravit; quae ab Agricola pri- 20 mum adsumpta in partem virium sequebatur egregia specie, cum simul terra, simul mari bellum impelleretur, ac saepe iisdem castris pedes equesque et nauticus miles mixti copiis et laetitia sua quisque facta, suos casus adtollerent, ac modo silvarum ac montium profunda, modo tempestatum ac flu- 25 ctuum adversa, hinc terra et hostis, hinc victus Oceanus
2 militari iactantia compararentur. Britannos quoque, ut ex captivis audiebatur, visa classis obstupefaciebat, tamquam aperto maris sui secreto ultimum victis perfugium claudere-

5 valentissimam *E*: volentissimam E^{2m} 9 differunt *Rhen.*: differt *E* in melius (*E*) *secl. Wex*: melius *vel* eius *Rhen.* 11 gentis *E*: gente E^{2m} 18 Bodotriam E^{2m}: uodotriam *E* 19 hostili exercitu E^{2m}: hostilis exercitus *E* 20 exploravit E^{2}: explorabit *E* 21 virium *E*: vinum E^{2m} 22 impelleretur *Rhen.*: impellitur *E* 23 mixti *E*: mixto E^{2m} 24 adtollerent E^{2}: adtollerant *E* 25 silvarum ac montium E^{2}: ac *om.* *E* fluctuum E^{2}: fluctum *E*

tur. ad manus et arma conversi Caledoniam incolentes 3
populi magno paratu, maiore fama, uti mos est de ignotis,
oppugnare ultro castella adorti, metum ut provocantes
addiderant; regrediendumque citra Bodotriam et ceden-
5 dum potius quam pellerentur ignavi specie prudentium
admonebant, cum interim cognoscit hostis pluribus agminibus inrupturos. ac ne superante numero et peritia 4
locorum circumiretur, diviso et ipse in tres partes exercitu
incessit.

10 Quod ubi cognitum hosti, mutato repente consilio universi **26**
nonam legionem ut maxime invalidam nocte adgressi, inter
somnum ac trepidationem caesis vigilibus inrupere. iamque in ipsis castris pugnabatur, cum Agricola iter hostium
ab exploratoribus edoctus et vestigiis insecutus, velocis-
15 simos equitum peditumque adsultare tergis pugnantium
iubet, mox ab universis adici clamorem; et propinqua luce
fulsere signa. ita ancipiti malo territi Britanni; et nonanis 2
rediit animus, ac securi pro salute de gloria certabant.
ultro quin etiam erupere, et fuit atrox in ipsis portarum
20 angustiis proelium, donec pulsi hostes, utroque exercitu certante, his, ut tulisse opem, illis, ne eguisse auxilio viderentur.
quod nisi paludes et silvae fugientes texissent, debellatum
illa victoria foret.

Cuius conscientia ac fama ferox exercitus nihil virtuti **27**
25 suae invium et penetrandam Caledoniam inveniendumque
tandem Britanniae terminum continuo proeliorum cursu
fremebant. atque illi modo cauti ac sapientes prompti
post eventum ac magniloqui erant. iniquissima haec
bellorum condicio est: prospera omnes sibi vindicant,
30 adversa uni imputantur. at Britanni non virtute se victos, 2

1 incolentes *E*[2]: incolentis *E* 3 oppugnare *E*[2]: oppugnase *E* castella *E*[2m]: castellum *E* 4 Bodotriam *E*[2]: Bodotria *E* et (cedendum) *E*[2m]: excedendum *E* 10 cognitum *E*[2]: incognitum *E* 15 peditumque *E*[2]: peditum *E* 17 nonanis *E*: Romanis *Guarnieri* 18 rediit *Wex*: redit *E* 19–20 portarum angustiis *E*[2]: partarum angustis *E* 20 hostes *E*[2m]: hostis *E* 24 conscientia *E*[2]: conscientiae *E* 26 proeliorum *Rhen.*: proelium *E*

sed occasione et arte ducis rati, nihil ex adrogantia remittere, quo minus iuventutem armarent, coniuges ac liberos in loca tuta transferrent, coetibus et sacrificiis conspirationem civitatum sancirent. atque ita inritatis utrimque animis discessum. 5

28 Eadem aestate cohors Usiporum per Germanias conscripta et in Britanniam transmissa magnum ac memorabile facinus ausa est. occiso centurione ac militibus, qui ad tradendam disciplinam inmixti manipulis exemplum et rectores habebantur, tres liburnicas adactis per vim gubernatoribus ascendere; et uno remigante, suspectis duobus eoque interfectis, nondum vulgato rumore ut miraculum praevehebantur. mox ubi aquam atque utilia raptum exissent, cum plerisque Britannorum sua defensantium proelio congressi ac saepe victores, aliquando pulsi, eo ad extremum inopiae venere, ut infirmissimos suorum, mox sorte ductos vescerentur. atque ita circumvecti Britanniam, amissis per inscitiam regendi navibus, pro praedonibus habiti, primum a Suebis, mox a Frisiis intercepti sunt. ac fuere quos per commercia venumdatos et in nostram usque ripam mutatione ementium adductos indicium tanti casus inlustravit.

(margin: 10, 2, 15, 3, 20)

29 Initio aestatis Agricola domestico vulnere ictus anno ante natum filium amisit. quem casum neque ut plerique fortium virorum ambitiose neque per lamenta rursus ac maerorem muliebriter tulit; et in luctu bellum inter remedia erat. igitur praemissa classe, quae pluribus locis praedata magnum et incertum terrorem faceret, expedito exercitu, cui ex Britannis fortissimos et longa pace exploratos addi-

(margin: 25, 2)

3 loca tuta E^2: locatura *E* 7 Britanniam *AB*: Britannias E^2: Brittanias *E* 9 inmixti *E*: inmixtis E^2: immixtis E^{2m} 11 remigante *E*: remigrante *Put.* 13 praevehebantur *E*: praebebantur E^{2m} mox ubi (*W. Heraeus*) aquam atque utilia (*Selling*) raptum exissent (*Till*) cum: mox adquam (adaquam E^2) adq. utillaraptis secum *E* 23 *ante* initio *add.* septimae *Brotier*: *post* initio *add.* insequentis *Koestermann* 26 in luctu *E*: inlustrans E^{2m}: *cf. c. 28, 3* 27 praedata E^2: praedatum *E*

derat, ad montem Graupium pervenit, quem iam hostis
insederat. nam Britanni nihil fracti pugnae prioris eventu 3
et ultionem aut servitium expectantes, tandemque docti
commune periculum concordia propulsandum, legationibus
5 et foederibus omnium civitatium vires exciverant. iamque 4
super triginta milia armatorum aspiciebantur, et adhuc
adfluebat omnis iuventus et quibus cruda ac viridis senectus,
clari bello et sua quisque decora gestantes, cum inter plures
duces virtute et genere praestans nomine Calgacus apud
10 contractam multitudinem proelium poscentem in hunc
modum locutus fertur:

'Quotiens causas belli et necessitatem nostram intueor, **30**
magnus mihi animus est hodiernum diem consensumque
vestrum initium libertatis toti Britanniae fore: nam et uni-
15 versi coistis et servitutis expertes, et nullae ultra terrae ac
ne mare quidem securum inminente nobis classe Romana.
ita proelium atque arma, quae fortibus honesta, eadem etiam
ignavis tutissima sunt. priores pugnae, quibus adversus 2
Romanos varia fortuna certatum est, spem ac subsidium
20 in nostris manibus habebant, quia nobilissimi totius Britanniae eoque in ipsis penetralibus siti nec ulla servientium
litora aspicientes, oculos quoque a contactu dominationis
inviolatos habebamus. nos terrarum ac libertatis extremos 3
recessus ipse ac sinus famae in hunc diem defendit: nunc
25 terminus Britanniae patet, atque omne ignotum pro magnifico est; sed nulla iam ultra gens, nihil nisi fluctus ac saxa,
et infestiores Romani, quorum superbiam frustra per obsequium ac modestiam effugias. raptores orbis, postquam 4

1 Graupium *E*: Grampium *Put.* 7 adfluebat *E*²: adfluebant *E* viridis *Guarnieri*: viris *E*: virens *E*²ᵐ, *fort. recte* 13 hodiernum diem consensumque vestrum *E*²: consensumque vestrum hodiernum diem *E* 15 coistis et *E*: *om. AB* 24 ac *E*: ad *E*²ᵐ: *cf. Hist. 3. 18, 2* 22 contactu *E*: conpactu *E*²ᵐ 24 sinus famae *E*: sinus fama (*abl.*) *Rhen.*: sinus (*gen.*) fama (*nom.*) *Boxhorn*: sinus a fama *Constans* nunc *E*² *s.l.*: tum *E* 25 patet *E*² *s.l.*: paret *E* atque . . . sed *post* defendit *transposuit Brueys*

cuncta vastantibus defuere terrae, mare scrutantur: si locuples hostis est, avari, si pauper, ambitiosi, quos non Oriens, non Occidens satiaverit: soli omnium opes atque 5 inopiam pari adfectu concupiscunt. auferre trucidare rapere falsis nominibus imperium atque ubi solitudinem 5 faciunt pacem appellant.

31 'Liberos cuique ac propinquos suos natura carissimos esse voluit: hi per dilectus alibi servituri auferuntur; coniuges sororesque etiam si hostilem libidinem effugerunt, nomine amicorum atque hospitum polluuntur. bona fortu- 10 naeque in tributum, ager atque annus in frumentum, corpora ipsa ac manus silvis ac paludibus emuniendis inter 2 verbera et contumelias conteruntur. nata servituti mancipia semel veneunt, atque ultro a dominis aluntur: Britannia servitutem suam cotidie emit, cotidie pascit. ac sicut in 15 familia recentissimus quisque servorum etiam conservis ludibrio est, sic in hoc orbis terrarum vetere famulatu novi nos et viles in excidium petimur; neque enim arva nobis aut metalla aut portus sunt, quibus exercendis reservemur. 3 virtus porro ac ferocia subiectorum ingrata imperantibus; 20 et longinquitas ac secretum ipsum quo tutius, eo suspectius. ita sublata spe veniae tandem sumite animum, tam quibus 4 salus quam quibus gloria carissima est. Brigantes femina duce exurere coloniam, expugnare castra, ac nisi felicitas in socordiam vertisset, exuere iugum potuere: nos integri et 25 indomiti et in libertatem non in paenitentiam †laturi, primo statim congressu ostendamus, quos sibi Caledonia viros seposuerit.

32 'An eandem Romanis in bello virtutem quam in pace lasciviam adesse creditis? nostris illi dissensionibus ac dis- 30 cordiis clari vitia hostium in gloriam exercitus sui vertunt;

1 terrae mare *E*: terram et mare *E*² 8 dilectus *E*: delectus *Guarnieri* 13 conteruntur *Jacob, Fröhlich*: conterunt *E* 23 Brigantes *E*: Trinobantes *Camden* 26 laturi *E*: bellaturi *Koch*: arma laturi *Mohr, Wex*: nati *Muretus*: educati *Richmond–Ogilvie*

quem contractum ex diversissimis gentibus ut secundae res
tenent, ita adversae dissolvent: nisi si Gallos et Germanos
et (pudet dictu) Britannorum plerosque, licet dominationi
alienae sanguinem commodent, diutius tamen hostes quam
5 servos, fide et adfectu teneri putatis. metus ac terror sunt 2
infirma vincla caritatis; quae ubi removeris, qui timere
desierint, odisse incipient. omnia victoriae incitamenta pro
nobis sunt: nullae Romanos coniuges accendunt, nulli
parentes fugam exprobraturi sunt; aut nulla plerisque patria
10 aut alia est. paucos numero, trepidos ignorantia, caelum
ipsum ac mare et silvas, ignota omnia circumspectantes,
clausos quodam modo ac vinctos di vobis tradiderunt. ne 3
terreat vanus aspectus et auri fulgor atque argenti, quod
neque tegit neque vulnerat. in ipsa hostium acie invenie-
15 mus nostras manus: adgnoscent Britanni suam causam,
recordabuntur Galli priorem libertatem, tam deserent illos
ceteri Germani quam nuper Usipi reliquerunt. nec quicquam ultra formidinis: vacua castella, senum coloniae,
inter male parentes et iniuste imperantes aegra municipia
20 et discordantia. hic dux, hic exercitus: ibi tributa et metalla 4
et ceterae servientium poenae, quas in aeternum perferre
aut statim ulcisci in hoc campo est. proinde ituri in aciem
et maiores vestros et posteros cogitate.'

Excepere orationem alacres, ut barbaris moris, fremitu **33**
25 cantuque et clamoribus dissonis. iamque agmina et armorum fulgores audentissimi cuiusque procursu; simul
instruebatur acies, cum Agricola quamquam laetum et vix
munimentis coercitum militem accendendum adhuc ratus,
ita disseruit: 'septimus annus est, commilitones, ex quo 2

3 dictu *E*²: dicto *E* 4 commodent *Put.*: commendent *E* 5 sunt *Beroaldus*: est *E* 9 exprobraturi *A*ᶜ: exprobaturi *E* 10 trepidos *E*: *add.* circum *E*²ᵐ (*sc.* circum trepidos) 16 itam (tam *E*ᶜ) deserent . . . quam *E*: deserent . . . tamquam *E*²ᵐ 17 necquicquam *E*²: nequicquam *E* 19 aegra municipia *E*: taetra mancipia *E*²ᵐ 26 procursu *E*²: procursus *E* 27 instruebatur *E*: instituebatur *E*²ᵐ 28 munimentis *E*²ᵐ: monitis *E* 29 septimus *Acidalius*: octavus *E*

virtute et auspiciis imperii Romani, fide atque opera nostra Britanniam vicistis. tot expeditionibus, tot proeliis, seu fortitudine adversus hostes seu patientia ac labore paene adversus ipsam rerum naturam opus fuit, neque me militum
3 neque vos ducis paenituit. ergo egressi, ego veterum 5 legatorum, vos priorum exercituum terminos, finem Britanniae non fama nec rumore sed castris et armis tenemus:
4 inventa Britannia et subacta. equidem saepe in agmine, cum vos paludes montesve et flumina fatigarent, fortissimi cuiusque voces audiebam: "quando dabitur hostis? quando 10 †animus?". veniunt, e latebris suis extrusi, et vota virtusque in aperto, omniaque prona victoribus atque eadem victis
5 adversa. nam ut superasse tantum itineris, evasisse silvas, transisse aestuaria pulchrum ac decorum in frontem, ita fugientibus periculosissima quae hodie prosperrima sunt; 15 neque enim nobis aut locorum eadem notitia aut commeatuum eadem abundantia, sed manus et arma et in his omnia.
6 quod ad me attinet, iam pridem mihi decretum est neque exercitus neque ducis terga tuta esse. proinde et honesta mors turpi vita potior, et incolumitas ac decus eodem loco 20 sita sunt; nec inglorium fuerit in ipso terrarum ac naturae fine cecidisse.

34 'Si novae gentes atque ignota acies constitisset, aliorum exercituum exemplis vos hortarer: nunc vestra decora recensete, vestros oculos interrogate. hi sunt, quos proximo 25 anno unam legionem furto noctis adgressos clamore debellastis; hi ceterorum Britannorum fugacissimi ideoque tam
2 diu superstites. quo modo silvas saltusque penetrantibus fortissimum quodque animal contra ruere, pavida et inertia ipso agminis sono pellebantur, sic acerrimi Britannorum 30

1 nostra *E*: vestra *Put.* 6 exercituum *E*²: exercitum *E*
9 montesve *E*: montesque *Urlichs*; *cf. Löfstedt, Peregr. Aeth., p. 201*
11 animus *E*: in manus venient *F. Walter* (veniet *Anderson*); cominus veniet *Anderson*: acies *Rhen.* 12 omniaque *E*²: omnia quae *E*
14 ita *Rhen.*: item *E* 15 periculosissima *E*²: riculosissima *E*
29 quodque *Laetus*: quoque *E* ruere *E*: ruebant *E*²ᵐ

iam pridem ceciderunt, reliquus est numerus ignavorum et timentium. quos quod tandem invenistis, non restiterunt, 3 sed deprehensi sunt; novissimae res et extremo metu torpor defixere aciem in his vestigiis, in quibus pulchram 5 et spectabilem victoriam ederetis. transigite cum expeditionibus, imponite quinquaginta annis magnum diem, adprobate rei publicae numquam exercitui imputari potuisse aut moras belli aut causas rebellandi.'

35 Et adloquente adhuc Agricola militum ardor eminebat, 10 et finem orationis ingens alacritas consecuta est, statimque ad arma discursum. 2 instinctos ruentesque ita disposuit, ut peditum auxilia, quae octo milium erant, mediam aciem firmarent, equitum tria milia cornibus adfunderentur. legiones pro vallo stetere, ingens victoriae decus citra Romanum san-15 guinem bellandi, et auxilium, si pellerentur. 3 Britannorum acies in speciem simul ac terrorem editioribus locis constiterat ita, ut primum agmen in aequo, ceteri per adclive iugum conexi velut insurgerent; media campi covinnarius eques strepitu ac discursu complebat. 4 tum Agricola superante 20 hostium multitudine veritus ne in frontem simul et latera suorum pugnaretur, diductis ordinibus, quamquam porrectior acies futura erat et arcessendas plerique legiones admonebant, promptior in spem et firmus adversis, dimisso equo pedes ante vexilla constitit.

36 25 Ac primo congressu eminus certabatur; simulque constantia, simul arte Britanni ingentibus gladiis et brevibus caetris missilia nostrorum vitare vel excutere, atque ipsi magnam vim telorum superfundere, donec Agricola quat-

1 reliquus E^2: reliquis E et timentium *Till*: dementium E: et metuentium E^{2m} 4 torpor *Ritter*: corpora E 6 quinquaginta E: quadraginta *W. Heraeus* 8 rebellandi E: bellandi E^{2m} 17 agmen in aequo ceteri *Bekker, cf. c. 36, 2*: agminae quoceteri E: quo steteri E^{2m} 18 conexi *Put.* (connexi): convexi E covinnarius, *cf. c. 36, 3*: convinnarus E eques E^2: eque E 20 in frontem simul et *Fröhlich*: simul in frontem simul et E 22 arcessendas E, *cf. Hist. 3. 71, 2, Ann. 2. 50, 1*: accersendas E^{2m}, *cf. Hist. 1. 14, 1, Ann. 4. 29, 1, et al.*

tuor Batavorum cohortes ac Tungrorum duas cohortatus est, ut rem ad mucrones ac manus adducerent; quod et ipsis vetustate militiae exercitatum et hostibus inhabile [parva scuta et enormes gladios gerentibus]; nam Britannorum gladii sine mucrone complexum armorum et in arto pugnam 5
2 non tolerabant. igitur ut Batavi miscere ictus, ferire umbonibus, ora fodere, et stratis qui in aequo adstiterant, erigere in colles aciem coepere, ceterae cohortes aemulatione et impetu conisae proximos quosque caedere: ac plerique semineces aut integri festinatione victoriae relinquebantur. 10
3 interim equitum turmae—fugere ⟨enim⟩ covinnarii—peditum se proelio miscuere. et quamquam recentem terrorem intulerant, densis tamen hostium agminibus et inaequalibus locis haerebant; minimeque equestris ea pugnae facies erat, cum aegre in gradu stantes simul equorum 15 corporibus impellerentur; ac saepe vagi currus, exterriti sine rectoribus equi, ut quemque formido tulerat, transversos aut obvios incursabant.

37 Et Britanni, qui adhuc pugnae expertes summa collium insederant et paucitatem nostrorum vacui spernebant, de- 20 gredi paulatim et circumire terga vincentium coeperant, ni id ipsum veritus Agricola quattuor equitum alas, ad subita belli retentas, venientibus opposuisset, quantoque ferocius
2 adcucurrerant, tanto acrius pulsos in fugam disiecisset. ita consilium Britannorum in ipsos versum, transvectaeque 25 praecepto ducis a fronte pugnantium alae aversam hostium

1 Batavorum *E*[2m]: uataevorum *E* 3 vetustate militiae *AB*: uetustatenniliae *E* inhabile *E*[2]: inabitabile *E*: it *ex praeced.* mil(it)iae *inrepserat* 3–4 parva . . . gerentibus *secl. Wex* 5 in arto *Fr. Medicis*: in aperto *E* 7 fodere *Gesner*: foedare *E* stratis *Ernesti*: tratis *E*: foede recti trates. vel traces *E*[2m] 11 enim *add. Wex*: ut *ante* fugere *add. Doederlein* 14 equestris ea pugnae facies *Rhen.*: equestris (*E*[2]: equestres *E*) ea (ei *E*[2m]) enim pugnae facies *E*: aequa nostris iam pugnae facies *Anquetil* 15 aegre in gradu stantes *Richmond–Ogilvie*: aegradiu aut stante *E*: aegre clivo instantes *Triller*: in gradu stantes *Lips.* 19 expertes *B*: expertis *E* 20 paucitatem *E*[2]: paucitate *E* 24 adcucurrerant *E*: adcurrerant *Put., fort. recte*; *cf. ad 23, 1* pulsos *E*[2]: pulso *E*

aciem invasere. tum vero patentibus locis grande et atrox spectaculum: sequi vulnerare capere, atque eosdem oblatis aliis trucidare. iam hostium, prout cuique ingenium erat 3 catervae armatorum paucioribus terga praestare, quidam 5 inermes ultro ruere ac se morti offerre. passim arma et corpora et laceri artus et cruenta humus; et aliquando etiam victis ira virtusque. nam postquam silvis adpropinquaverunt, 4 primos sequentium incautos collecti et locorum gnari circumveniebant. quod ni frequens ubique Agricola 10 validas et expeditas cohortes indaginis modo et, sicubi artiora erant, partem equitum dimissis equis, simul rariores silvas equitem persultare iussisset, acceptum aliquod vulnus per nimiam fiduciam foret. ceterum ubi compositos firmis 5 ordinibus sequi rursus videre, in fugam versi, non agminibus, 15 ut prius, nec alius alium respectantes: rari et vitabundi in vicem longinqua atque avia petiere. finis sequendi nox et satietas fuit. caesa hostium ad decem milia: nostrorum 6 trecenti sexaginta cecidere, in quis Aulus Atticus praefectus cohortis, iuvenili ardore et ferocia equi hostibus inlatus.

20 **38** Et nox quidem gaudio praedaque laeta victoribus: Britanni palantes mixto virorum mulierumque ploratu trahere vulneratos, vocare integros, deserere domos ac per iram ultro incendere, eligere latebras et statim relinquere; miscere in vicem consilia aliqua, dein separare; aliquando frangi 25 aspectu pignorum suorum, saepius concitari. satisque constabat saevisse quosdam in coniuges ac liberos, tamquam misererentur. proximus dies faciem victoriae latius aperuit: 2 vastum ubique silentium, secreti colles, fumantia procul tecta, nemo exploratoribus obvius. quibus in omnem par-30 tem dimissis, ubi incerta fugae vestigia neque usquam

4 praestare *E*: praebere *E*[2m] 7 nam *hic locavit Andresen*: nam *ante* primos *E*: iam *Hedicke* 9 gnari *Dronke*: ignari *E*: ignaros *Put.* 11 equis simul *E*: qui simulati *E*[2m] 12 equitem persultare *Rhen.*: equite persultari *E* 19 iuvenili *Guarnieri*: iuvenali *E* 24 aliqua *secl. Classen, Wölfflin* 28 ubique *E*[2]: ibique *E* 30 dimissis *E*[2]: demissis *E*

conglobari hostes compertum et exacta iam aestate spargi bellum nequibat, in fines Borestorum exercitum deducit.
3 ibi acceptis obsidibus, praefecto classis circumvehi Britanniam praecipit. datae ad id vires, et praecesserat terror. ipse peditem atque equites lento itinere, quo novarum 5 gentium animi ipsa transitus mora terrerentur, in hibernis
4 locavit. et simul classis secunda tempestate ac fama Trucculensem portum tenuit, unde proximo Britanniae latere praelecto omnis redierat.

39 Hunc rerum cursum, quamquam nulla verborum iactantia 10 epistulis Agricolae auctum, ut erat Domitiano moris, fronte laetus, pectore anxius excepit. inerat conscientia derisui fuisse nuper falsum e Germania triumphum, emptis per commercia quorum habitus et crines in captivorum speciem formarentur: at nunc veram magnamque victoriam tot mili- 15
2 bus hostium caesis ingenti fama celebrari. id sibi maxime formidolosum, privati hominis nomen supra principem adtolli: frustra studia fori et civilium artium decus in silentium acta, si militarem gloriam alius occuparet; cetera utcumque facilius dissimulari, ducis boni imperatoriam 20
3 virtutem esse. talibus curis exercitus, quodque saevae cogitationis indicium erat, secreto suo satiatus, optimum in praesens statuit reponere odium, donec impetus famae et favor exercitus languesceret: nam etiam tum Agricola Britanniam obtinebat. 25

40 Igitur triumphalia ornamenta et inlustris statuae honorem et quidquid pro triumpho datur, multo verborum honore cumulata, decerni in senatu iubet addique insuper opinionem, Syriam provinciam Agricolae destinari, vacuam tum

2 deducit *E*: reducit E^{2m} 8 trucculensem *E*: trutulensem E^{2m}: Rutup(i)ensem *Lips.* 9 prelecto A^{cm}: prelecta *E*: lecto E^{2m} omnis *E*: omni E^2 10 nulla E^2: ulla *E* 11 auctum *Lips.*: actum *E* Domitiano moris E^{2m}: Domitianus *E* 12 excepit *Put.*: excipit *E* 14 quorum E^2: quarum *E* crines *Put.*: crinis *E* 21 quodque E^2: quoque *E* saevae E^2: saevire *E* 23 praesens E^{2m}: praesentia *E* 28 addique *E*: additque *Muretus*

morte Atili Rufi consularis et maioribus reservatam. credi- 2
dere plerique libertum ex secretioribus ministeriis missum
ad Agricolam codicillos, quibus ei Syria dabatur, tulisse
cum eo praecepto ut, si in Britannia foret, traderentur;
5 eumque libertum in ipso freto Oceani obvium Agricolae, ne
appellato quidem eo ad Domitianum remeasse, sive verum
istud, sive ex ingenio principis fictum ac compositum est.
tradiderat interim Agricola successori suo provinciam 3
quietam tutamque. ac ne notabilis celebritate et frequentia
10 occurrentium introitus esset, vitato amicorum officio noctu
in urbem, noctu in Palatium, ita ut praeceptum erat, venit;
exceptusque brevi osculo et nullo sermone turbae servientium inmixtus est. ceterum uti militare nomen, grave inter 4
otiosos, aliis virtutibus temperaret, tranquillitatem atque
15 otium penitus hausit, cultu modicus, sermone facilis, uno
aut altero amicorum comitatus, adeo ut plerique, quibus
magnos viros per ambitionem aestimare mos est, viso aspectoque Agricola quaererent famam, pauci interpretarentur.

Crebro per eos dies apud Domitianum absens accusatus, **41**
20 absens absolutus est. causa periculi non crimen ullum aut
querela laesi cuiusquam, sed infensus virtutibus princeps et
gloria viri ac pessimum inimicorum genus, laudantes. et ea 2
insecuta sunt rei publicae tempora, quae sileri Agricolam
non sinerent: tot exercitus in Moesia Daciaque et Germania
25 et Pannonia temeritate aut per ignaviam ducum amissi, tot
militares viri cum tot cohortibus expugnati et capti; nec iam
de limite imperii et ripa, sed de hibernis legionum et possessione dubitatum. ita cum damna damnis continuarentur 3
atque omnis annus funeribus et cladibus insigniretur,
30 poscebatur ore vulgi dux Agricola, comparantibus cunctis
vigorem, constantiam et expertum bellis animum cum

1 Atili *E*[2]: Atilli *E* 2 *Post* missum *desinunt vetera codicis E folia; sed notae E*[2m] *in marg. fol. 69 et 76, quamvis erasae, interdum magna ex parte legi possunt* 4 eo *e*: *om. AB* Britannia *Put.*: Britanniam *codd.*: Britannia etiam *Halm* 15 hausit *Wex*: auxit *codd.* 24 Moesia *e*: Misia *AB* 26 tot (cohortibus) *e*[m]*AB*: totis *e* 31 cum inertia et formidine *eAB*: inertiae et formidini *E*[2m] *e*[m]

4 inertia et formidine eorum. quibus sermonibus satis constat Domitiani quoque aures verberatas, dum optimus quisque libertorum amore et fide, pessimi malignitate et livore pronum deterioribus principem extimulabant. sic Agricola simul suis virtutibus, simul vitiis aliorum in ipsam 5 gloriam praeceps agebatur.

42 Aderat iam annus, quo proconsulatum Africae et Asiae sortiretur, et occiso Civica nuper nec Agricolae consilium deerat nec Domitiano exemplum. accessere quidam cogitationum principis periti, qui iturusne esset in provinciam 10 ultro Agricolam interrogarent. ac primo occultius quietem et otium laudare, mox operam suam in adprobanda excusatione offerre, postremo non iam obscuri suadentes simul 2 terrentesque pertraxere ad Domitianum. qui paratus simulatione, in adrogantiam compositus, et audiit preces 15 excusantis et, cum adnuisset, agi sibi gratias passus est, nec erubuit beneficii invidia. salarium tamen proconsulare solitum offeri et quibusdam a se ipso concessum Agricolae non dedit, sive offensus non petitum, sive ex conscientia, 3 ne quod vetuerat videretur emisse. proprium humani 20 ingenii est odisse quem laeseris: Domitiani vero natura praeceps in iram, et quo obscurior, eo inrevocabilior, moderatione tamen prudentiaque Agricolae leniebatur, quia non contumacia neque inani iactatione libertatis famam 4 fatumque provocabat. sciant, quibus moris est inlicita 25 mirari, posse etiam sub malis principibus magnos viros esse, obsequiumque ac modestiam, si industria ac vigor adsint, eo laudis excedere, quo plerique per abrupta sed in nullum rei publicae usum ambitiosa morte inclaruerunt.

1 eorum *codd.*: aliorum *edd. Bipontini*: ceterorum *Grotius* 4 extimulabant *e*: existimulabant *E*[2m]*e*[2m] 13 iam *Rhen.*: tam *codd.* obscuri *e*[c]*AB*: obscuris *e* 15 simulatione *eB*: simulationis *E*[2m]*A* 16 annuisset *AB*: amnuiset *e* 17 proconsulare *e*: proconsulari *e*[c]*AB* 21 laeseris *e*[c]*AB*: laeserit *e* 28 excedere *codd.*: escendere *Lips.* plerique qui *J. Müller* 28–29 nullum re p̅ (*sc.* rei publicae) *E*[2m]: ullum rei post *eAB* enisi *post* abrupta *add. Heumann, post* usum *Schömann, pro* sed *substituit C. Heraeus*

Finis vitae eius nobis luctuosus, amicis tristis, extraneis **43**
etiam ignotisque non sine cura fuit. vulgus quoque et hic
aliud agens populus et ventitavere ad domum et per fora
et circulos locuti sunt; nec quisquam audita morte Agri-
5 colae aut laetatus est aut statim oblitus. augebat miseratio- 2
nem constans rumor veneno interceptum: nobis nihil comperti adfirmare ausim. ceterum per omnem valetudinem
eius crebrius quam ex more principatus per nuntios visentis
et libertorum primi et medicorum intimi venere, sive cura
10 illud sive inquisitio erat. supremo quidem die momenta 3
ipsa deficientis per dispositos cursores nuntiata constabat,
nullo credente sic adcelerari quae tristis audiret. speciem
tamen doloris animo vultuque prae se tulit, securus iam odii
et qui facilius dissimularet gaudium quam metum. satis con- 4
15 stabat lecto testamento Agricolae, quo coheredem optimae
uxori et piissimae filiae Domitianum scripsit, laetatum eum
velut honore iudicioque. tam caeca et corrupta mens adsi-
duis adulationibus erat, ut nesciret a bono patre non scribi
heredem nisi malum principem.

20 Natus erat Agricola Gaio Caesare tertium consule idibus **44**
Iuniis: excessit quarto et quinquagesimo anno, decimum
kalendas Septembris Collega Priscinoque consulibus. quod 2
si habitum quoque eius posteri noscere velint, decentior
quam sublimior fuit; nihil impetus in vultu: gratia oris
25 supererat. bonum virum facile crederes, magnum libenter.
et ipse quidem, quamquam medio in spatio integrae aetatis 3
ereptus, quantum ad gloriam, longissimum aevum peregit.
quippe et vera bona, quae in virtutibus sita sunt, impleverat, et consulari ac triumphalibus ornamentis praedito

5 oblitus *Muretus*: oblitus est *codd.*: oblitus. et *Wex* 7 *ante* adfirmare *add.* quod *Acidalius*, quodve (*vel* aut quod) *Ritter*, nec *Ernesti*, ut *Wex* 8 visentis $E^{2m}e^{c}A$: visentes *e*: viseritis *B* 11 constabat *e*: constabant *AB* 13 animo vultuque *codd.*: habitu vultuque *Ernesti* 14–15 constabat *eAB*: constat E^{2m} 20 tertium *Ursinus*: ter *codd.*: iterum *Nipperdey* 21 quarto *Petavius*: sexto *codd.* 22 Priscino *Harrer*, *cf. ILS 9059*: Prisco *codd.* 24 impetus *eA*: metus $E^{2m}e^{m}A^{m}$: metus et impetus *B*

4 quid aliud adstruere fortuna poterat? opibus nimiis non gaudebat, speciosae contigerant. filia atque uxore superstitibus potest videri etiam beatus incolumi dignitate, florente fama, salvis adfinitatibus et amicitiis futura
5 effugisse. nam sicut ei ⟨non licuit⟩ durare in hanc beatissimi 5 saeculi lucem ac principem Traianum videre, quod augurio votisque apud nostras aures ominabatur, ita festinatae mortis grave solacium tulit evasisse postremum illud tempus, quo Domitianus non iam per intervalla ac spiramenta temporum, sed continuo et velut uno ictu rem publicam exhausit. 10

45 Non vidit Agricola obsessam curiam et clausum armis senatum et eadem strage tot consularium caedes, tot nobilissimarum feminarum exilia et fugas. una adhuc victoria Carus Mettius censebatur, et intra Albanam arcem sententia Messalini strepebat, et Massa Baebius etiam tum reus erat: 15 mox nostrae duxere Helvidium in carcerem manus; nos Maurici Rusticique visus ⟨adflixit,⟩ nos innocenti sanguine
2 Senecio perfudit. Nero tamen subtraxit oculos suos iussitque scelera, non spectavit: praecipua sub Domitiano miseriarum pars erat videre et aspici, cum suspiria nostra 20 subscriberentur, cum denotandis tot hominum palloribus sufficeret saevus ille vultus et rubor, quo se contra pudorem muniebat.

3 Tu vero felix, Agricola, non vitae tantum claritate, sed etiam opportunitate mortis. ut perhibent qui interfuere 25

1–2 opibus . . . contigerant *post* peregit *transposuit Gudeman* 2 speciosae contigerant *Rhen.*: spetiosae contigerant $E^{2m}e^{m}$: spetiose non contigerant *e* (non *del. Guarnieri*) *AB*: *ita distinxit Rhen., post* superstitibus *AB*: *sine distinctione e* uxore *eAB*: uxoris $E^{2m}e^{m}$ 5 sicut ei non licuit *Dahl*: sicuti *codd.*: sicut iuvaret *Müller* hanc . . . lucem *Acidalius*: hac . . . luce *codd.* 8 grave $e^{m}A^{m}$: grande E^{2m} *eAB* 10 velut *eAB*: vel e^{m} exhausit *eAB*: hausit $E^{2m}e^{m}$ 14 Mettius $E^{2m}e^{m}A^{c}$: Mitius *eAB* arcem *eAB*: villam $E^{2m}e^{m}A^{m}$ 15 etiam tum *e*: iam tum *A*: tum *B* 17 Maurici Rusticique visus *eAB*: adflixit *suppl. R. Reitzenstein,* dehonestavit *Anderson, alii alia*: Mauric(i)um Rusticumque divisimus $E^{2m}e^{m}A^{m}$ 22 quo *Lips.*: a quo *codd.* 25 perhibent *Put.*: perhiberent *codd.* interfuere *e*: interfuerunt *AB*

novissimis sermonibus tuis, constans et libens fatum excepisti, tamquam pro virili portione innocentiam principi donares. sed mihi filiaeque eius praeter acerbitatem parentis 4 erepti auget maestitiam, quod adsidere valetudini, fovere
5 deficientem, satiari vultu complexuque non contigit. ex- 5 cepissemus certe mandata vocesque, quas penitus animo figeremus. noster hic dolor, nostrum vulnus, nobis tam longae absentiae condicione ante quadriennium amissus est. omnia sine dubio, optime parentum, adsidente amantissima
10 uxore superfuere honori tuo: paucioribus tamen lacrimis compositus es, et novissima in luce desideravere aliquid oculi tui.

Si quis piorum manibus locus, si, ut sapientibus placet, **46** non cum corpore extinguuntur magnae animae, placide
15 quiescas, nosque domum tuam ab infirmo desiderio et muliebribus lamentis ad contemplationem virtutum tuarum voces, quas neque lugeri neque plangi fas est. admiratione 2 te potius et laudibus et, si natura suppeditet, similitudine colamus: is verus honos, ea coniunctissimi cuiusque pietas.
20 id filiae quoque uxorique praeceperim, sic patris, sic mariti 3 memoriam venerari, ut omnia facta dictaque eius secum revolvant, formamque ac figuram animi magis quam corporis complectantur, non quia intercedendum putem imaginibus quae marmore aut aere finguntur, sed, ut vultus
25 hominum, ita simulacra vultus imbecilla ac mortalia sunt, forma mentis aeterna quam tenere et exprimere non per alienam materiam et artem, sed tuis ipse moribus possis.

5 excepissemus *Acidalius*: excepissem *codd.* 7 figeremus *EeAB*: pingeremus $E^{2m}e^{m}A^{m}$ tam $E^{2m}AB$, *Guarnieri*: tum eA^{m} 8 est *codd.*: es *Rhen.* 11 compositus $e^{m}A^{m}$: comploratus $E^{2m}eAB$ 15 nosque *codd.*: nosque et *Urlichs* 18 te potius $E^{2m}AB$: te *om. e*: te *add. Guarnieri s.l.* et *Muretus*: temporalibus et *Ee*: temporalibus *AB*: temporibus $E^{2m}e^{m}$: et immortalibus (*Lips.*) *Acidalius* similitudine *Grotius*: militum *EeA*: multum *B*: aemulatu *Heinsius* 19 colamus *Muretus*: decoramus *codd.*: decoremus *Ursinus* 22 formamque *T*, *Muretus*: famamque *eAB*

4 quidquid ex Agricola amavimus, quidquid mirati sumus, manet mansurumque est in animis hominum in aeternitate temporum, fama rerum; nam multos veterum velut inglorios et ignobiles oblivio obruet: Agricola posteritati narratus et traditus superstes erit. 5

3 fama *codd.*: in fama *Halm* 4 obruet *EeAB* (*et Decembrio*); obruit *Haupt* Cornelii Taciti de vita Iulii Agricolae liber explicit *E*: Cornelii Taciti de vita et moribus Iulii Agricolae liber explicit *e*

COMMENTARY

1–3. *Preface*

IT was customary to preface an historical work with a *prooemium* in which the author set out the scope and purpose of his book and attempted to win the goodwill of his readers. The themes of such prefaces were commonplace (cf. Lucian, *Quomodo Historia* 52–55; see Ogilvie's nn. on the *Praefatio* of Livy) and had been formulated by rhetoricians. The subject and aim of a biography required less formal introduction—neither Nepos nor Suetonius employs general prefaces—but the *Agricola* was Tacitus' first composition and afforded him an opportunity for explaining his reasons for writing at all. Thus besides many conventional reflections (c. 1, 1 n.; 1, 2 n.; 1, 3 n.; 1, 4 n.; 2, 3 n.; 3, 1 nn.; 3, 3 n.) there is heard a more personal note. Two themes dominate the Preface: Tacitus apologizes for the unfashionable nature of his subject-matter and for his own literary shortcomings and he seeks to excuse both by an appeal to the restrictions on freedom of expression which Domitian's régime had imposed. Within this framework he contrasts the demands of biography and autobiography, the individualism of the old Republic and the conformity of the principate, the decline of literature under Domitian and its revival under Nerva. These balanced themes weave a tightly knit unity. Underlying them is a fundamental conflict which Tacitus never satisfactorily resolved. The last years of the Republic showed that an autocracy was necessary if Rome was to have a stable government and to be saved from anarchy. Yet an autocracy was demoralizing, because it sapped the self-respect of free citizens, and could easily become tyrannical. Tacitus recognized that there could be no middle course, no mixed constitution or dyarchy (cf. *Dial.* 41, 4), but he hoped that if the ruler was good and the people well disposed, autocracy (*principatus*) and freedom of opinion (*libertas*) could be reconciled. As the years passed and his thought deepened he came to see that no reconciliation was possible, that *principatus* was the same thing as *regnum* (*A.* 1. 6, 1; 1. 9, 5; 3. 60, 1, etc.) or *potentia* (*H.* 1. 13, 1, etc.) and was incompatible with *libertas*. In the *Agricola*, however, the paramount emotion is one of relief at the downfall of Domitian; the profound pessimism of the *Annals* is still far away. Both in the introductory chapters and in the epilogue (see also nn. on c. 46) Tacitus employs conventional themes but infuses into them personal feelings in a way that is rare in Latin. [See also Büchner, *Tacitus*, pp. 23 ff.]

1, 1. Clarorum virorum : Tacitus echoes the opening words of the elder Cato's *Origines* (fr. 1 Peter = Cic. *Planc.* 66 'clarorum virorum atque magnorum non minus otii quam negotii rationem exstare oportere'). Similarly the introduction of the *Dialogus* recalls the introduction of Plato's *Symposium*, and *Annals* 1. 1, 1 'urbem Romam a principio reges habuere' is modelled on the opening of the narrative of Sallust's *Catiline* (6, 1). The first words of the *Germania* (*G. omnis* . . .) may be compared with Caesar, *B.G.* 1. 1, 1 *Gallia omnis* and stamp the work as a piece of ethnographical writing. This conceit, which became fashionable in later Latin (Wijkstrom, *Apophoreta . . . Lundström*, 135 ff.), makes it certain that Tacitus is recalling Cato directly and not at second-hand through Cicero, but he seems to combine the beginning of the *Origines* with a passage elsewhere in the same work where Cato speaks of the old custom of chanting at meals 'clarorum virorum laudes atque virtutes' (Cic. *Brut.* 75; *Tusc. Disp.* 1, 3; 4, 3; cf. Hor. *Ep.* 2. 1, 249). In emphasizing his relation to Cato, Tacitus is not so much proclaiming that he is reverting to an older style than that of Cicero, although his style does owe much to Sallust who looked to Cato for linguistic inspiration (Mendell; see Syme, *Sallust*, p. 267; and Introduction, p. 24), or endorsing Cato's distrust of demoralizing Greek influences (Paratore), as acknowledging his agreement with Cato's belief (cf. Nep. *Cato* 3, 3) that success in life is to be won by personal achievement (*virtus*) rather than by circumstances, birth, or position. This was one of the themes of the *Origines*: it was illustrated by the career of Agricola.

antiquitus usitatum, 'a custom of the past'. The use of a neuter past participle or adjective, in apposition to the object or subject, instead of a relative clause, is frequent in Tacitus. Cf. *G.* 31, 1; *H.* 4. 23, 3.

quamquam incuriosa suorum : the analogy of 'vetera extollimus recentium incuriosi' in *A.* 2. 88, 3, where the reference is to Arminius, suggests that *suorum* here is neuter, 'its own affairs', i.e. contemporary events. The thought is commonplace; cf. Pliny, *Ep.* 8. 20, 1, 'proximorum incuriosi longinqua sectemur'; Sen. *Tranq. Animi* 14, 4. Elsewhere Tacitus uses the substantival *sua* only in the accusative but the conventional character of the Preface makes it implausible to take *suorum* as masc. (sc. *clarorum virorum*). *Quamquam* is rarely used in classical prose, but often by Tacitus, with an adj. or part. without a finite verb: cf. c. 16, 2; 22, 1.

aetas : personified, as γενεά often is.

virtus, instance of merit in an individual, almost 'some great man'; for *magna ac nobilis* cf. Cic. *Dom.* 115.

vicit ac supergressa est, 'has overcome and surmounted'.

vitium : the thought is commonplace; cf. Sall. *Jug.* 10, 2; Vell. Pat. 1. 9, 6; Nep. *Chabrias* 3, 3, 'est enim hoc commune vitium magnis liberisque civitatibus ut invidia gloriae comes sit'. It is Greek in origin (cf. Dem. *Olyn.* 3, 24; Hdt. 7, 236).

ignorantiam recti et invidiam, 'ignorance of what is right and jealousy'. The use of the singular *vitium* implies that these are two aspects of one vice, related as cause and effect. The common sort cannot understand an exalted character, and hate its eminence. Contrast *boni intellectus* in *A.* 6. 36, 3.

1, 2. memoratu : supine, occasionally used by Sallust (*Cat.* 6, 2; 7, 3; *Jug.* 40, 3) and Livy.

pronum magisque in aperto, 'easy and with a freer field', metaphors virtually synonymous, and taken from a favourable course, in opposition to what is *arduum* and *impeditum*, 'uphill and full of obstacles'. Cf. c. 33, 4. Like *supergressa* above, the image is inspired by the comparison of virtue with a difficult, narrow, steep path which was first made by Hesiod (*W.D.* 289 ff.) and which subsequently became a commonplace (cf. Plato, *Rep.* 2, 364 c; Hor. *Odes*, 3. 24, 44, etc.).

celeberrimus = *clarissimus*, as in Livy 7. 21, 6; 26. 27, 16. The sense is 'in the olden days not only was it easier to *do* memorable deeds but also distinguished writers who recorded those deeds were motivated not by self-seeking but by *bona conscientia*'.

sine gratia aut ambitione, 'without partiality or self-seeking'. An author might falsify history to please friends or to gain some object. Cf. the contrast in *A.* 6. 46, 2, 'non perinde curae gratia praesentium quam in posteros ambitio'. The claim of disinterested impartiality was traditionally made by historians

(Livy, *Praef.* 5. with Ogilvie's note) but the remains of earlier Roman historical writing hardly confirm it. Cf. also Cic. *Brut.* 62.

bonae . . . conscientiae, 'consciousness of well-doing', i.e. of having fulfilled the historian's duty 'ne virtutes sileantur' (*A.* 3. 65, 1). Such expressions as *bona* or *mala conscientia,* in which the adjective has the force of an objective genitive, approach nearly to the modern 'conscience': cf. Sen. *Ep.* 43, 5 'bona conscientia turbam advocat, mala etiam in solitudine anxia atque sollicita est'. For *pretium* in the sense of *praemium,* cf. c. 12, 6, etc.

1, 3. ac plerique, 'indeed, many': cf. c. 36, 2. In the following words (as elsewhere) two sentences are combined, (1) that they wrote their own lives, (2) that they did not consider it arrogance, but confidence in their own worth to do so. They felt that they had a just claim on the appreciation of their hearers.

fiduciam potius morum : cf. Val. Max. 3. 3, ext. 2 'tanta fiducia ingenii ac morum suorum fretus'.

Rutilio : P. Rutilius Rufus was a *novus homo* who at an early date became a protégé of the Metelli and their *factio.* He had served under Scipio Aemilianus at Numantia and throughout his life showed a keen interest in military techniques. He held the praetorship probably in 119 B.C. when L. Metellus Delmaticus was consul, but although he stood with M. Metellus for the consulship of 115 B.C. he was defeated by M. Aemilius Scaurus (see next note). Rutilius and Scaurus counter-charged each other with bribery in connexion with this election (Cic. *Brut.* 113) but both prosecutions were unsuccessful and the relations between the two men do not seem to have been embittered. He went with Metellus Numidicus to Africa where he served against Jugurtha (109–107 B.C.). His friendship with the Metelli earned him the implacable hostility of Marius but also helped him to the consulship in 105 B.C. In 94 B.C. he went with Q. Mucius Scaevola, *pontifex maximus,* to Asia to reorganize the affairs of the province which had for some years suffered from the exactions of equestrian tax-farmers. On his return he was prosecuted in 92 B.C. for extortion by a certain Apicius—presumably one of the offended *equites* abetted by Marius (Dio fr. 97, 3; see Badian, *Studies in Greek and Roman History,* p. 106; Badian in *Latin Historians* (ed. T. A. Dorey, 1966), pp. 23–25)—and being convicted retired as an exile to the very province he was alleged to have plundered. He settled at Smyrna which adopted him as a citizen (*A*. 4. 43, 5). In his retirement he composed his memoirs (fragments in Peter, *Hist. Rom. Rel.* 1², 189 ff.; see G. L. Hendrickson (*C.P.* 28 (1933), 153 ff.)) in which he speciously defended his career, and a history in Greek. They were used by Sallust as a source for his *Jugurtha*. Rutilius was much admired by later generations as a man, an orator, and a philosopher of Stoic persuasion (Cic. *de Orat.* 1, 229; Vell. Pat. 2. 13, 2; Sen. *Ep.* 98, 12).

Scauro : M. Aemilius Scaurus came from an ancient patrician family which had for generations been in eclipse. He defeated Rutilius for the consulship of 115 B.C. and seems to have taken sides with the *factio* of the Metelli: he married a Metella and Metellan support must have been responsible for his election as *princeps senatus* (115 B.C.). He dominated the political scene for the next twenty years (president of the *quaestio Mamilia* 110 B.C.; censor 109 B.C.) and endeavoured to maintain the supremacy of his party against Marius: he supported the elder Q. Servilius Caepio after his defeat at Arausio in 103 B.C. and was the principal witness when Caepio's prosecutor, C. Norbanus, was himself prosecuted *c.* 95 B.C.; like the elder Caepio he deprecated any extension of citizenship (Cic. *de Orat.* 2, 257). He perhaps led an investigation into the affairs of Asia in 97 or 96 B.C. which was to result in the reorganization by Scaevola and Rutilius. Subsequently he supported the younger Drusus and was arraigned before the *quaestio Variana* in 90 B.C. He died soon afterwards. The details of his career explain the unfavourable judgement which Sallust makes of him (*Jug.* 15, 4 'avidus potentiae honoris divitiarum, ceterum vitia sua callide occultans') but he was much admired by Cicero. His memoirs, in three volumes, which may also have been used by Sallust, were mentioned by Cicero as valuable but no longer read (*Brut.* 112; fragments in Peter, *Hist. Rom. Rel.* 1², 185; see P. Fraccaro, *Opuscula* 2 (1957), 125 ff.; Badian in *Latin Historians* (ed. T. A. Dorey 1966), pp. 23–25). They are also cited by Val. Max. 4. 4, 11 and Pliny, *N.H.* 33, 21.

citra fidem, 'beneath credibility', i.e. 'unworthy of belief'. A contrast with *fiduciam* is perhaps intended. *Citra* is used as nearly equivalent to *sine* by Ovid, Seneca, Quintilian, and the elder Pliny, etc., and by Tacitus in his minor works only. Cf. c. 8, 3 *extra*, and the English idiom 'beyond belief'.

aut obtrectationi, 'or matter of censure', as contrary to good taste. By the phrase 'citra fidem aut obtrectationi' Tacitus controverts the opinion of Cicero who tried to persuade Lucceius to write an account of his consulship of 63 B.C. by arguing that if he wrote it himself it would command less belief and would incur disapproval (*ad Fam.* 5. 12, 8 'minor sit fides, minor auctoritas, multi denique reprehendant').

adeo, etc., 'so truly does the age most fruitful in excellence also best appreciate it'. Cf. the sentiment in *H.* 3. 51, 2, and 'simplex admirandis virtutibus antiquitas' (Sen. *Cons. ad Helv.* 19, 5).

1, 4. nunc, 'in these times', of the present age generally, contrasted with *apud priores* (§ 2), but with particular reference to the suppressive tyranny of the last years of Domitian's reign. Cf. *H.* 3. 72, 1; 83, 3 where *nunc* alludes to events of A.D. 69 contrasted with those of earlier times.

narraturo, etc., 'when about to relate the life of a man who was dead', one removed from the envy and jealousy of the present (cf. *A.* 4. 35, 1), in contrast to the examples of men who wrote their own lives, and published them in their lifetime.

venia opus fuit, 'I had to seek permission', for the choice of an unpopular subject (see below). The past *fuit* implies that Tacitus had sounded Domitian or his court for permission (*venia*) to compile Agricola's biography and had received a discouraging reply which he would not have received if he had proposed a work of invective. This, the natural interpretation of *opus fuit*, might help to explain both why Tacitus delayed writing the biography for four years and the particular intensity of feeling shown in it against Domitian.

quam non petissem incusaturus. *Incusaturus* must answer to *narraturo*, which is antithetical as implying a defence of Agricola, and must mean *si incusaturus fuissem*. Logic requires a stop after *incusaturus*, for the following words give the reason for the preceding statement; and there is an obvious balance between this and the previous sentence. It is, moreover, characteristic of Tacitus to end a section with a terse epigram (cf. c. 2, 3 fin.; c. 4, 3 fin.; *G.* 8, 3 with Anderson's n.). With this punctuation, the natural object of *incusaturus* is that of *narraturo* (*vitam defuncti hominis*): cf. Pliny, *Ep.* 7. 31, 6 (A.D. 107) 'cum plerique hactenus defunctorum meminerint ut querantur'. Tacitus had to ask indulgence for championing the career and record of a kinsman—which formerly was a recognized licence, indeed an expected duty (*Dial.* 10, 6; *H.* 4. 42,

1)—since under Domitian only the denunciation of the dead was appreciated. Alternatively *incusaturus* may be taken absolutely: 'had invective been my purpose'. The sentiment that invective secures a ready hearing is commonplace (cf. *H.* 1. 1, 2).

tam saeva, etc., sc. *fuere*: the times were hostile to merit (cf. preceding note). For *tam* so used at the beginning of a sentence, with the force of *adeo*, cf. Juv. 13, 75 'tam facile et pronum est', and Pliny, *Ep.* 5. 20, 4 'tam longas . . . periodos contorquent'. That the praise of others excites jealousy and hate is a commonplace (cf. e.g. Livy 3. 12, 8) and Tacitus' language echoes Cic. *Orator* 35 'tempora timens inimica virtuti'.

[Others interpret this concluding section differently, understanding *nunc* of A.D. 97–98, the time of publication, the period of renewed hope under Nerva and Trajan (cf. c. 3, 1 *nunc demum*), and taking *opus fuit* in the sense 'I have found it necessary to ask for indulgence', and supplying *sunt* with *tam saeva*, etc. The meaning then will be that Tacitus feels that even in the better times after the death of Domitian he has to justify writing biography; so hostile is the age to merit. The opening three chapters contain, therefore, a request for indulgence from the public (*veniae petitio*) and he concludes the introduction on the same note of excuse: 'hic interim liber . . . aut laudatus erit aut excusatus'. This interpretation, however, seems at variance both with c. 1, 1 (*nostris quidem temporibus*) which implies that the writing of biographies did not require permission or indulgence from the present public and also with the optimistic tone of c. 3, 1 which recognizes that Nerva and Trajan have introduced a new era in which no apology is needed for freedom of speech and which is no longer hostile to merit.]

2, 1. Legimus: the analogy of other passages where an example, intended to support a general proposition, is introduced by *legistis* (*Dial.* 18, 5) or *vidimus* (*G.* 8, 2; cf. Hor. *Odes*, 1. 2, 13) suggests that *legimus* here is the past tense. The plural will designate either Tacitus himself or the Romans as a whole, especially senators (cf. c. 45, 1 *nostrae . . . manus*; *nos*), 'we have read that eulogy of the dead was a capital offence'. Tacitus probably refers to the official publication of the sentence in the *acta senatus* or *acta diurna* (although Domitian, according to Dio 67. 11, 3, sometimes suppressed mention of trials in the latter) and it should be recalled that Tacitus was out of Rome in A.D. 93 and may not have witnessed all these events. The *acta diurna* circulated widely in the provinces (*A.* 16. 22, 3). Most editors understand *legimus* as present, 'we read', 'it is open for us to read', and think that Tacitus is emphasizing the definite and official charge against Rusticus and Herennius, but the ostensible grounds will have been *maiestas* whereas the real offence (the eulogy of Thrasea and Helvidius) would have had to be read outside the formal court records.

Aruleno Rustico : Q. Arulenus Junius Rusticus, brother of Junius Mauricus (c. 45, 1 n.) and probably son of Junius Rusticus mentioned by Tacitus as keeper of the senatorial minutes under Tiberius (*A.* 5. 4, 1), belonged to a family that came from the north of Italy. He had Stoic sympathies (Dio 67, 13; Pliny, *Ep.* 1. 5, 2) and was an adherent of Thrasea Paetus. As tribune in A.D. 66 he offered to interpose his veto at Thrasea's trial but was dissuaded by Thrasea from doing so (*A.* 16. 26, 4). He was praetor in A.D. 69 and suffect consul in A.D. 92. He published an account of Thrasea's death (see Introd., p. 14; Suet. *Dom.* 10 attributes to him 'Paeti Thraseae et Helvidii Prisci laudes' but is probably conflating him with Herennius) in which he referred to him as *sanctus* (Dio 67. 13, 2). It is not known when the work was written but it was the main charge against him when he was prosecuted during Domitian's purge in A.D. 93. His memory was attacked in a pamphlet by the time-serving orator M. Aquilius Regulus (Pliny, *Ep.* 1, 5). His book was probably used by Tacitus as a source in *Annals* 16. See O. Murray, *Historia* 14 (1965), 56 ff.

Paetus Thrasea : P. Clodius Thrasea Paetus belonged to an old family from Patavium (Padua) in the north of Italy (*P.I.R.*[2] C 1187). He had a distinguished public career under Claudius and Nero (consul in A.D. 56) and was on good terms with Nero during the early years of his reign. Although no fanatic but a man of gentle and humane temperament (Pliny, *Ep.* 8. 22, 3) he became, after the murder of Agrippina in A.D. 59, increasingly disgusted with the excesses of Nero's régime and felt obliged to protest against the servility and corruption of the senate (*A.* 13. 49; 14. 48; 15. 20; 15. 23, 4) until finally in A.D. 63–64 he withdrew altogether from public life (*A.* 16. 22, 1). His actions were determined not so much by doctrinaire ideology (he was a Stoic and, according to Plut. *Cato* 37, wrote a Life of Cato) or by nostalgia for the defunct Republic (he celebrated the birthdays of M. Brutus and Cassius, the regicides) as by a strong moral sense and personal dignity for which Padua was renowned (Strabo 3, 169; 5, 213). He was prosecuted for *maiestas* by Cossutianus Capito and Eprius Marcellus in A.D. 66 and committed suicide (*A*. 16. 35, 2). For his life and character see C. Wirszubski, *Libertas*, pp. 138 ff.; Syme, *Tacitus*, pp. 555 ff.

Herennio Senecioni: born in Hispania Baetica, Herennius pursued a public career as far as the quaestorship (Pliny, *Ep*. 7. 33, 5) but declined to seek higher office because of his disgust with the régime. He wrote an account of Helvidius Priscus' death based on *commentarii* supplied by his widow, Fannia (Pliny, *Ep*. 7. 19, 5). In association with his friend, the younger Pliny, he impeached Baebius Massa (c. 45, 1 n.), governor of Baetica, for corruption but was himself successfully prosecuted for *maiestas* (Pliny, *Ep*. 7. 33, 7) in revenge by Mettius Carus (c. 45, 1 n.) in A.D. 93. It was said that Domitian was angered by his refusal to hold office (Pliny, *Ep*. 3. 11, 3). A copy of his book was saved from destruction by Fannia, and Tacitus probably had access to it for the now lost portion of the *Histories*.

Priscus Helvidius: C. Helvidius Priscus, the son-in-law of Thrasea, came from Samnite stock. He was exiled when Thrasea was condemned but returned to Rome on the accession of Galba and was elected praetor in A.D. 70 (*H*. 4. 4, 3). He carried on a vendetta against Thrasea's prosecutor Eprius Marcellus (*H*. 4. 6; 4. 43) but was himself banished and later executed in A.D. 74 when Eprius was consul for the second time. Helvidius, like his father-in-law, was a Stoic (*H*. 4. 5, 1) but his fate was due not to his philosophical leanings but to his tactless and intransigent character. He flaunted his independence of opinion (*H*. 4. 6, 1) even to the extent of heckling Vespasian. It was well said of him that 'non aliter quam libero civitatis statu egit' (*Σ* Juv. 5, 36). For a different picture of Helvidius see Dio 66. 12; J. M. C. Toynbee, *Greece & Rome* 13 (1944), 43 ff.; Syme, *Tacitus*, pp. 557 ff.

The four people mentioned by Tacitus belonged to a group, connected by family ties and common interests, which had a long history of opposition to the Caesars in the first century. They were not opposed to the principate as such but attempted to assert the status and integrity of the senate. For Helvidius' son see c. 45, 1 n.; Thrasea was the son-in-law of A. Caecina Paetus (*cos*. A.D. 37) who met his end opposing Claudius (Pliny,

Ep. 3. 16, 13); Arulenus' wife may have been Verulana Gratilla who was exiled for defiance of Domitian (Pliny, *Ep.* 5. 1, 8) and Arulenus himself attended the lectures of Plutarch who was probably one of those expelled after A.D. 93 (see below). For the history of the purge in A.D. 93 see below and c. 45, 1 n.

saevitum : sc. *esse*. This fact would also be recorded in the *acta*.

triumviris : sc. *capitalibus*. These officers, who formed one section of the minor magistrates collectively called *vigintiviri*, superintended the infliction of capital punishment. They were executive officers acting upon instructions from the aediles (cf. *A*. 4. 35, 4; Livy 25. 1, 10).

in comitio ac foro. The *comitium* was the space at the north-western end of the Forum adjoining the Senate-house and separated by the Rostra from the Forum. It was the meeting-place of the old Comitia Curiata and the ancient place for trials and punishments (Livy 9. 9, 2, etc.; Pliny, *Ep.* 4. 11, 10), and for the burning of condemned books (Livy 40. 29, 14). The addition *ac foro* (cf. Pliny, *N.H.* 15, 77 'ficus arbor in foro ipso ac comitio Romae nata') emphasizes the public character of the place chosen.

2, 2. **vocem** : Tacitus implies that creative originality died under Domitian but this is not wholly true. There were poets (Statius, Silius Italicus, Valerius Flaccus) and prose-writers, above all Quintilian, who flourished under the régime and the names of several other authors are known. But it is true that they were essentially court-writers dependent on official patronage. Freedom of judgement and expression was suppressed.

libertatem : see c. 3, 1 n.

conscientiam, 'the moral consciousness of mankind', not merely their knowledge or remembrance (*memoria*). A similar judgement is expressed in *A*. 4. 35, 5 'praesenti potentia credunt extingui posse etiam sequentis aevi memoriam'. There, however, Tacitus is referring to the preservation of such books in spite of these precautions; here he is expressing the futility of the attempt to suppress the free moral judgement of men.

arbitrabantur. The subject (Domitian and his advisers) is supplied from the sense.

expulsis : aoristic abl. abs., adding another fact (cf. c. 14, 3; 22, 1; 23, 2, etc.). An expulsion of philosophers by Domitian, attested by several writers, is connected by Suet. (*Dom.* 10) and Dio (67. 13, 3) with the execution of Arulenus Rusticus (see above), and is stated by Pliny (*Ep.* 3. 11, 2) to have taken place in his praetorship, which is placed in A.D. 93. The cause was political. Domitian's relations with the Senate had reached breaking-point and, like Vespasian before him, he struck at the most articulate supporters of the opposition. In addition to the politicians (Arulenus, Mauricus, etc.) he attacked the intellectuals, chiefly Greeks and orientals, who were traditionally viewed with suspicion by the Roman authorities for their free thinking (Fuchs, *Der geistige Widerstand gegen Rom*). Among those forced to leave were Artemidorus, Epictetus, and, probably, Plutarch. Dio of Prusa's withdrawal from Rome in *c.* A.D. 84 was for quite different reasons: his patron had been murdered. [The statement in Eusebius, *Chron.*, that there had been a previous expulsion of philosophers and astrologers in the year Oct. 88–Sept. 89 (not elsewhere alluded to), if not due to confusion with the expulsion under Vespasian, would indicate that a certain number of them were charged with complicity in the revolt of Saturninus (see c. 41, 2 n.; Sherwin-White, *Letters of Pliny*, pp. 763 ff.).]

atque, etc.: following Sallust's usage (*Cat.* 10, 4; *Jug.* 1, 3; see D. C. Earl, *The Political Thought of Sallust*, pp. 10 ff.) Tacitus generally indicates by *bonae artes* not liberal studies (which he calls *artes civiles, ingenuae, liberales*, etc.) but moral qualities in action (*A.* 1. 9, 3; 2. 73, 3; 6. 46, 1 etc.). The sentence therefore amplifies what has gone before. Not only was philosophy banished but with it the practice of all the moral virtues (*fides, probitas*, etc.) was driven into exile.

2, 3. patientiae, 'submissiveness': cf. c. 15, 1; 16, 2.

ultimum, 'the extreme'. The times referred to are ancient only by comparison, the reference being to the lawlessness of the later Republic.

nos : sc. *vidimus.*

inquisitiones, 'espionage': cf. c. 43, 2. The description of the terror produced by such a system under Tiberius in *A.* 4. 69, 3 is probably coloured by reminiscences of this time.

loquendi . . . commercio, 'the intercourse of speech and hearing', the interchange of ideas. It was a crime not only to have spoken, but to have listened.

oblivisci quam tacere : a commonplace which goes back at least to a saying of Themistocles (Cic. *Acad.* 2, 2; cf. Cic. *Flacc.* 61).

3, 1. nunc : since Domitian's death (a narrower sense than in c. 1, 4). The renaissance under Nerva and Trajan is attested by Pliny (*Ep.* 1. 10, 1; 13, 1; 3. 18, 5) and besides Pliny and Tacitus the names of several other writers are known who were emboldened to publish: e.g. poets (Vestricius Spurinna, Arrius Antoninus, Pompeius Saturninus, C. Passennus), letter-writers (Voconius Romanus), biographers (Titinius Capito), and orators. Martial had retired to Spain, but it is significant that the bulk of Plutarch's writing belongs after A.D. 96 when he was over 50. He too must have felt constrained to silence under Domitian (see C. P. Jones, *J.R.S.* 56 (1966), 73.

redit = *redire incipit.* For *animus redire* cf. Livy 2. 43, 8; Ovid, *Ep. Her.* 13, 29.

et, 'and yet', as often: cf. c. 9, 3; 14, 3, etc. The correction to *set* is here extremely easy, but we have a parallel use of *et quamquam*, with the force of *quamquam autem*, and with *tamen* (as here) marking the apodosis, in c. 36, 3 and *H.* 2. 30, 2.

primo statim : coupled for emphasis.

saeculi ortu. Editors assume that this new period is imagined as rising like a star (cf. c. 44, 5), but the metaphor may rather be derived from birth (cf. Virgil, *Ecl.* 4, 5 *saeclorum nascitur ordo*). So Pliny uses *saeculum* of the Trajanic period (*Ep.* 2. 1, 6).

Nerva Caesar, etc. For the absence of the title *divus*, see Introd., p. 10. The language is complimentary to Trajan, who was in fact absent from Rome till A.D. 99.

olim, 'long since': cf. *H.* 1. 60, 1, etc.

dissociabiles, 'incompatible', as in Hor. *Odes*, 1. 3, 22. The word is not found elsewhere in Tacitus but *insociabilis*, which Novák would read here, is used in the same sense in *A.* 4. 12, 4; 13. 17, 1.

principatum ac libertatem, 'the principate and freedom of judgement'. An inscription dated on the day of Nerva's election (18 Sept. A.D. 96) was erected on the Capitol by *S.P.Q.R.* to 'Libertas Restituta' (*ILS* 274). Cf. the expression of Pliny (*Ep.* 9. 13, 4), 'primis diebus redditae libertatis'. What people meant by *libertas* at this period was not constitutional safeguards of citizen-rights nor the freedom of a citizen to determine his own and his country's destiny—

the principate was accepted as inevitable (*Dial.* 41, 5; *A.* 4. 33, 2) and republicanism, except once after the death of Caligula, was never seriously envisaged (Suet. *Claud.* 10, 3)—but the right of a senator to make his own contribution in the senate and in the service of the state (Wirszubski, *Libertas*, pp. 124 ff.). The abuse of absolute power by the emperors after Augustus imperilled the self-respect of the senate and the very lives of the senators. The position could only be rectified by the moderation of the emperor.

The conflict between *principatus* and *libertas* was a commonplace (cf. Lucan 7, 691 ff.; Plut. *Galba* 6) and Tacitus is here making use of contemporary slogans. The legend 'Libertas Restituta' had already appeared on coins after the death of Nero. See also on *fel. temp.* and *sec. publ.* below.

felicitatem temporum, a phrase used in *H.* 1. 1, 4 *rara temporum felicitate*; Pliny *Ep.* 10. 58, 7. On coins the goddess Felicitas is often mentioned and figured with the titles *F. temporum*, *publica*, etc. or more commonly *F. Augusti*, *Caesarum*, etc., the Emperor being the author and guarantor of the prosperity of the State (cf. Suet. *Aug.* 58). The legend *F. publica* starts with Galba (*The Roman Imp. Coinage*, 55) and is abundant under Vespasian but does not occur on the coins of Nerva or Trajan. It was probably being replaced as a popular slogan by *F. temporum* which is first found on the coins of M. Aurelius in A.D. 161.

nec spem, etc.: an obscurely phrased expression for 'our prayers for law and order are now in process of fulfilment and we are encouraged accordingly'. 'Public security has not merely framed hopes and prayers but has gained the assurance of her prayers' fulfilment and strength therefrom'. With the first clause some word like *conceperit* is supplied by zeugma. *Fiduciam ac robur*, balancing *spem ac votum* (a frequent pair; cf. *G.* 19, 2; *A.* 4. 39, 2) may be a hendiadys ('a strong assurance of her prayers' fulfilment') but *robur* seems rather to add the idea of strength gained from the confidence that the prayers are being fulfilled; cf. Sen. *Ep.* 94, 46 'duae res plurimum roboris animo dant, fides veri et fiducia'.

Securitas publica (*rei publicae*, *temporum*, etc.) or *Securitas Augusti*, a personification of the public and political security which the world owed to the imperial government, was deified like Felicitas and other abstract ideas. The legend *Securitas publica* first appears on the coins of Hadrian (*The Roman Imp. Coinage*, 221; A.D. 132–4) but was already a current conception

(Sen. *Ep.* 73, 2 'multum ad propositum bene vivendi confert securitas publica'). *Felicitas* and *securitas* are similarly juxtaposed by Pliny (*Ep.* 10. 58, 7). To Securitas vows were offered (*ILS* 2933, 3788), altars erected (*CIL.* xiv. 2899), and sacrifice made, e.g. on 10 Jan. A.D. 69 after Piso's adoption (*CIL.* vi. 2051, i. 30).

tardiora, 'slower to act'; so *tarda legum auxilia* (*A.* 6. 11, 2). The thought is a commonplace for which cf. Lucr. 1, 556 ff.

subit, 'comes over us', as in Livy 2. 42, 1.

3, 2. quid, si, etc.: a rhetorical formula introducing a new and stronger argument by means of an appeal to a circumstance which either is the case (so with the indicative, as here) or might easily have been (and perhaps yet can be) the case, though it was not (or is not) at the moment (so with the subjunctive). 'What if we have lost not only the inclination, but (by the destruction of the fittest and by disuse) even the power to write?' The answer is left to be supplied.

quindecim: the whole reign of Domitian, A.D. 81–96. His policy of repression is elsewhere noticed before his last and worst period (c. 39, 2); nor is this inconsistent with the generally good character of his early government as described in Suet. *Dom.* 9.

fortuitis, a word often used of natural in contrast to violent deaths; cf. *A.* 4. 8, 1, etc.

promptissimus: sc. *ingenio,* 'the most active minds', such as Rusticus and Senecio. Cf. Livy 5. 3, 1, etc.

pauci et, etc. The construction is 'pauci superstites sumus et (= *et quidem,* cf. c. 10, 3; 20, 3, etc.) non modo aliorum sed etiam nostri'. *Et* is necessary, for without it the words would imply that there were other survivors who had not outlived their faculties.

ut sic dixerim. *Uti dixerim* of the MSS. cannot be satisfactorily defended. Tacitus uses only the form *ut sic dixerim* for the classical *ut ita dicam* to qualify a strong expression (*Dial.* 34, 2; 40, 3; *G.* 2, 1; *A.* 14. 53, 4). The corruption is easy, resulting from haplography: *c* and *t* are constantly confused (c. 6, 4 n.).

nostri superstites: an expression used (also with a qualifying word) in Sen. *Ep.* 30, 5, 'vivere tamquam superstes sibi'. 'We have outlived ourselves, i.e., our faculties.'

exemptis, 'taken out', as in *A*. 3. 18, 1 (where it is perhaps used with simple abl.). Elsewhere Tacitus uses this verb with dative.

iuvenes, etc. By old Roman law a man passed from the *iuniores* to the *seniores* after his forty-fifth year (Aul. Gell. 10. 28, 1); after his fiftieth year he was not liable for military service (Sen. *de Brev. Vitae* 20, 4); after his sixtieth he was not required to attend the Senate (Sen., loc. cit.). Tacitus himself had passed from about his twenty-fifth to his fortieth year under Domitian.

exactae aetatis terminos, 'the limits of spent life', i.e. the end of life's course. Cf. *H*. 3. 33, 1.

per silentium : used in *A*. 4. 53, 1, etc., with merely the sense of *silens,* but here like *per cultum* in c. 4, 2; or possibly with instrumental force = *silendo,* to imply that they only saved their lives by silence: cf. the use of *per* in c. 6, 1; 40, 4; 46, 3. For the picture of enforced servility at this time cf. Pliny, *Ep*. 8. 14, 8.

3, 3. non tamen pigebit, 'yet (in spite of the difficulties which beset me) it will not be an unpleasant task' (= *iuvabit*). In *A*. 1. 73, 1, a possibly distasteful subject is prefaced by *haud pigebit referre*. The past tense of *composuisse* refers to the time of publication. It is a common idiom (cf. Virg. *Aen*. 7, 233; Livy, *Praef*. 3; Quint. 1. 1, 34; 3. 1, 22, 'non tamen pigebit . . . posuisse sententiam').

vel incondita ac rudi voce, 'even in an uncouth and rough style'. Tacitus refers to the neglect of his own literary powers during the last fifteen years rather than to a general dearth of historical writing. In fact he had cultivated oratory with great success and such self-depreciation was conventional (cf. Cic. *Orator* 230 (Caelius Antipater); Stat. *Silv*. 1, *pr*.; later examples in T. Jansen, *Latin Prose Prefaces*, p. 133).

memoriam, etc., 'a record of our past slavery'. The passage shows that, soon after Nerva's accession, Tacitus had formed the project of writing the *Histories,* not, however, quite in the form in which they appeared. The work is spoken of as intended to be, if not a monograph on Domitian, at least chiefly a history of his rule; and, though he could not at that early date have projected a history of Nerva, still less of Trajan, a *testimonium* of the happy change inaugurated was to come in as a climax and contrast. By the time the work was published, it had grown into a complete history from Galba to Domitian; and the great subsequent era of Trajan, with the career of conquest opened out by it, was relegated to a separate work (*H.* 1. 1), and ultimately abandoned.

interim. It was conventional (and often tactful) under the early Empire for authors in many fields to include in their prefaces a promise to write at a later date about the reigning emperor and the blessings of his age (cf. Virg. *Georg.* 3, 16–39 and esp. 40 'interea Dryadum silvas saltusque sequamur' (cf. *interim* here); Pliny, *N.H. praef.* 20; Stat. *Theb.* 1, 32; *Ach.* 1, 17; Tac. *H.* 1. 1, 4). Such promises need not have been intended seriously.

professione, etc. Cf. *H.* 2. 60, 2, *pietate . . . excusatus.* The context would seem to connect this with the previous apology for any want of finish in style, but the plea that his work is an act of dutiful affection is intended as a further deprecation (cf. c. 1, 4) of the jealousy roused by the praise of others. For the general dislike of a picture of exalted virtue, cf. *A.* 4, 33, 4, and Seneca, *de Vita Beata* 19, 2 'quasi aliena virtus exprobratio delictorum omnium sit'.

4–9. *The Early Career of Agricola*

Tacitus begins his biography proper with an account of the background and career of Agricola until his appointment as governor of Britain. The themes here treated (parentage, education, habits) are, naturally, of a kind conventional in biography; and it is instructive to compare the material of these chapters with, e.g., Horace's account of his early years (*Sat.* 1. 6, 64 ff.) where the same stress is laid on parental influence and on modesty. For details see Introd., pp. 2 ff.

4, 1. Cn. Iulius Agricola : the only certain case where Tacitus gives all the *tria nomina* of a person together (cf. *A.* 2. 1, 1; 12. 41, 1). This formal, and perhaps unique, introduction signals that Agricola is the subject of the work (for this device cf. *A.* 1. 1, 1 *urbem Romam,* and see Gow's note on Theocr. 1, 65). Agricola's full name also occurs on a lead pipe from Chester (*ILS* 8704a).

vetere, etc. This is a good example of the growth of a new aristocracy under the early Empire. See Syme, *Tacitus,* pp. 585 ff.

Foroiuliensium: Fréjus, *Octavanorum colonia, quae Pacensis appellatur et Classica* (Pliny, *N.H.* 3, 35), owing its foundation to Julius Caesar and its importance to the refounded *colonia* and naval station established there by Augustus (*A.* 4. 5, 1). On the date of Agricola's birth there, 13 June A.D. 40, see c. 44, 1 and note.

utrumque avum: nothing is known of either of these men.

Caesarum, i.e of more than one Caesar, doubtless Augustus and Tiberius. Cf. the career of C. Herennius Capito (*AE*, 1941, 105) . . . *proc. Ti. Caesaris Aug. proc. C. Caesaris Aug. Germanici.*

quae equestris nobilitas est, 'which is a noble equestrian office'. The tenure of the greater procuratorships, held by *equites* after serving as officers in the army, such as those carrying with them the government of lesser Imperial provinces, or control of finance in the greater provinces (see c. 9, 4 and note), or in grouped provinces, was considered to confer distinction on their holders, just as in Republican times the attainment of curule office by plebeians gave nobility to their families. Such *equites* are designated by Tacitus *equites inlustres* or *insignes* (in contrast to *equites modici*) and by the younger Pliny *equites splendidi*; Livy speaks of *equites nobiles* (23. 46, 12). The distinction was entirely unofficial. [For the careers and powers of procurators see A. H. M. Jones, *Studies in Roman Government*, 111 ff.; P.-G. Pflaum, *Les Carrières Procuratoriennes Equestres*; F. G. B. Millar, *Historia* 13 (1964), 180 ff.; id. 14 (1965), 362 ff.]

senatorii ordinis. The inference from a reference to his *ludi* in Sen. *de Ben.* 2. 21, 5, that Graecinus reached the praetorship, is confirmed by his tombstone (*Boll. Com.* 68 (1940), 178 = *AE*, 1946, no. 94) which cites him as *tribunus plebis* and *praetor*. The *cognomen* belongs also to the Pomponii (see *A.* 13. 32, 3).

studio, etc. He is called *vir egregius* in Sen., loc. cit., and *Ep.* 29, 6, and besides being an orator and philosopher, is mentioned in Col. 1. 1, 14, as author of a treatise 'de vineis', *composita facetius et eruditius.* See Introd., p. 2. It seems probable that the *cognomen* of 'Agricola' given to his son reflected his interest in agriculture.

namque: explaining the opportunity taken to gratify his spite. Seneca rhetorically says of Graecinus, 'quem C. Caesar occidit ob hoc unum quod melior vir erat quam esse quemquam tyranno expedit' (*de Ben.* 2. 21, 5). As Wex observed, the tense of *abnuerat* implies an interval between the order and the death of Graecinus; and as Agricola was born on 13 June A.D. 40 (c. 44, 1), his father cannot have died before September of the previous year, and probably not before Agricola's birth, since the son is not called *postumus*. Urlichs thought that Graecinus may have perished when Gaius returned from Gaul early in A.D. 40 (not later than 29 May, *CIL*. vi. 2030, 15), and this is very probable. The refusal to act against Silanus cannot have been the immediate cause of his death, but he was doubtless in disfavour and the ultimate pretext is unknown.

M. Silanum: it is usually assumed that this was M. Junius Silanus C. f., *cos. suff.* in A.D. 15, the father of the first wife of Gaius (*A*. 6. 20, 1) and an intimate of Tiberius. He became heartily disliked by Gaius after his accession and was forced to commit suicide (Suet. *Cal*. 23; Dio 59. 8, 4; Philo, *Gai*. 62–65). His death took place early in A.D. 38 since his successor in the Arval Brethren was co-opted on 24 May A.D. 38. There is, however, no record that this Silanus was threatened by prosecution and the interval between his death and that of Graecinus (see above), about 2½ years, is surprising. It is possible that M. Junius Silanus M. f., *cos*. in A.D. 19 is meant. He was proconsul of Africa for at least six years, probably A.D. 33–38 (*ILS* 2305, 6236), during which he incurred the suspicion and enmity of Gaius (*H*. 4. 48, 1, as here, *M. Silanum*). As a potential claimant to the throne, he may well have been menaced with impeachment on his recall. He is not heard of after his proconsulship.

4, 2. mater. On her death see c. 7, 1. On the name Procilla, which is frequent in Gaul, see Introd., p. 2.

rarae castitatis: the eulogy is conventional and is found on tombstones (*CIL*. vi. 8508; ix. 1893; see R. Lattimore, *Themes in Greek and Roman Epitaphs*, p. 296).

Tacitus' account of Agricola's early life is fulsomely rhetorical. Notice especially the accumulation of virtual synonyms (*bonam integramque*; *mixtum ac bene compositum*; *sedem ac magistram*; *incensum ac flagrantem* (cf. Cicero, *de Fat*. 3); *sublime et erectum* (cf. Quintilian 11. 1, 16); *magni excelsique* (Cicero, *de Off*. 1, 81, etc.). It should be compared with the picture in *Dial*. 28–30.

sinu indulgentiaque, best taken as hendiadys: 'under her loving care'. Cf. Sid. Apoll. 5, 16 'in indulgentissimo sinu nutritus'. *Indulgentia* has often a bad sense, but is used of parental tenderness in several places. For a mother thus to bring up her child herself instead of putting it out to nurse is spoken of as an old custom becoming uncommon: cf. *Dial.* 28, 4 'nam pridem . . . inservire liberis'. Cf. also Juv. *Sat.* 14. For the phrase cf. also Livy 1. 39, 3 *indulgentia nutriamus*.

per omnem, etc., 'by a complete training in liberal studies' (see c. 2, 2, note). On the liberal arts, as understood by Tacitus, see *Dial.* 30, 4, where five (geometry, music, grammar, dialectic, ethics) are expressly mentioned. *Per* often denotes the mode in which time is spent: cf. c. 3, 2 n.; 18, 5.

peccantium : substantival, '(the attractions) of evil-doers, i.e. of vice'. See c. 11, 2 note.

bonam integramque naturam, 'his good and upright disposition'. Cf. *Dial.* 28, 6 'sincera et integra et nullis pravitatibus detorta . . . natura'.

Massiliam. Cicero speaks strongly of the *disciplina* and *gravitas* of this city (*Flacc.* 63), and Strabo, in a very interesting description of its condition at his time (4, 179–81), says that the best Romans preferred it to Athens as a place of Greek culture (rhetoric and philosophy), which he ascribes to its greater simplicity of life. The young Octavian received part of his education there. Cf. also *A.* 4. 44, 3; Val. Max. 2. 6, 7. Its reputation went back to the time of Plautus (cf. *Cas.* 963).

comitate, 'courtesy', refinement of manners, opposed to *adrogantia* (*H.* 1. 10, 2), or roughness generally (cf. *A.* 4. 7, 1).

provinciali parsimonia. Cf. *A.* 3. 55, 3; 16. 5, 1.

mixtum, etc., 'presenting a blend and happy combination': the latter expression lays stress on *bene*, and such a concise use of *mixtus* for *in quo mixta sunt* resembles *H.* 1. 10, 2; *A.* 6. 51, 3.

4, 3. philosophiae. Tacitus, as Wölfflin notes, generally uses *sapientia* and *sapiens*, substituting *philosophia* or *philosophus* only here and in *A.* 13. 42, 4 (for variation in the same passage) and in *H.* 3. 81, 1: all three passages have a pejorative tinge.

acrius : to be taken as an adverb defined by the following words, which are in asyndeton; cf. *Vita Persi* 5, *acriter philosophantium*. Gronovius took it as an adjective qualifying *studium* (cf. *Dial.* 33, 2; *A.* 2. 5, 2, etc.) but the word-order is against this.

concessum. The old Roman antipathy to philosophy, noted apologetically by Cicero (*Off.* 2, 2, 'vereor ne quibusdam bonis viris philosophiae nomen sit invisum') still survived and rested on its drawing men away from a public life: cf. 'ut nomine magnifico segne otium velaret' (*H.* 4. 5, 2) and 'a philosophia eum (Neronem) mater avertit, monens imperaturo contrariam esse' (Suet. *Ner.* 52; so in later times, cf. *SHA. Alex.* 14, 5). The antipathy also existed in Athens (see the discussion in Plato, *Gorgias* 484 c). Cf. also note on c. 2, 2.

senatori: the phrase, which has been suspected since Agricola was not yet a senator, is rhetorical; cf. Livy 7. 13, 9 'ut viris ac Romanis dignum sit'; 7. 35, 8, etc. For a similar rhetorical inexactitude cf. c. 32, 4 n. Senators' sons belonged, like other *laticlavii*, to the equestrian order till the tenure of the quaestorship gave them a seat in the Senate, but they were described as belonging to the *ordo senatorius*, in the sense of 'senatorial class' (*A.* 13. 25, 2 compared with Suet. *Ner.* 26).

hausisse, 'would have imbibed', used figuratively. Cf. *Dial.* 30, 3 'omnes philosophiae partes penitus hausisse'. In direct speech *hauserat* would have been used (cf. *H.* 3. 27, 3 'incesserat cunctatio ni duces . . . monstrassent'), a dramatic substitute for *hausisset*. The idiom is common in Livy.

pulchritudinem ac speciem: 'the (inner) beauty and (outward) fine appearance of glory', i.e. 'the beauty and splendour'. For *speciem gloriae* cf. Cic. *ad Fam.* 10. 12, 5 'omnia quae habent speciem gloriae', there used pejoratively. A similar conjunction is found in c. 34, 3 'pulchram et spectabilem victoriam', and in Livy 1. 9, 12 'insignem specie ac pulchritudine'.

vehementius quam caute. We should expect *cautius* (cf. c. 44, 2, etc.); but we have a parallel in *H.* 1. 83, 3 'acrius quam considerate', if Walther's correction of *M*'s *considerat* is accepted.

ratio, 'discretion': cf. c. 6, 4.

modum: probably best taken in the sense of *moderatio*, 'a sense of proportion', a temperament preventing him from being carried into extremes of thought or action, like some members of the Stoic opposition. The thought, which is something of a paradox, was a rhetorical commonplace (cf. Doxopater in *Rhet. Graec.* (ed. Walz) 2, 429); cf. also the proverbial *optimus modus est* (Pliny, *Ep.* 1. 20, 20, etc.). Tacitus here strikes the key-note: moderation is throughout the prominent trait of Agricola's character.

5, 1. castrorum rudimenta, 'first lessons in army life': cf. *castrorum experimentis* (c. 16, 3) and the poetical use of *belli rudimenta* (Virg. *Aen.* 11, 156). He was *tribunus laticlavius*, the qualification for admission to the quaestorship and senate (Suet. *Aug.* 38). The duties were mainly administrative.

Suetonio Paulino : cf. cc. 14–16. C. Suetonius Paulinus as ex-praetor in A.D. 41 had held a successful command against the Mauretanians during which he became the first Roman to cross the Atlas Mts. (Pliny, *N.H.* 5, 14). He was rewarded with the suffect consulship, probably in A.D. 43. When Nero decided to persevere with the annexation of Wales, he selected Paulinus as an expert in mountain warfare to succeed Veranius. After his recall Paulinus is not heard of until A.D. 69 when he appears as one of Otho's leading generals and fought at Bedriacum. He was a cautious but determined commander (*moderatus*; cf. *H.* 2. 25, 2). He wrote memoirs which were probably used as a source by Tacitus.

adprobavit = *effecit ut probarentur*, 'he won approval for' a concise combination of two statements, that he performed his first service under Paulinus, and to his satisfaction. Cf. the use of *adprobare* in c. 42, 1.

electus = *nam electus est*, aoristic (timeless) participle, giving the proof of the preceding statement. The choice was an indication of the approval already won by the young officer and a means of testing his further capacity (see next note).

quem contubernio aestimaret. Agricola was picked out by Paulinus 'to be tested on Headquarters staff' (*contubernio* instrum. abl.). The construction is similar to 'delectus cui . . . Antonia . . . in matrimonium daretur' (*A.* 4. 44, 2), and for the abl. cf., e.g., Val. Max. 4. 7, 3, 'si amicitiae fido pignore aestimetur (L. Reginus)'. A tribune would not ordinarily be attached to the headquarters staff: he would be subordinate to the *legatus legionis*. Agricola's selection was due to his proved efficiency in routine duties (cf. Pliny, *Ep.* 7. 31, 2). He had come to Paulinus' notice and was seconded for probationary service at headquarters. The position must be distinguished from that of a *contubernalis* on a proconsul's staff: for such *comites* (not all young men) were civilians; in Imperial times, whether chosen by senatorial governors or nominated by the Emperor they assisted in administration, especially as judicial assessors, and were paid a salary (*Dig.* 1. 22, 4; W. T. Arnold, *Roman Provincial Administration*[3] (1914), p. 68).

nec Agricola, etc.: equivalent to *et Agricola neque*, as in c. 8, 3 and in 18, 6 to *et . . . non*. In the construction of the following words, the supposition that *egit* is to be supplied with *licenter* (cf. c. 19, 2; *H*. 1. 84, 1) is inadmissible when it has to stand in contrast with another verb, and its insertion or the omission of *neque segniter* are very violent methods of procedure. It is quite possible to refer *rettulit*, etc., to both clauses, and to take the whole to mean 'and Agricola did not either irresponsibly ("at his own sweet will"), like young men who turn military service into self-indulgence, or indolently (i.e. did not either from love of amusement or dislike of work) regard his title (rank) of military tribune and his inexperience as a ground for taking pleasure and furlough'. *Voluptates* seems to correspond to *licenter* (explained by *in lasciviam*)—since *lascivia* means 'frivolity', as in *A*. 11. 31, 3; *H*. 2. 68, 2, etc.—and *commeatus* to *segniter*; *et* being used for *aut* (as in c. 22, 2 *ac fuga*) because *voluptates* and *commeatus* are parts of one idea, pleasure taken on the spot and on leave. Others take *et* as explanatory. In any case the thought is somewhat confused, for *licenter—in lasciviam* involves *segnitia*, and *segnitia* implies love of amusement (*lascivia, voluptates*). Alternatively it is possible to understand (by zeugma) *tribunatum exercuit* (or the like) with *licenter*: 'Agricola did not either use his tribunate irresponsibly like many young men, who turn their military service into having a good time, or indolently take advantage of his inexperience and rank to lead an idle life of pleasure'.

Referre ad, 'to regard as a means to pleasure', is similar to Cicero's use (with *voluptatem* in *Lael*. 32, and often in his philosophical writings), 'to judge by the standard of', 'to regard as the end'. *Titulus* does not imply that the office was 'titular', but means 'rank': cf. Livy 7. 1, 10; 28. 41, 3. The demoralization of the service by the constant purchase of leave and exemptions is dwelt upon in *H*. 1. 46, 2–3.

noscere . . . agere. The infinitives are historical.

in, 'for the sake of' (cf. c. 8, 3, and note). **ob,** 'by reason of'; *propter* in this sense was avoided by Tacitus, as being a popular word. Notice the variation of prepositions.

simulque : coupling *agere* to the other verbs.

et anxius et intentus, 'both with caution and alertness'. The former word denotes that he did not despise his enemy, the latter that he was alert to seize an opportunity. Cf. Val. Max.

8. 7, 7, 'attenti et anxii et numquam cessantis studii praemia'. Agricola is depicted as doing everything that a model young officer should do (cf. Xen. *Cyrop*. 5. 3, 48; cf. also Thuc. 2. 11, 4 ff.).

5, 2. **exercitatior,** 'more troubled', is probably what Tacitus wrote here, though elsewhere he uses the word in a very different sense (c. 36, 1; *A*. 12. 12, 1; 14. 59, 2). *Exercitatus* has the meaning of 'troubled' in Cicero, *de Rep*. 6, 26 (with *curis*) and Hor. *Epod*. 9, 31, 'exercitatas . . . petit Syrtis Noto'. It seems therefore unnecessary to substitute *excitatior* (used by Livy 4. 37, 9, etc., but only of sounds and similar sensations).

in ambiguo, 'in uncertainty', its possession trembling in the balance.

coloniae : probably a rhetorical plural, referring only to Camulodunum (Colchester). Cf. c. 32, 3 note. It is most unlikely that the word is used to include other towns, not colonies (as London and Verulamium), which suffered also. The *veterani* were the colonists. On the events see *A*. 14. 32 and Introd., p. 53.

intersaepti : 'isolated, beleaguered'. The reference is to the Ninth Legion (*A*. 14. 32) which attempted to go to the aid of Camulodunum but after a bloody repulse was forced to take refuge behind the defences of a camp. *Exercitus* is a rhetorical plural, like *coloniae*. The emendation *intercepti* (as in c. 28, 3 and 43, 2), which would mean 'destroyed', is unnecessary. For *intersaepti* cf. *H*. 3. 21, 2; 53, 1.

de salute, etc. Cf. c. 26, 2. The antithesis is conventional: cf. Caes. *B.C*. 3. 111, 5.

5, 3. **alterius.** This genitive is constantly used for *alius* to avoid the ambiguity of that form: cf. c. 17, 2.

summa rerum. Tacitus uses *summa rerum* to mean 'the supreme direction of affairs' (as in *H*. 2. 33, 2; 4. 25, 4: so commonly in Livy) or 'the general plan (situation)' (as in *H*. 2. 81, 3; 3. 50, 3) or 'the whole issue' (as in *H*. 3. 70, 3; cf. Vell. Pat. 2. 68, 'in acie de summa rerum dimicat'). Here Tacitus means that the overall responsibility lay with Paulinus, some such meaning as 'devolved upon' being supplied by zeugma from *cessit in*. For *cessit in* 'passed to', 'fell to', cf. *A*. 1. 1, 3, etc.

artem et usum, 'skill and experience'. Cf. Caes. *B.G*. 2. 20, 3 'scientia atque usus militum'; Veg. *de Re Mil*. 3, 9 'scientes artem bellicam an ex usu temere pugnantes'. *Addere stimulos* is not a new phrase (Lucan 1, 263; Sen. *Med.* 833; Stat. *Theb.* 10, 629; Sen. *Ep.* 34, 2; *de Ben.* 1. 15, 2; Quint. 10. 7, 16).

cupido. In the minor works of Tacitus this word occurs here only, the more popular *cupiditas* four times; the latter is rare in *Hist.* and never found in *Ann.*, while *cupido* is very common in both. On this tendency see Syme, *Tacitus*, App. 51.

temporibus, abl. The later years of Nero are referred to (c. 6, 3 *sub Nerone temporum*), and the chief instance in the writer's mind is no doubt that of Corbulo.

sinistra, 'unfavourable'. Cf. *H.* 1. 51, 5, *sinistra fama,* etc.

erga, 'against', or 'in relation to', a sense common in Tacitus and very rare before him.

ex magna . . . ex mala. Notice the alliteration. The epigram, as usual closing a section, is conventional; cf., e.g., Pindar, *Nem.* 8, 21–22; Livy 6. 11, 7.

6, 1. degressus (cf. c. 18, 2) is used of departing from a place, as *digredi* of parting from a person. As Urlichs suggested, Agricola probably left with his chief towards the end of A.D. 61, and may have held the *vigintiviratus* in 62.

natalibus : used of ancestry in the silver age by Tacitus, the younger Pliny, Juv., etc. Her father [T.?] Domitius Decidius, probably (like his son-in-law) a native of Gallia Narbonensis, is shown by an inscription (*ILS* 966) to have been one of the first *quaestores aerarii* chosen by nomination of Claudius (in A.D. 44: *A.* 13. 29, 2; Dio 60. 24, 1), and to have been afterwards praetor (by the same ordinance). He must have been a relation of T. Decidius Domitianus known as a procurator of Augustus in Spain (*P.I.R.*[2] D 22; cf. D 143). Probably the marriage took place in 62, and the son (§ 2) was born in time to enable Agricola to gain a year under the provisions of the *lex Papia Poppaea,* so as to stand for the quaestorship at the end of 63, in his twenty-fourth year. See Appendix I.

decus ac robur, 'gave distinction and substantial help'. The marriage brought Agricola both distinction (*decus*) and more solid advantages (*robur*) from the support of so illustrious a family and its connexions. The words are similarly paired by Stat. *Ach.* 1, 787.

vixeruntque mira concordia, etc. The thought and language of this sentence is sincerely, if conventionally, eulogistic.

Tacitus pays a fine compliment to his mother-in-law who was still alive (c. 46, 3). The theme of conjugal affection is common on epitaphs; cf. *CIL*. iii. 4592 'vixit concorditer' (see R. Lattimore, *Themes in Greek and Roman Epitaphs*, pp. 280 ff.). For *per mutuam caritatem* cf. Sen. *Ep*. 9, 11. See also next note. *Concordia* is modal abl., and *per . . . anteponendo* express the instrumentality by which the concord was maintained.

in vicem se anteponendo, 'putting each other first', each putting the other before self. For the thought cf. St. Paul, *Ep. ad Rom*. 12, 10. As a rule, *in vicem* is used, without *se*, for the classical *inter se*, 'each other' (which is also found in Tacitus, *A*. 3. 1, 3, etc.), e.g. c. 37, 5 *vitabundi in vicem*; *A*. 13. 2, 1 *iuvantes in vicem*. But *se* is sometimes added, e.g. *Dial*. 25, 5 'in vicem se obtrectaverunt'; Pliny, *Ep*. 3. 7, 15.

nisi quod, 'except that', 'only'; an expression often used to qualify something that has been stated (cf. c. 16, 5, etc.), and sometimes, as here (cf. *A*. 1. 33, 3, etc.), to qualify something implied in a previous statement. Here the implied thought is difficult to supply, but appears to be: Domitia deserved more credit since a good wife is a rarity and is, therefore, as praiseworthy as a bad wife is detestable. The thought is proverbial (cf. Eur. *Melan*. fr. 494 Nauck; Semonides 6; Stobaeus); for the form cf. Hesiod, *W.D*. 346 πῆμα κακὸς γείτων ὅσσον τ' ἀγαθὸς μέγ' ὄνειαρ.

6, 2. sors quaesturae, etc. One of the quaestors of the year was allotted to the proconsul of each senatorial province. L. Salvius Otho Titianus, the brother of the emperor Otho, and prominent in the first two Books of the Histories, had been consul A.D. 52, and it appears that his proconsulship of Asia fell in the year A.D. 63–64. (Cf. Appendix I.) Agricola was quaestor in the latter year; but it is not certain whether his quaestorship extended also into the proconsulate of Titianus' successor, L. Antistius Vetus, a most upright man (*A*. 16. 10, 2).

neutro = *neutra re*. So *nullo* in *A*. 3. 15, 2.

parata peccantibus (sc. *esset* from below), 'lying ready for wrong-doers', through the temptations of its wealth and works of art, and the facility of finding tools for iniquity. Cf. *H*. 1. 6, 2, 'materia . . . audenti parata'. Cicero (*ad Q.F*. 1. 1, 19) congratulates his brother, who had held that proconsulship for three years, on having abstained from all plunder and preserved his integrity 'in tanto imperio, tam depravatis moribus, tam corruptrice provincia', and he uses similar language elsewhere.

quantalibet: here alone in Tacitus; first in Livy and Ovid.

facilitate, here in a bad sense, 'complaisance', as, e.g., in *A.* 2. 65, 2. In a good sense, c. 9, 3.

redempturus esset, etc., 'was ready to purchase a mutual concealment of misdeeds': cf. *H.* 4. 56, 3.

auctus est: so used of the growth of a family in *A.* 2. 84, 2; it is a rather formal expression, cf. Plaut. *Truc.* 384; Cic. *ad Att.* 1. 2, 1. Here the marginal reading of *E* is obviously the better. On the daughter see c. 9, 6. His wife, therefore, accompanied him to the province (cf. also c. 29, 1), a custom which had been attacked (*A.* 3. 33–34), but was not forbidden.

in subsidium: Tacitus may refer, as Furneaux thought, to the privileges of a parent (see note below, § 4) which Agricola thereby acquired, but the sense is probably more general. A family needed heirs for its support and continuity; cf. Suet. *Cal.* 12, 1 'deserta desolataque reliquis subsidiis aula . . . ad spem successionis paulatim admoveretur'; *A.* 1. 3, 5; Cat. 68, 119 ff.

solacium, 'consolation', for the loss of his son.

6, 3. inter . . . tribunatum plebis: best taken as an adjectival phrase qualifying *annum*, understood from the following *annum* (= *eum qui erat inter*, etc.), like *sub Nerone temporum*; although it might mean 'the interval between', as in *G.* 28, 2 'inter Hercyniam silvam Rhenumque et Moenum amnes', 'the country between the forest and the rivers'. In either case a somewhat harsh construction is preferred to the repetition of *annum*.

quiete et otio: modal abl. These synonyms recur in c. 21, 1; 42, 1. Agricola's tribunate fell in the year A.D. 66, in which Arulenus Rusticus, one of his colleagues, with less discretion contemplated exercising his veto in the trial of Thrasea (see on c. 2, 1). *Transiit* for the normal *transegit*.

sub Nerone, equivalent to an adjective: cf. *inter quaesturam*, etc., above and c. 16, 1.

inertia pro sapientia. In those times Memmius Regulus was *quiete defensus* (*A.* 14. 47, 1), and Galba made his real indolence pass for prudence (*H.* 1. 49, 3). Pliny speaks similarly of his own times 'suspecta virtus, inertia in pretio' (*Ep.* 8. 14, 7).

6, 4. praeturae. This office might be held after the interval of a year, and fell in Nero's last year (A.D. 68): cf. *sequens annus* (c. 7, 1). The normal age for the praetorship was 30, but another year would be remitted to Agricola on account of the birth of a daughter; cf. note on § 1.

tenor. This emendation of Rhenanus (for the MSS. *certior*) has not been improved upon by later editors; and, given a script like that of *E*, *tenor* might be corrupted to *cerior* and then to *certior* (*c* and *t* are easily confused, cf. c. 14, 1 *Togidumno* for *Cogidumno*; c. 38, 1 *notare* for *vocare*; also c. 3, 2 and c. 34, 3). The word is often qualified by *idem* (Pliny, *Pan.* 84 'idem tenor vitae'; Livy 7. 32, 16, etc.) and is especially used of the character of a magistracy: cf. Livy 4. 10, 9 'quinque consulatus eodem tenore gesti'; 7. 40, 9. Other conjectures (*torpor, otium, gestio, et ius, rector*) are less attractive and transposition ('inertia ⟨certior et⟩ pro sapientia fuit': so Peerlkamp) is weak in that it destroys the characteristic hendiadys and produces an unsatisfactory sense.

nec enim, etc. *Iurisdictio*, strictly speaking, belonged only to the *praetor urbanus* and *peregrinus*, though in a less technical sense to several others. But under the Julio-Claudian dynasty the whole number amounted sometimes to eighteen, some of whom had no judicial duties of any kind (Dio 60. 10, 4). Urlichs suggested that he probably was one of those who had charge of a city region.

ludos. The *cura ludorum*, in old times partially devolving on praetors, was wholly assigned to them by Augustus in 22 B.C. (Dio 54. 2, 3), and became one of their most prominent functions.

et inania honoris, 'and (other) vanities of office': cf. c. 13, 1; *A.* 1. 5, 3, where a general expression is similarly added to a particular.

medio . . . duxit. The meaning is that in giving his games Agricola steered a middle course between thrift and lavishness. *Duxit* for *edidit* is a rhetorical experiment in the art of expression, on the analogy of *ducere choros* or *funus*, and *medio* is an abl. of direction, as Andresen explains, comparing Virg. *Aen.* 4, 184 'volat caeli medio terraeque'. 'He conducted in a course midway between.'

Tacitus evidently intended to say that Agricola avoided two extremes—meanness and extravagance—so that his games were fine enough to win him popular recognition (*fama*) without incurring criticism for reckless expense (*luxuria*) but, as if reluctant even to entertain the possibility that Agricola could be guilty either of meanness or extravagance, Tacitus clouds his thought by choosing two favourable words to express the extremes—*ratio*, i.e. 'judicious management (of money)' (not 'parsimony'; cf. c. 18, 4; c. 20, 3 *tanta ratione curaque*) and *abundantia* 'generosity' (not 'extravagance'; cf. Cic. *Phil.* 2, 66 'non illa quidem luxuriosi hominis sed tamen abundantis'). [Emendation is unnecessary: *medio moderationis* (Gudeman) sounds ill and is unconvincing because *moderatio*, far from being an extreme, is the very quality of steering between extremes; *modo* (for *medio*; Puteolanus, Lipsius, Duff) is not used by Tacitus to mean 'by means of'.]

uti . . . ita = 'while . . . yet', as often.

longe a = *procul a*, c. 9, 4, 'far removed from'.

famae propior, 'coming near to (popular) distinction'. *Propior* is common in this sense (= *iuxta*), the contrasted idea being sometimes unexpressed (*A.* 6. 42, 2; 16. 35, 1; *G.* 30, 3).

6, 5. tum, etc. This commission was given to him while he was still praetor. After the fire Nero had repaired the loss of works of art in Rome by the pillage of temples throughout the empire (see *A.* 15. 45, Suet. *Ner.* 32), which is the *sacrilegium* here referred to. But no restoration of this plunder took place (as the concluding words of this sentence show). It is clear, therefore, that Agricola was commissioned to inquire into other misappropriations of temple treasure by individuals during the fire or afterwards. In early times we hear of a special board appointed for such a *conquisitio* (cf. Livy 25. 7, 5 'triumviri sacris conquirendis donisque persignandis'). In the imperial age the temples and their property were under the care of a board of two *curatores aedium sacrarum et operum publicorum* (the office is carefully studied by A. E. Gordon, *Univ. of Cal. Publ. in Class. Arch.* 2 (1952), pp. 279–304). Urlichs suggested that this was the office to which Agricola was appointed by Galba, but it is much more probable that his commission was a special one since the curatorship seems always to have been held after the consulship.

ne = *ut non*, as in *A.* 14. 11, 1; 28, 2, etc., and in classical Latin. The MSS. reading *fecit ne* may be compared with Livy 2. 45, 12 'velle ne scirem, ipsi fecerunt' etc. (see R. G. Nisbet, *A.J.P.* 44 (1923), 27 ff.). *Effecit*, a very easy emendation, accords with the usage of Tacitus but is unnecessary. For the genit. *alterius* cf. c. 5, 3 n.

sensisset. The force of the pluperf. is 'that it was as though the State had never felt'. A very similar instance is cited from Pliny, *Pan.* 40, 3 'effecisti ne malos principes habuissemus', implying that Trajan had blotted out the memory of past misgovernment.

7, 1. Sequens annus : the famous year of the four emperors, A.D. 69. Tacitus often thus personifies *annus* (c. 22, 1), *dies*, etc.

classis. The dispatch of this fleet, probably about March, is described in *H.* 1. 87, and its raid upon Liguria in *H.* 2. 12–15.

Intimilios. It is clear both from the narrative in *H.* 2. 13, which mentions the attack on the town but tells of a heroic Ligurian lady who was not Procilla, and from the fact that Agricola's mother is said to be living in her country house (*praediis*) that the sailors from the fleet ravaged not only the town of Intimilium but also the district round about. It is therefore preferable (and perhaps palaeographically easier) to read *Intimilios*, the people (*Ἰντεμέλιοι* in Strabo 4, 202) rather than *Intimilium*, the town. For the parenthesis cf. Livy 28. 46, 9 'Ingauni (Ligurum ea gens est)'; 36. 11, 9 'Leucadios (quod Acarnaniae caput est)'. For *populari* applied to a people cf. *A.* 14. 29, 1. *Hostiliter populari* is Livian (28. 24, 4; 37. 17, 3; 18, 3).

Mommsen (*CIL.* v, p. 900) showed the correct form of the name to be *Intimili-*, not *Intemeli-*. The town, the modern Ventimiglia, 17 miles east of Nice, is called Albintimilium in *H.*, loc. cit., and *Ἄλβιον Ἰντεμέλιον* by Strabo.

causa caedis, 'motive for murder'; the narrative in *H.*, loc. cit., says that the naval troops sated their greed by the ruin of the innocent. Elsewhere *causa caedis* usually means 'pretext for murder' (cf. Livy 3. 36, 5; Cic. *ad Att.* 2. 24, 4 with Shackleton Bailey's note).

7, 2. sollemnia pietatis, not the actual funeral which must have taken place before Agricola's arrival but 'the ceremonies required of filial duty', i.e. the erection of a tombstone and so on.

adfectati . . . imperii, 'aiming at the empire'. Vespasian's *primus principatus dies* was 1 July, on which day the legions at Alexandria took the oath in his name, as did those of Judaea in his presence on the 3rd (*H.* 2. 79, 1).

deprehensus, 'was overtaken': cf. c. 34, 3. One of the early acts of Vespasian's party was to send letters to Gaul (*H.* 2. 86, 4).

in partes, etc. Forum Iulii was occupied for Vespasian by the procurator, Valerius Paulinus, about October (*H.* 3. 43, 1).

initia, etc. C. Licinius Mucianus, governor of Syria since A.D. 67, entered Rome at the end of the year, just after the death of Vitellius (who died on 20 Dec.), when the city was in a state of anarchy: see *H.* 4. 11, 1. He held no formal magistracy at that time, but was a powerful champion of Vespasian's interests. His character is sketched in *H.* 1. 10 (see Syme, *Tacitus*, pp. 195–6, and App. 85).

iuvene admodum. He was eighteen years old. Tacitus uses the same expression of himself at the professed date of the *Dialogus* (1, 2). The profligacy and licence of Domitian at this time are described in *H.* 4. 2, 1; 39, 2. He was made praetor at the beginning of A.D. 70, Vespasian being then in Egypt, Titus in Palestine.

fortuna, 'imperial rank' (cf. c. 13, 3, and note).

7, 3. is must refer to Mucianus who in *H.* 4. 11 is portrayed as taking the initiative throughout the crisis (cf. also Suet. *Dom.* 1, 3). Paratore, Desideri, and others take it as referring to Domitian but this is against both the language and the facts. If it did refer to Domitian it would imply that Agricola was much more in Domitian's debt than Tacitus liked to admit.

ad dilectus agendos : early in A.D. 70 and probably in Italy, primarily to fill up the newly constituted *legio II Adiutrix* (Introd., p. 77). When a *dilectus* was held in Italy (which was rare), commissioners of senatorial rank were appointed (cf. *ILS* 1098). In senatorial provinces the duty was discharged by the governor or by senatorial officers appointed by him; in the Imperial provinces by the governor or by equestrian officers. These officers were known as *dilectatores*. See G. Forni, *Il Reclutamento delle Legioni da Augusto a Diocleziano*, pp. 22 f.

integre, 'with rectitude', allowing no one to buy exemption from service. Cf. Cicero, *Leg. Man.* 2.

vicesimae : one of the legions engaged in the first invasion of Britain, and probably stationed at Wroxeter (see p. 77). It is strange that the province is not mentioned till the next chapter; but it is difficult to suppose (with Ritter) that *in Britannia* has dropped out after *transgressae*. Agricola's appointment as *legatus legionis* was made later in A.D. 70.

tarde. It appears from *H.* 3. 44, that the only British legion forward to accept Vespasian was the Second, which he had commanded in the Claudian expedition.

sacramentum : in the early Republic troops took an annual oath of loyalty to the consul or general. Marius replaced this by an oath which was binding on all recruits for the whole of their military service ('se esse facturos pro republica nec recessuros nisi praecepto consulis post completa stipendia'). Under the Empire the oath was made to the princeps and was renewed annually on 1 Jan. or on the anniversary of his accession. One of the first acts of a new reign was to exact the oath from the army (*A.* 1. 8, 4, etc.).

ubi decessor, etc. *Ubi* = *apud quam*, a Tacitean usage, cf. *A.* 1. 40, 1 and *H.* 3. 31, 1; and *decessor* is used, as here, of a retiring official in correlation to *successor* in Cic. *pro Scauro*, 33. The retiring legionary legate was Roscius Coelius, and the *legati consulares* were the governors Trebellius (A.D. 63–69) and his successor Vettius Bolanus; see c. 16, 3–5. Tacitus gives the report which reached Rome, and modifies it: the legion, indeed, was too much even for consular governors, and its commander, a man of praetorian rank, was unable to restrain it, whether his inability was due to his own or to the soldiers' character. Tacitus suspends judgement—Roscius, who was consul in A.D. 81, may still have been alive—but in c. 16 he ascribes the outbreak of mutiny to the demoralizing effects of idleness, and the continuance of the mutinous spirit under Bolanus to the same cause. *H.* 1. 60 gives an account of the mutiny under Trebellius, in which auxiliaries were also involved. Feeling against Trebellius was inflamed by Coelius, who had long been his enemy, on the ground that the legions were stripped of men and deprived of pay. The troops took Coelius' part and drove Trebellius from Britain. Events under Bolanus are not recorded except that now Coelius could not control his men. His attitude towards Vespasian, to whom the Twentieth Legion was slow to swear allegiance, must have been loyal, since in A.D. 81 he attained the consulship under Titus.

legatis . . . consularibus. The governors of the major Imperial provinces containing two or more legions were always of consular rank, in order to control the praetorian legates.

nimia : as in English 'too much for', 'too strong'; so in Vell. Pat. 2. 32, 1 Pompeius is called 'nimius liberae reipublicae'. Cf. *A.* 2. 34, 4.

legatus praetorius. The commanding officer of a legion (*legatus legionis*) was regularly one who had been or was qualified to be praetor.

potens: cf. Livy 9. 26, 16.

successor . . . ultor, as also in *H.* 1. 40, 2.

moderatione: 'mildness', 'clemency'. The prospect of an efficient commander determined to restore discipline was sufficient to bring the troops back to their loyalty. On his arrival, therefore, Agricola was not obliged to adopt punitive measures but could exercise a clemency that was unusual in such a situation. For the thought cf. Sall. *Jug.* 45, 3 'prohibendo a delictis magis quam vindicando exercitum brevi confirmavit'. *Moderatio* is often linked with *clementia* (cf. Curtius 5. 3, 15; Suet. *Jul.* 75, 1; note also *A.* 12. 49, 2).

[**videri . . . fecisse,** recalls the formula with which judgement was pronounced in Roman law (cf. Cic. *Verr.* 2. 93; Livy 2. 54, 10; 8. 15, 6; see Daube, *Forms of Roman Legislation*, pp. 73–77). Tacitus adapts it into an epigram that gives Agricola's own verdict on his conduct.]

8, 1. Vettius Bolanus: sent out by Vitellius after Trebellius fled to him (*H.* 2. 65, 2). He had been *legatus legionis* in the East under Corbulo in A.D. 62 (*A.* 15. 3, 1), and was *cos. suff.* about A.D. 66. His government of Britain is similarly represented as inactive in c. 16, 5; *H.* 2. 97, 1. It should be noted that *legio XIV* was absent from Britain in A.D. 69 except for a few months, and was withdrawn finally in 70, being replaced by *legio II Adiutrix* in 71; cf. Introd., p. 77. Statius, in a poem written to Bolanus' son Crispinus (*Silv.* 5. 2, 143–9), credits him with founding *castella* and dedicating trophies won in battle from a British king. It is possible that as part of the aftermath of the Boudiccan revolt he engaged in successful operations against the Brigantes and that some of the Yorkshire forts were planted by him, but it is unlikely that he penetrated further. The picture given by Tacitus may be taken as just. He was subsequently proconsul of Asia in A.D. 76.

feroci, 'warlike': cf. *H.* 1. 59, 1 *ferox gens.*

dignum, '(than a warlike province) requires'. Normally Tacitus omits the copula (*esse*) with *dignus*, but it is designedly inserted here: without it *fuit* would have to be supplied, and this would limit the sense unduly.

ne incresceret: sc. *ardor*, 'so that it should not become too strong'; cf. Livy 1. 33, 8 'ad terrorem increscentis audaciae'. The verb, found only here in Tacitus, is sparsely used throughout Latin (cf. Varro *ap*. Aul. Gell. 16. 12, 7; Furius Ant. 3; Virgil, Livy, etc.). *Vis* and *ardor* are similarly associated in *Dial*. 24, 1, etc.

peritus . . . eruditus: here alone with inf. in Tacitus. But the former is so used in Virg. *Ecl*. 10, 32, etc., the latter in Plin. *N.H*. 33, 149. Ritter's emendation *obsequii*, for *obsequi*, would be in accordance with c. 42, 1, etc., but is needless.

utilia honestis miscere, 'to combine interest with propriety' (honourable conduct); not so to push his own reputation as to forget due subordination to his superior. The sentiment is commonplace: contrast Livy 23. 14, 3 'cum honesta utilibus cedunt'.

8, 2. Petilium Cerialem. He had commanded the Ninth Legion in its disaster in Britain during the rising of Boudicca in A.D. 60 (*A*. 14. 32, 3), and in the civil war he took up the cause of Vespasian, who was closely related to him (*H*. 3. 59, 2), was *cos. suff*. probably for a short time in A.D. 70, and was immediately afterwards sent to suppress the rising of Civilis (*H*. 4. 68). After his government of Britain (A.D. 71–74: see c. 17, 2; Introd., pp. 53–55), he was again *cos. suff*. in May A.D. 74. His full name is Q. Petillius Cerialis Caesius Rufus (*ILS* 1992). He may well have been Vespasian's son-in-law.

habuerunt virtutes, etc., 'good qualities now had room for display'; *exempla* are deeds worthy of being taken as examples, as in *A*. 13. 44, 4, etc. Cf. the sentiment on the appointment of Corbulo, 'videbaturque locus virtutibus patefactus' (*A*. 13. 8, 1). The image is common in Cicero; cf. *Mur*. 18 'nullum vobis sors campum dedit in quo excurrere virtus cognoscique posset'.

in experimentum, 'to test him': cf. *in famam*, below.

ex eventu, 'on the strength of his success'. For *eventus* in the sense of successful result, cf. c. 22, 3, etc.

8, 3. in, 'with a view to', as often in Tacitus. Cf. c. 5, 2 *in iactationem*; 10, 1, etc.

ad auctorem, etc. For the adversative asyndeton cf. c. 37, 5, etc. Agricola is represented as speaking, not of his achievements (*gesta* = *res gestae*, as, e.g., in Livy 6. 1, 3; 8. 40, 5), but of the success (*fortunam*) attending plans due to the originator and leader, whose lieutenant he had been. A similar principle of loyalty is noted among the Germans, *G.* 14, 2.

ut minister must mean 'as a subordinate should, (which he was)'.

obsequendo is unexpected since outstanding merit (*virtus*) is hardly evidenced by mere obedience (*obsequendo*) but Tacitus picks up *peritus obsequi* in § 1 above and implies that Agricola's strict compliance with Cerialis' orders gave him opportunities to display his abilities as well as earning him his general's favour. Cf. *A.* 6. 8, 4 *obsequii gloria.* [*Exsequendo* (Voss) is no improvement.]

extra : cf. the use of *citra* in c. 1, 3, etc.

nec = *nec tamen*: cf. c. 19, 3, and the use of *et* for *et tamen* (c. 3, 1, etc.).

9, 1. Revertentem. Agricola returned from Britain either in A.D. 73 or with Cerialis early in 74. Probably the present tense implies that the elevation to the patriciate took place immediately on his return: cf. *ingredienti* (c. 18, 5); *respondens* (*H.* 2. 4, 2), etc. In other places it has a more aoristic force.

inter patricios adscivit : for the technical *adlegit* (cf. Cic. *de Rep.* 2, 35; Livy 6. 40, 4). The old power to co-opt new patrician *gentes* into the *curiae* had been long obsolete, and the patriciate became a gradually diminishing body, from which a few very ancient priesthoods (those of *rex sacrorum,* the Salii, and the three *flamines maiores*) had still to be filled up. Partly to provide for these, partly to widen the prestige attaching to the oldest Roman nobility and to pay a compliment to distinguished men and families, the patriciate had been granted to individuals by Julius Caesar and Augustus, under special enactment, and by Claudius and Vespasian, as a censorial function analogous to that of choosing senators. Vespasian increased the patriciate to 1,000.

Aquitaniae. The part of Gaul originally so called lay between the Garonne and the Pyrenees (Caes. *B.G.* 1. 1, 2), but the province as constituted by Augustus extended northwards to the Loire. Its capital was Bordeaux (Burdigala): for its history see R. Étienne, *Bordeaux Antique* (1962), pp. 80 ff.

splendidae inprimis dignitatis : concise genit. of quality, with abl. of respect added. All the *tres Galliae* were Imperial provinces under ***legati*** of praetorian rank, and were among the most important of that class. Galba had held Aquitania just before his consulship (Suet. *Galb.* 6), and several others are known to have done so.

administratione, 'in respect of its functions'.

spe consulatus cui destinarat: sc. *eum*, an omission characteristic of Tacitus (cf. c. 42, 2). Tacitus means that Vespasian had marked him out as a future consul. The language is non-technical, although consulships were often decided six months or more in advance (cf. *H.* 1. 77, 2; provincial governors sometimes record the title '*cos. des.*' (*ILS* 1055; *IGR* iii. 840)) and *destinare* has a technical meaning elsewhere denoting the guiding vote given by special centuries (*praerogativae*) at the actual election (A. H. M. Jones, *Studies in Roman Government and Law*, pp. 29 ff.).

9, 2. plerique. The belief that soldiers lack depth and versatility was (and is) a commonplace. The Spartans were unfavourably contrasted in this respect with the Athenians (Thuc. 1. 141, 3), and Scipio Africanus is regarded as exceptional, whereas Mummius, the uncultured conqueror of Corinth, is typical (Vell. Pat. 1. 13, 4). Cf. also Livy 2. 56, 8; [Quint.] *Decl.* 310, p. 210. 11 'contemnunt hominem militarem nihil minus quam litibus idoneum'.

subtilitatem: here 'judicial discrimination', capacity for drawing fine distinctions.

secura et obtusior, 'irresponsible and somewhat blunt'. *Securus*, lit. 'free from care', implies that a general did not have to fear that his judgement would be subject to popular criticism or higher appeal. Camp justice is satirized in Juv. 16, 13 ff.

manu, 'by the strong hand', summarily: cf. *G.* 36, 1 'ubi manu agitur'. *Ac* after *et* couples the following words closely with *obtusior*. A similar contrast is made by Sallust, *Jug.* 31, 18.

exerceat, 'bring into play'.

naturali prudentia, 'with native good sense', either modal abl. or abl. of quality.

togatos, 'civilians', as often, in contrast to soldiers. As no troops were quartered in Aquitania, the duties of its *legatus* would be judicial and administrative only.

facile iusteque agebat, 'dealt readily and equitably'. Seneca speaks of 'ingenia facilia, expedita' (*Ep.* 52, 6), and Pliny of 'ingenium facile, eruditum in causis agendis' (*Ep.* 2. 13, 7).

9, 3. iam vero, 'furthermore', so used in transitions, sometimes with emphatic force: cf. c. 21, 2.

curarum remissionumque, 'of business and relaxation': cf. *Dial.* 28, 5. Africanus was credited with the same ability to make the fullest use of his time; cf. Vell. Pat. 1. 13, 3 'intervalla negotiorum otio dispunxit'.

ubi . . . poscerent. This subjunctive of repeated action, with *ubi, quoties,* etc., very frequent in *Hist.* and *Ann.*, and adopted chiefly from Livy, is perhaps found here alone in the minor works. Cf. the indicative below and in c. 20, 2.

conventus ac iudicia, 'assize courts'. *Conventus* denotes the meetings of provincials at appointed places where the governor administered justice on circuit, and *iudicia* is added to define the terms more exactly, just as *conventus* in the extended meaning of 'assize district' is often defined by the adjective *iuridicus.* Some editors interpret *iudicia* as trials held in the capital of the province, where the governor normally resided.

severus et saepius misericors. In all judicial business he was 'serious and earnest (earnestly attentive), strict and yet more often merciful'. *Et* = *et tamen* (see c. 3, 1 note). *Severus* and *misericors* are opposites (cf. Cic. *Mur.* 6), and where opposite ideas are thus coupled by *et* (or *ac*) *saepius, aliquando* or *modo* is implied with the first, e.g. *H.* 2. 62, 2 'pecunia et saepius vi': cf. c. 38, 1 'aliquando . . . saepius'; c. 19, 3 'nec semper . . . sed saepius'; *A.* 11. 16, 2 'modo . . . saepius'. Sometimes the opposite is not expressed, and in such cases *saepius* or *et saepius* means 'more often than not', 'generally', as in *G.* 22, 1; *A.* 12. 7, 3; 46, 3. The interpretation of Furneaux and others, according to which the meaning is 'tempering strictness with compassion', would seem to require *etiam,* or some such word, with *misericors.*

ubi . . . nulla ultra potestatis persona, 'when his duty was discharged, the official pose (mien) was no longer kept up', he could lay aside the demeanour of the official and be affable (*facilitas*). The thought is a commonplace. Plutarch (*Cato mi.* 21) attributes the same combination of a stern public image with an affable private manner to the younger Cato.

tristitiam . . . exuerat. The difficulties in this sentence are these. *Exuerat* must mean 'discarded', continuing the metaphor of *persona.* The metaphor is common (cf. Cic. *Off.* 3, 10 'ponit enim personam amici cum induit iudicis', etc.; see Wendland, *Hermes* 51 (1916), 481 ff.) but at first sight *exuerat* seems to imply that in his public life Agricola displayed *tristitia*, *adrogantia*, and *avaritia*, which he discarded when he reverted to a private capacity. These three qualities are in Tacitus uniformly vicious and cannot be stretched, as Gudeman, Persson, Forni, and others wish, to cover the impassive demeanour of a stern, dignified, and thrifty governor (cf. *H.* 1. 38, 1 'tristitia et avaritia sui simillimum'; *H.* 1. 51, 4, 'avaritiam et adrogantiam, praecipua validiorum vitia'; Livy 33. 11, 8, etc.). The deletion (*et avaritiam* secl. C. Heraeus) or emendation (*amaritiem* for *avaritiam*, Lipsius, Duff) of one of these three qualities does not touch the problem. Most editors (Peerlkamp, Anderson, Till, Ernout), feeling that Tacitus could not have implied that his father-in-law in his public capacity ever wore the mask of such distasteful failings, follow the suggestion of Wex and delete the whole sentence as a gloss on *persona*, but this is unlikely. Tacitus' subsequent account of Germanicus seems to recall at many points his picture of Agricola (see c. 18, 4 n.; c. 22, 4 n.; c. 43, 1 n.; 43, 2 n.) and in *A.* 2. 72, 2 he writes of Germanicus, 'cum magnitudinem et gravitatem summae fortunae retineret, invidiam et adrogantiam effugerat'. *Exuerat* cannot, however, be used as the equivalent of *effugerat* 'avoided, eschewed'. The true interpretation may lie in the force of the pluperfect tense. Normally when governors assumed the public 'front' (*persona*) of office, the elements of that *persona* did appear, if only in the eyes of provincials and of sensitive critics such as Tacitus (cf. c. 30, 4), to comprise *tristitia*, *adrogantia*, *avaritia* as well as *gravitas*, *auctoritas*, etc. Unlike other governors, when Agricola came to assume the *persona* of office he had already removed from it these three traditional but undesirable features. The point would be more clearly expressed by *avaritiam* ⟨*iam*⟩ *exuerat* (Dihle) but the addition is not essential.

facilitas : his affability in private life (cf. *A.* 2. 65, 2, etc.), as contrasted with his strictness (*severitas*) in official duties.

9, 4. integritatem atque abstinentiam, 'uprightness and self-restraint', Cicero's *integritas et continentia* (*ad Q. F.* 1. 1, 18). *Abstinentia* means self-control as opposed to *libido*, with special reference to freedom from *avaritia* (Val. Max. 4, 3). Tacitus rhetorically apologizes for mentioning what should be taken for granted. So Velleius (2. 45, 5) speaks of Cato as one 'cuius integritatem laudari nefas est'.

famam . . . cui . . . indulgent: cf. 'etiam sapientibus cupido gloriae novissima exuitur' (*H.* 4. 6, 1); Cic. *Arch.* 2; Hor. *Sat.* 1. 6, 23.

per artem, 'by intrigue', such, for instance, as governors often used to procure addresses of thanks from subjects (*A.* 15. 20–21). Notice how the variation of construction *ostentanda . . . per artem.*

collegas: governors of neighbouring provinces: so in *H.* 1. 10, 2. Such rivalries are often mentioned. In Republican times they endangered the very existence of the State.

procuratores. Imperial procurators charged with collecting sums due to the *fiscus* existed in all provinces. In Caesarian provinces, governed by *legati,* there was also a chief procurator who was charged with the financial administration, and corresponded to the quaestor of a senatorial province. These officers had received a more independent position and jurisdiction from the time of Claudius (see c. 15, 2; *A.* 12. 60), and were frequently hostile to the governor and a check upon him (cf. *A.* 14. 38, 3); thus in Spain Galba was powerless to curb their rapacity (Plut. *Galb.* 4). But they were far below him in rank (hence superiority over them conferred no *gloria*). The plural here might refer to successive procurators, but more probably is rhetorical (cf. c. 5, 2), balancing *collegas.*

atteri sordidum, 'ignominious to be worsted', by defeat (*vinci eoque atteri*). *Attero* literally means 'wear down' (cf. *G.* 29, 1, etc.; Justin 41. 6, 4, 'Eucratides bellis attritus'); it is degrading to be forced to submit by an inferior. The ancients were sensitive about their personal honour, and the distaste for unworthy opponents was commonplace: cf. Livy 5. 27, 5; Sen. *de Ira* 2. 34, 1, 'cum inferiore (contendere) sordidum est'; Eur. *El.* 189 with Denniston's note.

9, 5. minus triennium detentus. He was recalled early in A.D. 77 (see next note). Caesarian provinces were not held for a fixed term, but usually for from three to five years (Dio 52. 23, 2), except in case of misconduct. *Detentus,* implying that the term was shortened in order to accelerate Agricola's appointment to Britain, is an illustration of the panegyrist's art, which is also revealed in what follows.

statim ad spem cos. revocatus est. The prospect of the consulship, held out to him in his appointment (§ 1), had now become immediate. *Consules suffecti* were probably designated on 9 Jan. (cf. Mommsen, *Gesamm. Schriften,* iv, p. 428).

Agricola's recall followed his designation: praetorian governors were regularly designated during their governorship (see above). In what month his consulship actually began is unknown. Cf. Appendix I.

dari : i.e. was virtually already given (cf. *A.* 2. 34, 1 *abire*): the consulship was but a stepping-stone to it. The popular *opinio* was based merely on the fact that he had served with distinction there as military tribune and as legionary legate.

nullis . . . sermonibus : concise abl. abs., 'not that his conversation was ever directed to this end', i.e. he did not intrigue to secure the province of Britain or drop hints that he desired it.

par : taken by Andresen as neut., but better of Agricola: cf. 'par negotiis' (*A.* 6. 39, 3), 'par oneri' (*A.* 6. 28, 5). Britain was one of the most important military commands, and the only province where a forward movement was then in progress.

haud semper . . . eligit. Rumour is often said *destinare aliquem*: here *fama* (public opinion) 'sometimes makes (determines) the choice', leads to the appointment. The power of *fama* was proverbial; cf. Hesiod, *W.D.* 701; Aeschin. *in Tim.* 127. The sentence is an iambic line and its structure recalls similar aphorisms (cf. Sen. *H.F.* 328, 'quem saepe transit casus aliquando invenit') but it is most improbable that Tacitus is using a quotation. Such accidental iambic lines are not uncommon (see note on c. 10, 3); cf., e.g., Cic. *ad Att.* 2. 24, 1 'quam sit amor omnis sollicitus atque anxius'.

9, 6. egregiae tum spei, 'then a girl of excellent promise'. Cf. such expressions as *egregiae famae* (*A.* 12. 42, 2, etc.), and the frequent use of *spes* in this sense by Virgil, etc. Note also Livy 43. 17, 4 'egregia spe futuri status'. She would be then about thirteen years old (cf. c. 6, 3, note), and marriage of girls at the age of twelve was not unusual. See Dio 54. 16, 7; *CIL.* ix. 1817; and many instances collected in Friedländer, *Roman Life and Manners*, i, pp. 234 ff. An interesting letter of Pliny (5. 16) speaks of the mature qualities of a girl who had died before marriage in her fourteenth year. [Büchner, following Acidalius, takes the phrase with *consul* rather than *filiam* but this is against both sense and word-order.]

iuveni mihi : he was probably about twenty-two years old. Introd., p. 8.

statim should mean that the appointment followed immediately after the marriage of his daughter, not immediately after his consulship; but the interval need not have been long.

sacerdotio: added to distinguish it from civil magistracies (*honores*). The pontiffs, and members of the other great priesthoods, were formally chosen by the senate (representing the old *comitia*) from a list furnished by the college; but candidates were in fact usually 'commended' by the princeps. See *A*. 3. 19, 1; the pontificate and augurship were not often given to persons below consular rank. For priesthoods just before or just after the consulship cf. *ILS* 1005; 1036, etc.

10–12. *The Ethnography of Britain*

Tacitus breaks off the narrative of Agricola's life to give an account of the ethnography (c. 10–12) and history (cc. 13–17) of Britain. The information was necessary for a reader to have a clear understanding of the nature of Agricola's achievement, but Tacitus inserts it at this point in his biography rather than earlier in order to mark off the culmination of Agricola's career, the governorship of Britain. In the same way the digression on the Usipi (c. 28) marks off the climax of Agricola's campaigning, the Battle of Mons Graupius (c. 29–38). This technique of using digressions to separate the principal stages of the narrative is taken from Sallust, who uses it to the same effect in the *Jugurtha* and the *Catiline*. In particular these chapters show close affinity with the digression on the 'Africae situs et gentes' in *Jugurtha* 17–19 but both belong to a larger and older genre of ethnographical writing whose roots go back to Herodotus (cf. the account of Scythia in Hdt. 4, 5 ff.). This genre, in which one should distinguish both technical treatises and more literary and superficial essays such as Tacitus' *Germania*, had fixed rules and a vocabulary of its own. It was regular to divide the consideration of a country into five topics: (1) the physical geography (*situs*); (2) the origins and features of the inhabitants (*gentes*); (3) climate; (4) mineral resources, agricultural products, etc.; (5) political, social, and military organization. The order of topics might vary (cf. c. 12, 3 n.) although *situs* usually precedes *gentes*, but all five topics, however cursorily, are treated. They can be seen, for example, in the passages of Herodotus and Sallust cited above and in Chapters 1–7 of the *Germania*. Besides developing a schematic method of treating the subject, the ethnographers evolved a technical phraseology. Tacitus' language in this section has many parallels from earlier writers (cf. c. 10, 1 n.

rerum fide; c. 10, 2 n. *spatio ac caelo*; c. 10, 2 n. *obtenditur*; c. 10, 2 n. *oblongae scapulae vel bipenni*; c. 10, 3 n. *transgressis*; c. 10, 5 n. *causa ac materia*; c. 11, 1 n. *qui mortales . . . coluerint*; c. 11, 3 n. *in universum . . . aestimanti*; c. 11, 4 n. *accepimus*; c. 12, 1 n. *in pedite robur*; c. 13, 1 n. *ipsi Britanni*, etc.). They also evolved a stock of conventional curiosities about individual countries. It is clear, for instance, that the *mare pigrum* (c. 10, 5 n.), the mild climate (c. 12, 3 n.) and the shortness of summer nights (c. 12, 4 n.) were regularly discussed by writers on Britain. In the same way they built up a stock of commonplaces about the characteristics and customs of the people. Tacitus compares the Caledonians with the Germans and singles out their *rutilae comae* and *magni artus* (c. 11, 2), just as Strabo (4, 200) had earlier written of the Britons that they were εὐμηκέστεροι τῶν Κελτῶν καὶ ἧσσον ξανθότριχες. Again, the description of the agricultural wealth of Britain (c. 12, 5 n.) is given in the same terms as that of Germany (*G.* 5, 1).

But it would be wrong to infer from this that Tacitus' account of Britain is a purely literary exercise and that his claim that it is based on actual observation is untrue. Tacitus was, of course, refining on the work of his predecessors in a formal genre. He cites Livy and Fabius Rusticus (c. 10, 3 n.) and in addition there survive accounts in Caesar (*B.G.* 5. 12–14), Strabo (4, 199–200) and Pomponius Mela (3. 6), and there were others, such as that written by Posidonius. The earliest was compiled by a Greek from Marseilles, Pytheas, *c.* 325 B.C., which was utilized by Timaeus and Eratosthenes and so is probably the original source of most of the commonplaces about Britain. All these authors retailed conventional material under the conventional heads, and Tacitus in his turn repeats them, but Tacitus adds the fruits of the explorations commissioned by Agricola (see Introd., p. 32), which were subsequently to be incorporated in Ptolemy's *Geography*. In particular the shape of Highland Scotland (c. 10, 3 n. *in cuneum*), the proof that Britain was an island (c. 10, 4), the conditions of the Pentland Firth (c. 10, 5 n.), the character of the western sea-lochs (c. 10, 6 n.) all derive from first-hand knowledge transmitted to Tacitus by Agricola himself, whom he last saw in A.D. 88, or members of his staff. [The section is discussed by P. Couissin, *Rev. de Phil.* 6 (1932), 97–117; G. Walser, *Rom, das Reich und die fremden Völker* (Historia, Einzelschriften 1, 1951), pp. 25 f. For the ethnographical genre see K. Trüdinger, *Studien zur griechisch-römischen Ethnographie* (Diss. Basel, 1918); J. J. Tierney, *Proc. Royal*

Irish Acad. 60 (1960), 189 ff. 'The Celtic Ethnography of Posidonius'.]

10, 1. situm populosque, 'the position and peoples'. Cf. Sall. *Jug.* 17, 1 'Africae situm'; Amm. Marc. 29. 5, 18, etc.

multis scriptoribus, dat. of agent: cf. c. 2, 1. On earlier writers on this subject cf. Introd., pp. 36–37.

in, 'with a view to': cf. c. 5, 2; 8, 3, etc.

curae, 'study', here probably not industry in collecting material but literary elaboration; *ingenium*, 'talent', i.e. literary ability. The two terms are often linked (cf. *Dial.* 16, 1; Stat. *Silv.* 1. 5, 64; Pliny, *Ep.* 3. 5, 3; 3. 7, 5; Quintilian 7. 1, 40; 3. 1, 22 'ubicumque ingenio non erit locus, curae testimonium meruisse contentus'). It was a commonplace to claim that one was either improving on one's predecessors' style or advancing new evidence: cf. Livy, *Praef.* 2 'novi semper scriptores aut in rebus certius aliquid allaturos se aut scribendi arte rudem vetustatem superaturos credunt'. Tacitus asserts that the complete subjugation of Britain has brought accurate knowledge for the first time.

tum primum perdomita. So in *H.* 1. 2, 1, 'perdomita Britannia et statim missa'.

nondum comperta . . . percoluere, 'where my predecessors have worked up with fine language things that had not yet been investigated'. *Percolere* is elsewhere used of putting the finish on a work (Pliny, *Ep.* 5. 6, 41 *incohata percolui*).

rerum fide, 'with truth of facts': the phrase is common in ethnographers; cf. Mela 3. 6, 49 (Claudius returning from Britain) *rerum fidem declaraturus*; Sall. *Jug.* 17, 7, etc.

10, 2. spatio ac caelo, 'as regards its extent and situation'. *Caelum* is the region of the sky (*plaga caeli*) under which the island lies as marked out by astronomical geography, i.e. the belt of latitude. Cf. c. 11, 2 *positio caeli*; Virg. *Ecl.* 3, 40–41; Livy 5. 54, 3.

in orientem Germaniae. Germany began at the mouth of the Rhine, and extended to and included Scandinavia.

in occidentem Hispaniae. Cf. c. 11, 2. The idea that Britain lies opposite to Spain on the west is found also in Caesar (*B.G.* 5. 13, 2), Strabo, and the elder Pliny (4, 102). This erroneous view of the orientation of Spain was prevalent among geographers of the Roman period till Ptolemy (*c.* A.D. 150) or his source Marinus (*c.* A.D. 100). The Gallic coast from the Rhine to the Pyrenees was imagined to be parallel to that of southern Britain (Strabo 2, 128); the western point of Britain lay opposite the Pyrenees, which were thought to run due north and south; and the Spanish coast beyond was supposed to run in a westerly, or north-westerly, direction (Strabo 3, 137; 4, 199, etc.). The deep recess formed by the Bay of Biscay was unknown. The true orientation of the Pyrenees and the Spanish peninsula had been given by Eratosthenes (*c.* 250 B.C.) on the basis of the reports of the explorer Pytheas of Marseilles (*c.* 325 B.C.). The error goes back to Polybius and was aggravated by Caesar who made the south coast of Britain parallel with the French coast. See J. J. Tierney, *J.H.S.* 79 (1959), 132 ff.

obtenditur, 'faces'. For this geographical sense cf. *G.* 35, 1; Mela 1. 4, 20; 2. 2, 23; Pliny, *N.H.* 5, 77; Solinus 10, 23.

inspicitur, 'is within sight of'. Tacitus thought that the two countries were parted all along by a narrow channel. See note above.

nullis contra terris, 'there being no land opposite'. For the abl. abs., cf. c. 9, 5; for the adjectival use of *contra*, cf. *in vicem* (c. 24, 1), *ultra* (c. 25, 1), etc.

vasto atque aperto, a stock phrase; cf. Caesar, *B.G.* 3. 9, 7; 3. 12, 5; Ovid. *Met.* 8, 165; Lucan 2, 619.

10, 3. Livius : nowhere else cited by Tacitus as an authority, but praised in the speech put into the mouth of Cordus (*A.* 4. 34, 3). The description would have come in Book 105, where he speaks of Caesar's expedition. For the use of a single name co-ordinate with a double name, cf. 'Lucio Sulla . . . Cinna' (*H.* 3. 83, 3).

Fabius Rusticus was an historian from Spain, patronized by Seneca. He was still alive when Quintilian was writing (10. 1, 104 f.) and may have survived at least until A.D. 108 (*CIL.* vi. 10229, 24). The extent of his history is not known but he certainly dealt with the reign of Nero (*A.* 13. 20, 2; 15. 61, 3) and perhaps carried his history on to the death of Vespasian or later. He is conjectured to have been a primary source for the *Histories* and for Plutarch's lives of Galba and Otho. It is uncertain in what context he discussed the geography of Britain but it may belong to his narrative of the rising of

Boudicca or, if he also dealt with the reign of Claudius, to his account of the invasion of Britain in A.D. 43. See Syme, *Tacitus*, pp. 289–94.

eloquentissimi. Tacitus selects Livy as the best-known historian of the past and Fabius Rusticus as the leading authority of the present.[1] The omission of any mention of Caesar, who discussed the ethnography of Britain in *B.G.* 5. 12–14, has surprised scholars and has led some (e.g. Couissin) to suppose that the omission is deliberate and intended to belittle Caesar's achievements in comparison with Agricola's. The supposition is most improbable in view of Tacitus' estimate and use of Caesar in the *Germania* (c. 28, 1 *summus auctorum*) and his reference to him in c. 13, 1. Nor is the omission to be explained on the hypothesis that *B.G.* 5. 12–14 is a later interpolation. Caesar's account is likely to have been extensively drawn on by Livy and Livy's history was in Tacitus' day a classic while Caesar's commentaries were little read. Tacitus may also refer to Caesar in c. 11, 4 (see n.).

oblongae scapulae vel bipenni. The MSS. read *oblongae scutulae*. *Scutula*, defined as a rhombus (Ps.-Cens. fr. 7, 4 'quod latera paria habet nec angulos rectos'), denoted various objects of more or less rhomboidal shape: a dish or tray (Martial 8. 71, 7; 11. 31, 19), a piece of tessellated pavement (Vitruvius 7. 1, 4; Palladius 1. 9, 5; often in Papyri), a patterned check in clothing (? tartan) (Prudentius, *ham.* 289; cf. Juv. 2, 97), a patch over the eye (Plaut. *Mil. Gl.* 1178), a sore on a mule (Chiron 185). An *oblonga scutula* would therefore have an elongated rhomboidal form, viz.:

A *bipennis* is usually taken to be a double-axe, resembling two axe-heads joined back to back (the shaft being inserted at the point of junction), viz.:

[1] E. Koestermann (*Athenaeum* 43 (1965), 170 ff.), who thinks that Tacitus originated from Transpadane Gaul, argues that he cites the two historians who had connexions with his home-land. But although Livy was a Paduan, there is nothing substantial to link Fabius with that region.

But it is clear from the following sentence ('et est ea facies . . .') that Tacitus regarded the two images, doubtless used by Livy and Fabius Rusticus respectively, as describing a single shape and *vel* commonly links virtual synonyms (cf. *G.* 6, 1; *Dial.* 9, 4, etc.). In fact, however, as Lacey has shown (*Proc. Camb. Phil. Soc.* 183 (1954/5), 16 ff.), the double-headed axe was not currently in use at Rome and the word *bipennis* was merely a fashionable and poetical synonym for *securis*, single axe, viz.:

The earlier ethnographers all agreed that the shape of Britain was triangular (Caesar, *B.G.* 5. 13; Strabo, 4, 199; Mela 3. 6, 50) and it is, *a priori*, very unlikely that this view would have been modified before Agricola's exploration of Scotland. The head of a Roman axe was roughly triangular. It follows that under the words *oblongae scutulae* there must lurk the meaning 'an elongated (triangle)'. Lacey suggests that Livy (and so presumably also Tacitus) wrote *oblongo scutulo. Scutula* (plural of *scutulum*) occurs once in the medical writer Celsus (8. 1, 15), meaning shoulder-blades. A shoulder-blade is exactly the right shape, viz.:

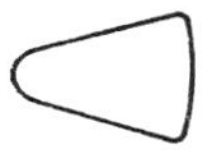

Scutulum is such a rare word (it occurs elsewhere,[1] in the sense of a little shield, only in Cic. *de Nat. Deorum* 1, 82; Vulg. 1 *Macc.* 4, 57; Cassian. *Conl.* 1. 5, 1) that its corruption would be easy. Its rarity, however, argues against Livy having used it and we prefer the even simpler correction *scapulae* already partly anticipated by *B. Scapulae* (plural) is applied to the whole shoulder structure of which the *scutulum* (shoulder-blade) is a part (Celsus 3. 22, 12) but it is also used loosely to mean just the shoulder-blades (Pliny, *N.H.* 21, 155). *Scapula* (sing.) could therefore naturally be used of a single shoulder-blade, and it is glossed by ὠμοπλάτη in *Corp. Gloss. Lat.* (Goetz) II. 179, 38; II. 482, 9; III. 351, 1. Comparisons of this kind are not uncommon in ancient geographers, e.g. Spain is likened to an outstretched ox-hide (by Posidonius; Strabo 3, 137), the

[1] Cet. Fav. 19, p. 302, 9 'aut tesserae aut scutula aut trigona aut favi' closely imitates Vitruvius 7. 1, 4 and *scutula* should be emended to *scutulae*.

Peloponnese to the leaf of a plane-tree (ibid. 2, 84), the inhabited world to a cloak (2, 113). Such comparisons were merely rough aids to popular conception.

The older ethnographers maintained that Britain as a whole was triangular in shape. Tacitus, on the basis of Agricola's exploration, modifies this by saying that Britain as far as the Forth–Clyde isthmus is triangular, but that north of the isthmus there is a huge tract which tapers towards the north like a wedge. This description tallies very closely with the shape of Britain given by Ptolemy (see Fig. 1) although he has distorted the actual alinement of Scotland (see Introd., p. 40). [Our interpretation, based on the assumption that the two shapes given by Fabius Rusticus and Livy are similar and, in fact, triangular, does still leave some difficulty in the meaning of *oblongae*. We take it to mean 'longer than normal'.]

adsimulavere, 'have compared': cf. *A*. 1. 28, 1; 15. 39, 3.

et est ea facies . . . : sed transgressis. 'And that is its shape below Caledonia, whence the statement is applied to the whole; but when you have crossed (into Caledonia), a huge and shapeless tract', etc.

For *citra* cf. note on c. 1, 3. *In universum* has the force of *universe* as in c. 11, 3; *G*. 5, 1, etc. *Transgressis* is dat. of point of view, cf. *H*. 3. 71, 1 *dextrae subeuntibus*; 5. 11, 3 *procul intuentibus*. The archetype evidently had the variants 'unde et in universum fama est' and 'unde et universis fama sed'. While *in universum* is clearly preferable to *universis*, *est* is weak and the adversative *sed*, although not necessary in so concise a writer as Tacitus, eases the connexion of the sentences. [The form *Caledoniam* is given here by *AB* and at c. 11, 4 by *A*ᶜ*B*: *e* and *E* throughout read *Calyd-*, but *Caled-* is the correct form, guaranteed by inscriptions and by the Greek transcription *Καλη-*, and is printed throughout. Variations in the spelling of the name are common in the manuscripts of many authors.]

inmensum, etc. The sense of this difficult sentence seems to be: When you have reached the Forth–Clyde isthmus (the division between Britain and Caledonia) you think that you have already reached the edge of the northern sea: but if you cross you find that a vast, irregular tract of land (for *spatium terrarum* cf. c. 23, 1; *G*. 35, 1; *H*. 3. 8, 3) stretches out from the point which you thought to be the furthermost coast (*extremo iam litore*) and eventually tapers as it were into a wedge. Tacitus' description agrees with the shape plotted by Ptolemy (see Fig. 1). (*Extremum litus* usually means the edge of the shore where it meets the sea (Virg. *Georg.* 3. 542; Livy 32. 32, 12; 32. 35, 7, etc.) but here it must apply to the most distant shore, the shore after which you think that there is only sea (cf. the unusual sense in c. 10, 6 *litore tenus*).) There is a contrast intended between what up till now had been thought to be the final shore of Britain (*extremo iam litore*) and the real north coast discovered by Agricola's fleet (*hanc oram novissimi maris*). *Extremo litore* may be abl. of motion from, although elsewhere *procurro* is used with the preposition *a*(*b*) and the abl. (cf., e.g., Florus 2. 2, 19 'Clupea procurrit a Punico litore') and in a comparable passage of Ovid (*Fasti* 4, 419 'terra tribus scopulis vastum procurrit in aequor') *scopulis* is an attendant abl., 'stretches out with three reefs'. The position and sense of *iam* may be compared with *G.* 32, 1 'certum iam alveo Rhenum Tencteri colunt'. *Enorme* means 'irregular', as in *A.* 15. 38, 3 rather than merely 'vast'. Alternatively, if it is felt that this interpretation puts too much weight on *iam*, *litore* could be regarded as an attendant or local abl. and the phrase be understood as 'with or at what is now really the last stretch of shore'. Britain as far as the Forth–Clyde isthmus is triangular but the remaining coastline thereafter encloses an irregular mass of land which eventually tapers into a wedge.

Litore . . . tenuatur forms a hexameter. Such, probably accidental, rhythms are censured by Cic. and Quintilian but are not uncommon in Tacitus (cf. *G.* 39, 1 *auguriis . . . sacram*; *A.* 1. 1, 1; 6. 37, 3, fin.). They are found even in Cicero (*ad Att.* 15. 4, 1 'eius concilium ad bellum spectare videtur') and there are some striking instances in the Greek New Testament (cf. *Philip.* 2. 1–2; *Galat.* 1. 10). [See Norden, *Ant. Kunst.*, p. 53, n.3.]

10, 4. **novissimi,** 'the remotest': cf. *A.* 2. 24, 1, etc. On the circumnavigation cf. c. 38, 4.

adfirmavit, 'established the fact'. Here (as in *A.* 14. 22, 4; *H.* 4. 73, 1) to prove by facts; usually to affirm in words. It was generally asserted to be an island by earlier writers (Caesar, *B.G.* 4. 20, 2; Cic. *ad Att.* 4. 16, 7; Livy; Mela; Pliny, *N.H.* 4, 102) but there was no proof and the question was sometimes debated in *suasoriae* (Quint. 7. 4, 2). Cf. also c. 24, 3 n.

incognitas . . . invenit. The Orkneys were already discovered and known (Pliny, *N.H.* 4, 103; Mela 3. 6, 54). Tacitus exaggerates but Agricola's fleet may have been the first to land on and over-run the islands. Cf. Juv. *Sat.* 2, 160–1 'et modo captas

Orcadas' (written after A.D. 100). The emperor Claudius had claimed that they had submitted to him (Eutrop. 7. 13, 2–3: see C. E. Stevens, *C.R.* 1 (1951), 7–9), and Tacitus' comment may be an implied rejection of that claim. Orcas is a name derived from Celtic **orci*, 'the young pigs', the name of the inhabitants, with a Greek termination.

domuitque. The fleet probably received some formal submission.

dispecta, emphatic, 'was thoroughly viewed, but no more'. *Dispicere* means 'to see clearly' (cf. Lucr. 6, 647; Catullus 66, 1; Caes. *B.G.* 7. 36, 2) not 'to see from afar'. The fleet viewed the Shetlands but did not land on them because their orders only went as far as viewing them and because winter was approaching. The fleet must have sailed on to the Shetlands since even if *dispecta* were to mean 'seen from afar' the only land likely to be visible from a boat fairly close to the Orkney coast would be Fair Isle. For *hactenus iussum* cf. *A.* 12. 42, 3; *G.* 25, 1.

Thule, first mentioned by Pytheas as lying six days' sail north of Britain, near the frozen sea (Strabo 1, 63, etc.). Six days' sail was between 500 and 700 miles (cf. Hdt. 7. 183, 3), and it is probable that Pytheas' Thule was Iceland which is about 600 miles from the north of Britain. The sea now does freeze round the north-east of Iceland but not round the Faroes and Shetland and, even if the sea was 3° colder then than now, the Faroes and Shetland would still be excluded. Agricola's Thule was, however, not Iceland but Shetland, as Camden, Pennant, and other antiquaries realized (cf. S. Hibbert, *A Description of the Shetland Islands* (Edinburgh, 1822), p. 586: the pretended etymology of Foula from Thule must be rejected, since Foula is Norse, Fowl Island), and it is evident that Tacitus, following earlier Roman writers, has transferred the name to the furthest land of which they had knowledge and has also reinterpreted Pytheas' observation of the frozen sea (see note on *pigrum mare* below).

10, 5. perhibent is used by Tacitus with a definite subject either understood or expressed, and not generally, 'men say': cf. c. 45, 3; *H.* 2. 3, 1, etc. The subject here is the Roman explorers who brought back first-hand experience of the conditions.

pigrum, etc., 'is sluggish and heavy, and is not', etc. In *G.* 45, 1, Tacitus gives a similar account of the sea in the far north beyond Scandinavia. The idea of the immovable and windless character of the outer Ocean was widely spread, and dates back to Pytheas, who surrounded his Thule with a coagulate of sea, land, and air (perhaps frazil or sludge ice), beyond which was the frozen sea, the *mare concretum* of Pliny, *N.H.* 4. 104.

Pytheas was certainly alluding to the freezing sea round Iceland but Tacitus describes a different phenomenon. The North Atlantic Drift Current passes close to the western shores of Shetland through the Faroe–Shetland channel. It is at its maximum intensity in Oct.–Nov. and Jan.–Feb. Roman ships coming from the shorter North Sea waves into the long oceanic rollers of the Current would notice the difference and, if faced by head-winds and the strong tidal streams, would be immobilized for long periods. See also A. R. Burn, *C.R.* 63 (1949), 94. [Deman (*Latomus* 17 (1958), 364) suggests 'black frost' but that is very different from what Tacitus describes.]

ne ventis quidem : i.e. still less by oars. Tacitus similarly describes the Dead Sea in *H.* 5. 6, 2.

perinde, 'as much as other seas'. The expression is so used in several places where the comparison is left to be supplied, so that it comes to mean 'less than would be expected'. The correction from *proinde* is supported by the general usage of Tacitus.

causa ac materia. The words are often paired (cf. Cic. *Fin.* 1, 18; Livy 6. 31, 2; 39. 1, 8, etc.) and represent the efficient and material causes of Aristotle's distinction. The origin of the winds was a subject of much discussion. Seneca ascribes whirlwinds to the resistance offered by high ground to the natural course of the wind which would otherwise expend itself (*N.Q.* 5. 13, 2). There are similar explanations in Lucretius and in *Aetna*, 300 ff.

tardius impellitur : falsely argued from the analogy of heavy solid bodies.

10, 6. neque . . . ac. For this very rare combination, cf. Suet. *Vesp.* 12, where *ac* bears the same sense of 'and moreover'; Colum. 3. 4, 1.; Cic. *ad. Q.F.* 3.3,4.

multi, especially Aristotle and Posidonius whose *περὶ 'Ωκεανοῦ* was the standard work on the subject.

multum fluminum . . . ferre. *Ferre* is naturally taken absolutely: 'set in various directions', cf. *A.* 2. 23, 4 'aestus . . . ferebat'; Caes. *B.G.* 3. 15, 3; Quint. 10. 3. 7. The change of subject is indicated by the chiasmus 'dominari mare, multum fluminum ferre'. For *flumina* meaning 'tidal currents' cf. Vitruv. 5. 12, 2; Curtius 9. 9, 9. The whole passage may be compared with Mela 3. 3, 31: 'mare . . . aquis passim interfluentibus ac saepe transgressis vagum atque diffusum facie amnium spargitur'.

nec litore, etc., 'nor does the sea flow and ebb (only) as far as the coast but flows inland deep and circuitously, making its way among highlands and mountains, as if in its own domain'. This description is obviously drawn from Agricola's experience of the western sea-lochs, based upon fleet explorations and land reconnaissance from high points.

velut in suo, cf. Cic. *ad Att.* 13. 2a, 2 ('Ariarathes) pedem ubi ponat in suo non habet', with Shackleton-Bailey's note.

11, 1. Ceterum, returning to the chief subject after the digression on tidal phenomena: cf. c. 25, 1.

qui . . . coluerint. It was a regular feature of ethnography to investigate the history of the population and to inquire whether the present inhabitants were indigenous or immigrant: cf. *G.* 2, 1 'ipsos Germanos indigenas crediderim', etc.; Sall. *Jug.* 17, 7 'sed qui mortales initio Africam habuerint, quique postea adcesserint aut quo modo inter se permixti sint'. The indigenous origin of the Britons was maintained by several historians represented by Diodorus 5. 21, 5: Caesar (*B.G.* 5. 12, 1) held that the inland tribes were indigenous but the coastal tribes were immigrant. See note on § 3 below.

ut inter barbaros, 'as might be expected', 'as is natural', (or 'as is usual') where barbarians are concerned: cf. c. 18, 5; *G.* 2, 2, etc.

parum compertum. The refusal by an author to commit himself to the solution of a disputed problem is especially characteristic of the ethnographical style: cf. *G.* 46, 4 'ego ut incompertum in medium relinquam'; Sall. *Jug.* 17, 2 'haud facile compertum narraverim'; Livy 5. 33, 4.

habitus corporum, here 'the physical types', as in § 2 and *G.* 4, 1; 46, 1. In c. 44, 2; *A.* 4. 57, 3, etc., the term is used of the physical characteristics (personal appearance) of individuals.

ex eo, 'from that variation', sc. *petuntur,* which is expressed in *H.* 5. 2, 1.

11, 2. rutilae . . . comae, etc. Cf. *G.* 4, 1 'truces et caerulei oculi, rutilae comae, magna corpora'. To Roman eyes the physical difference between German and Celt was slight: both had large frames and red (or fair) hair, but the Germans had the larger frames and the redder hair (cf. Strabo 7, 290; Manilius 4, 710–11). Hence Tacitus assigns to the Caledonians a Germanic origin. In fact they were evidently a mixed population, in which immigrant Celtic elements were blended with the old neolithic stock.

habitantium: substantival, 'of those inhabiting', a usage following that of the Greek participle with the article (cf. c. 4, 2; 32, 4, etc.). The transitive use of *habitare* is mostly found in the passive but cf., e.g., Livy 5. 51, 3.

Silurum. These inhabited Glamorgan, Monmouthshire, and south Brecknockshire, and maintained a resolute independence until the time of Frontinus (c. 17, 2 and note; cf. *A.* 12. 32). After their conquest a civic capital was founded at Caerwent (Venta Silurum) and in the second century villas were being built. Military control was still exerted over the wilder parts of their territory at least until the end of the second century but Romanization, though limited, was further advanced among them than among the other Welsh tribes.

colorati, 'swarthy', not here 'sunburnt' (Quint. 5. 10, 81, 'sol colorat: non utique, qui est coloratus, a sole est'). So used of Indians (Virg. *G.* 4, 293) and other dark races. [Gudeman interprets it as 'painted', with reference to the well-known practice of painting themselves with woad, but Tacitus is referring to natural characteristics.] The asyndeton *torti crines*, 'curly locks', is part of the same argument; *et* adds another from geography.

contra, adverbial.

Tacitus' geographical argument about the ethnical affinities of the Silures is based on the false notion that Spain was opposite (*posita contra*) and near to Britain (see note on c. 10, 2). The physical resemblances to the Spanish Iberians indicate that among the Celts of south Wales there was a strong strain of native British blood resulting from fusion with the descendants of older stock, who may be called Iberian in the sense in which the term is used by ethnologists to describe the short, dark race of non-Aryan stock which was widely spread over the Mediterranean lands and beyond in the neolithic period. Some slight support for Tacitus' theory is provided by a distinctive type of hill-fort, distinguished by a defensive feature known as *chevaux-de-frise*, which is found only in two Welsh sites (Craig Gwrtheyrn, Pen-y-gaer) and in north-central Spain (A. H. A. Hogg, in I. L. Foster and G. Daniel, *Prehistoric and Early Wales*, p. 121) but this is not found in Silurian country. It is, however, possible that there was contact between Wales and Spain in Iron Ages B and C.

eas: explained by the context, as *ea provincia* in *A*. 4. 56, 3.

proximi, etc., 'those nearest to the Gauls are also like them'. The reference is to the inhabitants of the south-east of England. Caesar had already noted the similarity between the customs of the Gauls and those of the people of Kent, the most civilized of the Britons (*B.G.* 5. 14, 1).

procurrentibus, etc., 'projecting in opposite directions' (north and south), and so approaching each other (see on c. 10, 2). *Diversus* has often the force of 'opposite', as in c. 23, 1; *A*. 2. 17, 2, etc.

positio caeli = *situs caeli*, their situation under a particular tract of the sky, involving uniform climatic conditions (see note on c. 10, 2). Here practically 'climate'. So also Columella 3. 4, 1. The celestial divisions determined the character of the terrestrial (cf. Virg. *G*. 1, 233 ff.); and *climata*, the term used by the astronomer Hipparchus (*c*. 140 B.C.) for belts of latitude (so 'climate' in old English), came to denote belts of temperature, whence the modern sense of the English word. The effects of climate were discussed, e.g., by Aristotle and Posidonius, and already formed a conventional topic in Herodotus.

11, 3. in universum . . . aestimanti, 'to form a general judgement' (so in *G*. 6, 3). The phrase belongs to the ethnographical style.

Tamen indicates that of the alternative explanations just suggested the former is the more probable. It is, of course, the true one. The first Celtic invaders were Goidelic Celts (or Celts who spoke Q-Celtic, the language from which Goidelic was evolved): from their dialect Gaelic, Irish, and Manx are descended. Later, apparently from about 600 B.C., came successive invasions of Brittonic Celts (P-Celtic), from whose dialect Welsh, Cornish, and Breton are derived. The first of these, bringing a late-Hallstatt type of culture, are known as Iron Age A, into which an element of early La Tène subsequently entered; these people had established themselves widely in Britain by 300 B.C., when the succeeding culture, Iron Age B, the British equivalent of developed La Tène on the continent, began to arrive, becoming widely established in the course of the third century. Iron Age C is largely though not entirely represented by the Belgae, whose first invasions probably came towards the end of the second century B.C., other immigrants following in several waves down to the time of Caesar's conquest of Gaul.

sacra, 'you would find (among the Britons) their (the Gaulish) rites'. Archaeological evidence shows that there was no essential difference between the religion of the insular and continental Celts. Gallo-Roman and Romano-British iconography is very close although in some cases it is doubtful how much came into Britain before the Roman conquest and how much with it. The Irish literary and philological evidence also shows that there was a common religious tradition among the Celts in pre-Roman times. In particular Druidism was current both in Britain and in Gaul (cf. Caesar, *B.G.* 6. 13) and the Catuvellauni brought over with them the cults of Camulos and Toutates amongst others. On Celtic religion generally see M. L. Sjoestedt, *Gods and Heroes of the Celts* (London, 1949); Anne Ross, *Pagan Celtic Britain* (1967).

superstitionum persuasionem, 'their religious beliefs'. The manuscript text (*deprehendas*) *superstitionum persuasione,* which is defended by Persson, Lenchantin, and Forni among others, is taken to mean 'because of their adherence to (Gallic) religious beliefs' but the abl. can grammatically only go with *deprehendas,* which is nonsensical. The sense should be: 'you may find the same rites and religious beliefs among the Britons as among the Gauls'. This requires *persuasionem* or *persuasiones.* The former is palaeographically easier and, although the plural is found (e.g. in Sen. *Ep.* 94, 30), Tacitus uses *persuasio* only in the singular (*G.* 45, 1; *H.* 5. 5, 3). The addition of *ac* is necessary because Tacitus never separates two objects in asyndeton by the predicate.

Superstitio, contrasted with *religiones* in *H.* 5. 13, 1, is used often of foreign religions other than Greek, whether barbarian (*G.* 39, 2; 43, 3, etc.), or Jewish (*H.* 5. 8, 2, etc.), or Christian (*A.* 15. 44, 3).

sermo, etc. The language in southern Britain at this time was derived from Gaulish and was probably sufficiently close, at least among the educated classes, to be mutually intelligible.

In northern Britain, south of the Forth, there were people who also spoke British and who had recently come from more southern territories. Further north the Caledonii probably spoke a form of Celtic, related to but not identical with British, which may have been contaminated with a non-Indo-European language. See K. H. Jackson, *Language and History in Early Britain* (Edinburgh, 1953); *The Pictish Language*, in *The Problem of the Picts* (ed. F. T. Wainwright, London, 1955), pp. 136–7, 148, 155–6.

in deposcendis, etc. Similarly Caesar (*B.G.* 3. 19, 6) and Livy (10. 28, 4).

11, 4. praeferunt, 'display'; cf. *A.* 4. 75, etc.

nam explains an unexpressed thought: '(as is the case with the Gauls), for . . .'.

pax emollierit. The demoralizing effect of peace was a commonplace; cf. Xen. *Cyr.* 3. 1, 26; Plato, *Laws* 698 b ff., Polybius 6. 57, 5; 31. 25, 3; Catullus 51; Sall. *Cat.* 10, 1; *Jug.* 41, 1; *Hist.* fr. 11 M.; Livy 1. 19, 4; 1. 22, 2, etc.; Tac. *G.* 14, 2; 36, 1. See below, c. 21, 2.

accepimus. The reference is probably to Caesar, *B.G.* 6. 24, 1, who is cited in *G.* 28, 1. On their subsequent unwarlike character, cf. *A.* 3. 46, 2–4; 11. 18, 1, etc. *Accepimus* is another ethnographical idiom (cf. Livy 5. 34, 1, etc.).

pariter, 'at the same time' (ἅμα): cf. *A.* 6. 18, 1; 13. 37, 2, etc.

olim : in the time of Claudius, taken closely with *victis.*

ceteri : such for instance as the Brigantes, and those of the north and west generally.

12, 1. In pedite robur. The same is said of the Germans (*G.* 6, 3) and of the Chatti in particular (*G.* 30, 3 'omne robur in pedite'). That the Britons had also cavalry is shown by Caes. *B.G.* 4. 24, 1; 32, 5, etc. The language is conventional and the structure of the sentences conforms to a standard pattern for which cf. *H.* 5. 8, 1 (another ethnographical section), 'magna pars Iudaeae vicis dispergitur; habent et oppida; Hierosolyme genti caput'.

nationes : here (as in *G.* 2, 3, etc.) of separate tribes; in *G.*, loc. cit., opposed to *gens*, but in c. 22, 1, below, interchanged with it. Under Roman rule the country was divided for administrative purposes into *civitates* roughly corresponding to the earlier tribal regions.

et curru, 'also with the chariot'. These warriors are the *covinnarii* of c. 35, 3, the *essedarii* of Caesar, who describes their skill and tactics (*B.G.* 4, 33). That these chariots were scythed is affirmed by Mela 3. 6, 52 and Silius Italicus 17, 417, but the silence of Caesar and Tacitus, who describe battles in which chariots take part, is against the supposition that they were generally scythed, and there is so far no archaeological evidence for it: cf. Powell, *The Celts*, p. 109. The use of chariots in battle was probably obsolete in Gaul by Caesar's time (cf. Livy 10. 28) but survived in Britain.

honestior auriga, etc. The general use of *propugnator* of one fighting from a place of vantage (as a ship, wall, etc.) seems to show that here the driver is opposed to those who fight from the chariot, and that the meaning is that (contrary to the rule in Homer and among the Gauls) the driver is the higher, the fighters the lower in rank. There was probably only one *propugnator*, chosen by the *auriga* from among his clients. So in Caesar's time the British chariot carried a driver and one warrior, like the Gallic (Diod. 5. 29). Caesar (*B.G.* 4, 33) describes the chariots as carrying the fighters among the troops of cavalry (their own, apparently), and then, while they alight and fight, taking position in rear to rescue them if pressed. The minor role played by chariots in Agricola's campaigns (cf. c. 35) is to be explained by the fact that Agricola was fighting against tribes with whom Caesar had not come into contact and whose terrain did not lend itself to such a mode of fighting. *Clientes* is used as in the case of a Gaulish (Caes. *B.G.* 1. 4, 2, etc.) or German (*A.* 1. 57, 3) chief, and presumably the same system obtained among the Britons.

propugnant, 'fight for' rather than 'fight in front of' (cf. Caes. *B.C.* 3. 45).

olim. In Caesar's time there were four kings in Cantium (*B.G.* 5. 22, 1); and monarchy was evidently general (cf. Diod. 5. 21, 6), with instances of pre-eminent kings ruling several tribes, like Cassivellaunus (Caes. *B.G.* 5. 11, 9) and afterwards Cunobellinus, described by Suetonius as *rex Britannorum*. Dio describes the Britons (60. 20, 1) as ἄλλοι ἄλλοις βασιλεῦσι προστεταγμένοι in the time of Claudius, and Claudius himself cites *reges Brit*[*annorum*] *XI* [*devictos*] (*ILS* 216). Some were established or continued as vassals of Rome, like Cogidumnus, Prasutagus, and Cartimandua.

nunc per principes, i.e. when Tacitus was writing no king remained. [The correction *distrahuntur* seems necessary and is easy after *studiis*. The meaning is that the Britons are divided by partisan factions (*factionibus et studiis* is a hendiadys) under rival chieftains. But no adequate parallel can be adduced for the simple verb *trahuntur*, and although Tacitus sometimes uses the simple verb instead of the compound the sense is always, unlike here, made clear by the context (as perhaps in *H*. 4. 20, 2, *portis rumpunt* (M: *prorumpunt* Ritter)). In Caes. *B.G.* 6. 38, 4, 'aegre per manus tractus servatur', *tractus* is literal, 'dragged'. In *A*. 3. 54, 5, 'haec (cura) omissa funditus rem publicam trahet', *funditus* defines the image.] For the sense cf. *H*. 1. 13, 1; 5. 12, 4; Justin 15. 4, 23 'in duas factiones diducuntur'. For *distrahi* cf. *A*. 12. 42, 1 'distrahi cohortes ambitu duorum'.

12, 2. pro nobis, 'on our behalf'.

in commune, 'in common', 'together': cf. 'in commune consultare' (*H*. 4. 67, 2), 'in medium consulere' (*H*. 2. 5, 2), etc. The thought should be compared with *G*. 33, 2 'maneat . . . si non amor nostri, at certe odium sui, quando urgentibus imperii fatis nihil iam praestare fortuna maius potest quam hostium discordiam'. The whole passage recalls Hermocrates' argument in Thuc. 4. 64, 4. Tacitus' estimate, although true as a generalization, is not wholly accurate. Cunobellinus seems to have ruled over a confederation of tribes during his lifetime and there was some form of union among the Brigantes. The evidence for common cult-centres in Britain which would have formed the nucleus of such confederations is still tenuous but the votive deposits at Llyn Cerrig Bach, which are widely diverse in age and place of origin, indicate the existence of such a centre in Anglesey rather than a last desperate rallying-ground for British refugees before its capture in A.D. 60 (Sir Cyril Fox, *A Find of the Early Iron Age from Llyn Cerrig Bach*).

duabus tribusve, 'two or (at most) three' (cf. Cic. *de Orat.* 1, 28; *ad Att.* 6. 1, 3; *Phil.* 14, 16; Ovid, *Trist.* 1. 5, 33, etc.); *tribusque* would mean 'two and (even) three', and would suggest a considerable number. *Civitas* is used of tribes, as the Brigantes (c. 17, 2), etc., and often of Gaulish and German tribes.

conventus, 'meeting together': cf. *A*. 2. 35, 2 *conventus Italiae.*

singuli, etc., 'they fight individually, and the whole are vanquished' (in detail).

12, 3. caelum, etc. The strange interposition of this account of the climate and products between two passages treating of the character of the people, has led to the supposition of some error on the part of a transcriber, which it is thought might be corrected by inserting c. 12, 3–6 at the end either of c. 10 (Reifferscheid: see Wölfflin, *Philol.* 26 (1867), 144–5) or of c. 11 (Baehrens). But the opening sentence of c. 13 (n.) leads into the account of the history of Roman dealings with Britain before the arrival of Agricola as governor (cc. 13–17). The order in which the topics in an ethnography were treated was not fixed, but the arrangement here corresponds to that in the *Germania.*

foedum, 'gloomy': cf. 'foedum imbribus diem' (*H.* 1. 18, 1), 'nubes foedavere lumen' (Sall. *Hist.* fr. 4. 80 M.).

asperitas frigorum abest. So Caesar says (*B.G.* 5. 12, 6), 'loca sunt temperatiora quam in Gallia, remissioribus frigoribus', his comparisons being, no doubt, between southern Britain and northern Gaul. Strabo also speaks of the weather as rainy and misty rather than snowy (4, 200).

dierum spatia, etc. Tacitus, like Juvenal (2, 161), speaks only of the long summer, not of the short winter, days. Caesar (*B.G.* 5. 13, 3), Strabo (2, 75, quoting Hipparchus), and Pliny (*N.H.* 2, 186) have some information as to both. Caesar, when in Britain, had verified the greater length of the day by a water-clock. Pliny comes very near accuracy in giving the longest day as fourteen hours at Alexandria, fifteen in Italy, seventeen in Britain (which would be about a medium between London and the north of Scotland).

nostri orbis, 'our world': cf. *G.* 2, 1, etc. So *nostrum mare* for the Mediterranean sea (c. 24, 2). *Dierum* is omitted for conciseness: cf. c. 24, 2.

ut . . . internoscas, 'so that you can hardly distinguish between evening and morning twilight' (the one passes into the other). Potential subj., cf. c. 22, 4.

12, 4. solis fulgorem is the 'glow' which may be seen long after sunset even in the English midlands on fine summer nights. Cf. *G.* 45, 1 'extremus cadentis iam solis fulgor in ortum edurat adeo clarus ut sidera hebetet'. The short summer nights of the north were already known to Homer (*Od.* 10, 83 ff.).

occidere et exsurgere. *Solem* should probably be supplied, as it was by Eumenius and Jordanes (see below). Peter keeps *solis fulgorem* as subject, but *occidere* is properly used of the sun and is inappropriate to *fulgor*. The actual sun is below the horizon but only casts a low shadow. *Et* after *nec* (cf. c. 1, 3) couples two parts of one idea, 'set, and then rise again'. The statement here reported is very loose: even at the north end of Shetland the sun rose at 2.23 and set at 21.41 on 21 June 1966. The night-long glow is assigned with more correctness to the extreme north of Scandinavia in *G.* 45, 1 (quoted above).

transire, 'passes along (the horizon)': cf. Jordanes' paraphrase (next note).

scilicet, etc. 'In fact the flat extremities of the earth, with their low shadow, do not project the darkness, and night falls below the level of the sky and the stars.' The theory implied is that night is cast by the shadow of the earth but that at its northernmost edge the earth is itself so flat, there being nothing but ocean, that the sun is hardly occluded by it. Cf. Pliny, *N.H.* 2, 47. Tacitus is followed by Eumenius (4th cent.), who says of Britain (*Pan.* 9), 'nullae sine aliqua luce noctes, dum illa litorum extrema planities non attollit umbras, noctisque metam caeli et siderum transit aspectus, ut sol ipse, qui nobis videtur occidere, ibi appareat praeterire'. Cf. also Jordanes (6th cent.), *Getica*, 3, 21 'et quod nobis videtur sol ab imo surgere, illis (sc. Scandzam insulam incolentibus) per terrae marginem dicitur circuire'. *Sidera* is more specific than *caelum*, and the two are thus coupled frequently in poetry and prose. [See Steinmetz, *Philol.* 111 (1967), 233 ff.]

12, 5. praeter, 'except': cf. Caes. *B.G.* 5. 12, 5 'praeter fagum atque abietem'.

oriri sueta, a phrase used by Sall. *Hist.* fr. 1, 11 M.

patiens frugum pecudumque fecundum. That *pecudumque fecundum* is the true reading is indicated by comparison with the similar passage describing the products of Germany, 'terra . . . frugiferarum arborum [im]patiens, pecorum fecunda, sed plerumque improcera' (*G.* 5, 1), as Lundström has pointed out (*Mélanges Vising*, 1925). Caesar (*B.G.* 5. 12, 2) and Strabo (4, 199) had already emphasized the abundance of cattle in Britain, and it is inconceivable that Tacitus should have omitted to mention such an important item of British economy. Trees, crops, and cattle comprise the three traditional aspects of husbandry (cf. Virg. *G.* 1–3). In both passages Tacitus added to the formula about crops and cattle an observation marked by harshness of style, but the harshness of *tarde mitescunt* (sc. *fruges* only) in the *Agricola* is somewhat greater than that of *sed plerumque improcera* (sc. *pecora sunt*) in the *Germania*. [It can be assumed that in the archetype *fecundum* had been omitted by haplography after *pecudumque* and was replaced in the margin whence *AB* substituted it for *pecudumque*.]

12, 6. fert . . . metalla. See Appendix 4. Caesar says nothing of precious metals, and Cicero had heard that there were none in Britain (*ad Fam.* 7. 7, 1; cf. *ad Att.* 4. 16, 7): but Strabo speaks of gold, silver, and iron (4, 199), Caesar of tin found in the interior, and a little iron on the coasts (*B.G.* 5. 12, 5); and an account of the tin trade from Belerion (Land's End) to the island of Ictis is given in Diod. 5. 22, 1 (cf. also Pliny, *N.H.* 4, 104). It is shown in Appendix 4 that there are traces of gold-mining, and that silver was extracted from lead ores, which were extensively worked. Copper was mined in Anglesey and elsewhere in N.W. Wales. Iron was extensively mined throughout the country. The working of Cornish tin which flourished in earlier centuries seems to have been disrupted by Caesar's wars: it may have been resumed in the third century. Cornish tin was less important than that of N.W. Spain.

pretium victoriae. For *pretium* = *praemium*, cf. c. 1, 2. These words should not be pressed to mean that Britain was invaded for the sake of these metals (Introd., p. 51). When Rome conquered a country the land became the public property of the Roman state; when the new province was formally organized, the ownership of the land was restored to the local communities and inhabitants but Rome could reserve for herself specific areas such as mineral deposits (A. H. M. Jones, *J.R.S.* 31 (1941), 26 ff.). These were either leased to contractors or were worked by the state itself, usually as extraterritorial domains under the management of a procurator (O. Davies, *Roman Mines in Europe* (1935), pp. 2–4, 140 ff.). Both systems are known from Britain. Private lessees can be inferred from Derbyshire (*CIL.* vii. 1214–16) whereas elsewhere, e.g. in the Mendips (*CIL.* vii. 1202) and Northumberland, direct military control is proved by inscriptions. The tendency was increasingly for the profitable mines to be worked by the state with forced labour (Florus 2, 25) while the poorer ones were leased (cf. Suet. *Tib.* 49). In either case the revenue accrued to the state.

gignit, 'produces'. Tacitus is not here alluding to the scandalous rumour that Caesar invaded Britain from financial greed (Suet. *Jul.* 47 'spe margaritarum'; cf. Catullus 29, 1–4; Cic. *ad Att.* 4. 16, 7). Mela (3. 6, 51) speaks of some British rivers as 'gemmas margaritasque generantia', and medieval writers give exaggerated accounts (Elton, *Origins of English History*, p. 220).

subfusca, etc. Pliny (*N.H.* 9, 116) calls the British pearls 'parvos atque decolores', instancing the breastplate dedicated by Caesar in the temple of Venus Genetrix, made of pearls professedly brought from there. The whole subject of pearls is treated by Pliny at length in §§ 105 ff. British pearls are also mentioned by Juba (Jacoby 275 F 70).

artem, i.e. the skill to dive for them.

rubro mari, the Indian Ocean. The phrase is sometimes used, as here, of the Indian Ocean as a whole (cf. *A.* 2. 61, 2: Syme, *Tacitus*, p. 768) and sometimes with a more limited reference to the Persian Gulf (Mela 3. 8, 72) or the Red Sea (Pliny, *N.H.* 6, 107). Pearls were in fact found in the Indian Ocean (off Malacca and between Indian and Ceylon), in the Persian Gulf (especially near Bahrein), and in the Red Sea (near the coast of Abyssinia).

viva ac spirantia: the adjectives are also paired in Cic. *de Dom.* 134.

saxis. The case here is generally taken to be dat., as in Virg. *Aen.* 2, 608, also in *A.* 1. 44, 4 (*avellerentur castris*), but Virgil has *complexu avulsus* (*Aen.* 4, 616), and such ablatives depending on the force of a prep. in composition are often found.

expulsa, 'cast on shore'.

naturam, 'quality', i.e. that of the best pearls. If, gathered alive, they were as good as others, greed would have found a way to get them alive. *Deesse* is taken by zeugma with *avaritiam* in a sense nearer to *abesse*: cf. Cic. *Brut.* 276. As usual Tacitus ends a section with an epigram.

13–17. *The History of Britain*

Tacitus gives a short account of the history of the Roman occupation of Britain before the governorship of Agricola. Although imprecise (c. 17, 1 n.) the account affords a just appreciation of the achievements of Agricola's predecessors.

(For details see Introd., p. 47.) Artistically Tacitus stresses their shortcomings in order to contrast Agricola's abilities. Didius' love of glory (c. 14, 2) and Trebellius' dilatoriness (c. 16, 3) are the very opposite of Agricola's modest and energetic government.

13, 1. Ipsi Britanni. These sentences describing the character of the Britons as subjects lead up to the account of their subjugation, and *ipsi* is similarly used in *G.* 2, 1 in a transition from the country to the people.

et iniuncta imperii munia: a general expression added to particular terms (c. 6, 4 n.): 'obligations enjoined by the government', including the provision of requisitioned corn (c. 19, 4), of labour for road-building (c. 31, 1), etc. Normally the form *munia* means the functions or duties of an office or calling (and elsewhere *munia imperii* in Tacitus means the duties of government; cf. *H.* 1. 77, 1, etc.), *munera* the duties imposed by a public authority, i.e. public burdens. Here *iniuncta munia* (for which cf. Livy 3. 35, 7, etc.) is equivalent to *munera.* There is no need to change the text. Tacitus never uses *munera obire*: he always writes *munia obire.* The use of *munia* for *munera* is typical of Tacitus' liking for variation. In the same way he unexpectedly writes *militiae munera* in *H.* 3. 13, 1 and *suprema munera* in *A.* 3. 2, 1. Livy uses the two forms interchangeably.

obeunt is sufficiently applicable by zeugma to the two more special terms to make correction (e.g. to *subeunt*) needless.

si iniuriae absint. The subjunctive is best taken as potential: cf. *H.* 3. 86, 2 'quae, ni adsit modus, in exitium vertuntur'.

igitur, here noting the beginning of a relation of the state of things already indicated by *domiti.* Cf. c. 29, 2, etc.

ostendisse . . . non tradidisse, 'to have pointed it out, not handed it over, to posterity'. Ancient writers vary in their estimates of the results of Caesar's two raids on Britain, but generally they depreciate them. If their objective was, as he alleged, to prevent British powers from aiding the Gauls, then it might be considered amply achieved. Furthermore he showed that Britain could be invaded, and that (as in Gaul) the disunity of its tribes could be turned to account. *Litore potitus,* if it implies lasting conquest, would be obvious exaggeration, like the 'bis penetrata Britannia' of Vell. 2. 47, 1, and the 'Caledonas secutus in silvas' of Florus, 1. 45 (3. 10), 18.

13, 2. et, carrying on the idea of *bella civilia*; *ac* adding another cause. *Principum*, 'leading men', as in *Dial.* 36, 4, and often in Cicero. Cf. c. 12, 1; 21, 2.

consilium, 'policy'; *praeceptum* 'an injunction'. That Tiberius regarded the practice of Augustus in this light is acknowledged by himself (*A.* 4. 37, 3; cf. also 1. 77, 3). Augustus more than once professed an intention to invade Britain, but really preferred to gain influence there by diplomacy, and dissuaded his successors from extending the empire (*A.* 1. 11, 4). The proposed invasion figures often in poetical propaganda (cf. Virg. *G.* 1, 30; 3, 25; Hor. *Odes*, 1. 35, 29). In 24 B.C. he seems to have gone so far as to claim that Britain now belonged to the Romans (cf. Livy, fr. Bk. 135 Weiss.; Serv. *ad Georg.* 3, 25).

agitasse : sc. *animo*, 'had entertained thoughts'. The great army collected by Gaius in Gaul is stated to have been marched to the coast as if to embark, and then to have been led back, after being told to pick up shells[1] as spoils of the ocean; a lighthouse was built to commemorate the occasion (Suet. *Cal.* 46; Dio 59. 25, 1). The lighthouse, a useful aid to cross-Channel shipping, may have been built in connexion with the invasion, and the abandonment of the invasion was probably caused by the refusal of the troops to embark (J. P. V. D. Balsdon, *The Emperor Gaius*, pp. 88 ff.; cf. the similar reluctance among Claudius' troops in A.D. 43, reported by Dio 60. 19).

ni, i.e. he had planned the invasion, and would have executed it but for the swift changes of plan through his fickle mind and the immense efforts against Germany being all in vain (cf. c. 4, 3; 37, 1).

velox . . . paenitentiae. *Velox* goes with *paenitentiae* (gen.), and *ingenio mobili* is a causal abl., 'through natural fickleness swift to change'. Such a genitive as *paenitentiae* expressing the thing in point of which a term is applied to a person, though nowhere else used with *velox*, is found with many adjectives, such as *pervicax*, *procax*, etc. *Velox ingenio* would imply praise, as in Quint. 6. 4, 8, etc.

[1] For a different explanation of the order to pick up *musculi* see Balsdon, *J.R.S.* 24 (1934), 18. R. W. Davies (*Historia* 15 (1966), 124 ff.) argues that Gaius' concentration of troops was no more than an extended training manœuvre and that he never intended to invade Britain but Claudius evidently believed that he had planned an invasion. See also Bicknell, *Historia* 17 (1968), 124 ff.

frustra fuissent, 'had failed', a construction first found in Ennius (*Sat.* 60–62V.) and then used by Sallust (*Cat.* 20, 2, etc.) and Livy (2. 25, 2, etc.). The expedition into Germany, of which Tacitus elsewhere speaks still more severely (*H.* 4. 15, 2; *G.* 37, 4), is described in a similar spirit by Suetonius (*Cal.* 43 f.), who speaks of a sham fight, in which the emperor's German bodyguard represented the enemy, and of Gauls dressed up to resemble German prisoners. Compare the similar (false) tale told of Domitian in c. 39, 2. For Gaius' German campaigns see Balsdon, *The Emperor Gaius*, pp. 58 ff. They seem to have been principally training exercises.

13, 3. auctor tanti operis. This correction of *E*'s *auctoritate*, made by Bezzenberger and Peerlkamp, is preferable to Wex's *auctor iterati*, read by most modern editors. In this phrase *operis* is not elsewhere qualified by a participle (cf. Val. Max. 1. 5, 2 'huius tam praeclari operis auctor'; Pliny, *Paneg.* 10; Albinov. *carm.* 1, 236) nor is *iterare opus* found. Claudius could not properly be said to have repeated Caesar's undertaking, because Caesar only made brief raids whereas Claudius planned the complete annexation of the island. On the other hand, it is natural to speak of the full-scale invasion of Britain as 'a great undertaking'. [The corruption (*auctor tāti* → *auctoritate*) is very easy. The simple *auctor* read by the Toledo MS. is an emendation by the scribe Crullus (see Introd., p. 88).]

legionibus, etc., four: see Introd., p. 76. **Transvectis** and **adsumpto** are aoristic.

in partem rerum, 'to share the undertaking'. For similar uses of *in partem*, cf. c. 25, 1; *A.* 1. 11, 1; 14. 33, 1. If this passage stood alone, we might suppose that Claudius commanded the first invasion in person, with Vespasian as his chief of staff, and that Plautius was sent out afterwards to govern the province: whereas our only narrative, that of Dio (60. 19–23), makes Plautius command the invading force, and barely mentions Vespasian, while Claudius arrives later and stays only a few days, to take the credit of the capture of Camulodunum. But Tacitus elsewhere (*H.* 3. 44, 1) states that Vespasian was *legatus* of the Second Legion (cf. Suet. *Vesp.* 4), and in the *Annals* he doubtless agreed with the account preserved by Dio. Here he is speaking loosely and rhetorically.

fortunae. He had been previously obscure; but his service in Britain advanced him to the consulship and *triumphalia*, and led Nero afterwards to select him to deal with the Jewish rebellion; which position led to his designation as emperor. *Fortuna* is used specially of the imperial rank (*A*. 6. 6, 2; 11. 12, 3, etc.), and of that of Vespasian in particular (c. 7, 2).

domitae gentes, capti reges. These asyndeta form one idea, distinct from *et* (cf. c. 11, 2). The syntax and the language recall the style of official communiqués: cf., e.g., *Inscr. Ital.* 13 Elogium no. 81 *domitis Liguribus*; Livy 2. 19, 2 'his consulibus Fidenae obsessae, Crustumeria capta'.

Tacitus seems to allude here to the triumphal arch of Claudius (*ILS* 216; 217; cf. Suet. *Cl.* 25) which was dedicated in A.D. 51–52, and recorded the conquest of eleven kings. (For a possible list of the kingdoms concerned see Introd., p. 51. Caratacus and his family were captured in A.D. 51: cf. *A*. 12. 36.)

fatis, best taken as abl.: cf. Virg. *Aen.* 6, 869, 'ostendent terris hunc tantum fata'; Livy 5. 15, 4 'interpres fatis oblatus'. By his achievements here destiny made him conspicuous as the coming man, a more rhetorical repetition of the fact stated above (*quod initium*, etc.).

14, 1. Consularium, i.e. governors (of consular rank).

Aulus Plautius, the leader of the original expedition in A.D. 43, who remained in Britain till A.D. 47, and received an ovation (*A*. 13. 32, 2), an honour usually reserved, like the full triumph, for the imperial family (see note on c. 40, 1). Plautius probably fixed the frontier of the Roman province from the Exe to the Humber along the limestone escarpment that runs from the Cotswolds to Lincoln Edge and built the Fosse Way to serve the frontier zone. Such a frontier proved too vulnerable to sudden inroads and necessitated the subsequent subjugation of Wales. He had been *cos.* in A.D. 29 (from July) and *legatus* of Pannonia from about 41 to 43 (*CIL*. v. 698). From Pannonia he brought the Ninth Legion with him to Britain.

subinde, 'in succession' (*A*. 6. 2, 4).

Ostorius Scapula (Publius), *legatus* A.D. 47–52. He must have been suffect consul about A.D. 44. His problem was to crush the marauding British tribes under Caratacus that threatened the Roman frontier zone. To that end he summoned the Twentieth Legion from Camulodunum where it was stationed as a watch on the Iceni and replaced it by a veteran colony. He drove a wedge to the sea at Chester to isolate Wales and, perhaps, to threaten Anglesey, and protected his northern flank by encouraging friendly relations with the Brigantian kingdom (G. Webster, *Arch. Journal* 115 (1958), 49–98). Although by cautious advances he penetrated into Wales, defeating and eventually capturing Caratacus (for a fuller account see *A.* 12. 31–39; Dudley and Webster, *The Rebellion of Boudicca*, p. 34; M. G. Jarrett, *Arch. Journal* 121 (1965), 26), he failed to subdue the Silures. He died in the province.

proxima pars. In *A.* 12. 31, 2 Tacitus defines the extension of the Roman province planned by Ostorius as *⟨cis Tris⟩antonam et Sabrinam* (Syme, *Tacitus*, p. 394 n. 4), that is, as far as the Trent and the Severn. These limits fit Ostorius' strategy.

veteranorum colonia : that of Camulodunum (Colchester). The site lay south-east of Cunobellinus' headquarters at Sheepen. This area and parts of the *colonia* have been excavated (see C. F. C. Hawkes and M. R. Hull, *Camulodunum* (1947); M. R. Hull, *Roman Colchester* (1958).

civitates : it is not known which these cantons were but, apart from west Sussex where his capital was situated (Noviomagus Regnensium—Chichester), east Sussex and central Surrey form distinct areas that could conveniently have been incorporated.

Cogidumno regi. Either 'certain cantons were given to king Cogidumnus' or 'certain cantons were given to Cogidumnus to be king over'. The former would imply that he was already king of the Regnenses at the time, although there is nothing to link him with the old reigning dynasty of the Commii. The latter would imply that he was imposed as a king from outside. Latin usage, however, supports the former (cf. *H.* 4. 4, 1) and the name is compounded from a common Celtic root **dubno* or **dumno*, 'deep, powerful'; cf. Dumnorix and Dubnorex, etc. Tacitus speaks as if he was still living in his own time or that of Agricola. His name occurs in a famous inscription from Chichester (*RIB* 91):

> [N]eptuno et Minervae templum [pr]o salute do[mus] divinae
> [ex] auctoritate [Ti.] Claud. [Co]gidubni r(egis)
> legat[i A]ug(usti) in Brit(annia), [colle]gium
> fabror(um) et [q]ui in e[o sun]t d(e) s(uo) d(ant),
> donante aream [ ]ente Pudentini fil(io).

The explanation of the inscription is beset with some difficulties: *R* is an unexampled abbreviation of *regis*, though *R.N.* occurs for *regnum Noricum*; and the title of *legatus Augusti* is unknown among vassal princes, but an analogy may be provided by the position of the client-king Herod the Great who in 20 B.C. was given some official status in the province of Syria (Josephus, *A.J.* 15, 360 = *B.J.* 1, 399, says that he was made procurator, ἐπίτροπος). For the very rare formula *pro salute domus divinae* cf. *RIB* 897, 919: *domus divina* first occurs in the Paris inscription of Tiberius' reign (*CIL.* xiii. 4635). The very fine lettering

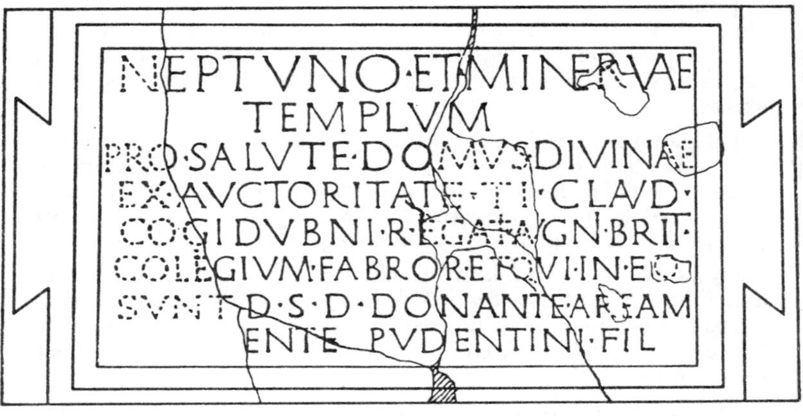

10. Inscription of Cogidubnus at Chichester

(with punctuation dots) is hardly later than the early Flavian age. Confirmation of Cogidumnus' devotion to the imperial house may be provided by a dedication to Nero found at Chichester (*RIB* 92), and the large palace at Fishbourne erected after the supply-base was dismantled *c.* A.D. 75 may well have been his (B. Cunliffe, *Antiquity* 39 (1965), 174 ff.).

vetere . . . consuetudine is a modal abl. (cf. *H.* 2. 71, 1) or abl. abs., and *ut haberet* explains *consuetudine* (cf. Cicero, *ad Fam.* 13. 76, 1, etc.). [With the MSS. reading, which places *ut* before *vetere*, it is very awkward to supply the subject of *haberet* from *populi R.* and the sense is inferior. *Ut* might easily drop out, like *et* in other places, and be inserted in the wrong position before *uet-*: and the erasure in *E* (see crit. note) perhaps points in this direction.]

In Livy 44. 24, 2 the Romans are said 'regum viribus reges oppugnare'. Among the instances in old times were Massinissa, Attalus, Eumenes, Herod, etc. In Britain other vassal royalties were king Prasutagus of the Iceni and queen Cartimandua of the Brigantes, and under the early Empire there were many such in the East (Thrace, Asia Minor, Syria, etc.).

instrumenta is predicative.

et reges, 'even kings', in ironical contrast to *servitutis*.

14, 2. mox . . . continuit, i.e. *secutus est et continuit*. There is a similar condensation below, 'Suetonius hinc . . . habuit'.

Didius Gallus (Aulus), *cos. suff.* in A.D. 36, had perhaps accompanied Claudius on the original invasion of Britain (*AE*, 1947, p. 76). He had gained experience in troublesome frontier politics as legate in Moesia about A.D. 46, winning the *triumphalia ornamenta* for restoring Cotys to the kingdom of Crimea (*ILS* 970) and campaigning against Mithridates of Bosporus (*A*. 12. 15, 1). He was legate of Britain in A.D. 52–58 and must have been preoccupied with consolidating the province in the face of the Silures and Brigantes. His activities are recorded briefly in *A*. 12. 40, where he is spoken of as 'senectute gravis et multa copia honorum', and said to have left all action to subordinates and contented himself with standing on the defensive. Elsewhere Tacitus says of him 'neque . . . nisi parta retinuerat' (*A*. 14. 29, 1). This was probably Imperial policy. During his five years' rule ground already won was secured by forts, roads, etc. He seems also to have held the governorship of Asia (*ILS* 970). See J. H. Oliver, *A.J.P.* 69 (1948), 219–22.

castellis . . . promotis is aoristic, as also *subactis* and *firmatis* below: c. 2, 2, etc. On the meaning of *castellis* see note on c. 16, 1 and Fig. 2. These forts have not been identified but may include some on the Welsh marches such as Usk and Weston-under-Penyard which had a long pre-Flavian occupation.

aucti officii. *Augere officium* elsewhere means 'to increase one's services (to someone)'; cf. Cic. *ad Att.* 13. 20, 1; Nep. *Att.* 2, 6. But here the sense is clearly that Didius is anxious for the glory that comes from achieving the conventional ambition whereby men aim not merely to retain (*parta continuit*) but to increase their inheritance. For this theme cf., e.g., Thuc. 2. 62, 3; Livy 4. 2, 4. *Officium* must, therefore be used of a sphere of duty and so almost of the province for which he was responsible (cf. c. 19, 3).

Veranius : Q. Veranius was the son of the legate and friend of Germanicus. He received rapid promotion under Tiberius and Claudius, being *triumvir monetalis*, quaestor in A.D. 37, and praetor (probably) in A.D. 42. He was then appointed to the newly created province of Lycia and Pamphylia which he held for five years, conducting successful campaigns against the bandits of Cilicia Aspera. His achievements won him an ordinary consulship and the patriciate in A.D. 49. His appointment to the governorship of Britain in A.D. 57 shows that Nero had decided on an active policy to rationalize the province by reducing and annexing Wales, and chose Veranius as a man whose talents and experience fitted him for the task. Veranius died before the policy could be completed (he boasted in his will (*A*. 14. 29) that he could have subdued the province (i.e. Wales) in three years) and the revolt of Boudicca set the whole subjugation of Britain back by several years and forced Nero to appoint governors who could secure the loyalty of the provincials rather than advance the conquest of the island. Veranius' career is documented by several inscriptions (*IGR* iii. 703; *AE*, 1953, p. 251; see Birley, *Roman Britain and the Roman Army*[2], pp. 1 ff.; A. E. Gordon, *Q. Veranius, Consul A.D. 49* (Univ. of California Publications in Class. Archaeology 2 (1952), 231 ff.)).

14, 3. Suetonius . . . Paulinus: see on c. 5, 1. His *biennium* would be A.D. 58 and 59. The revolt of Boudicca began in A.D. 60 and Paulinus was superseded by Turpilianus in A.D. 61 (*A*. 14. 39, 3: see Syme, *Tacitus*, App. 69). It is likely that Paulinus spent the *biennium* campaigning against the Silures in S. Wales and against the tribes of N. Wales.

hinc, 'after this'; so often in Tacitus.

firmatisque praesidiis. *Firmare praesidia* is regularly used to mean 'to strengthen garrisons' (cf. *A*. 13. 41, 2; Livy 32. 40, 10; 43. 20, 4; 44. 8, 1, etc.) and this must be the meaning here too. Didius had established some advanced forts: Paulinus consolidated them. There is no parallel for translating 'firmatis praesidiis' as 'praesidiis firmis positis', 'establishing strong forts' (Peter). To take *firmatis* and *nationibus* as abl. abs., and *praesidiis* as instrumental abl., 'securely held by forts', though supported by c. 23, 1; *H*. 2. 83, 2; 4. 55, 4; Livy 1. 33, 4; 5. 12, 4, etc., would here give an awkward construction.

quorum: probably 'which things' (cf. *A*. 3. 63, 1), not only the *praesidia*. *Fiducia* is causal abl.

Monam: Anglesey, mentioned by Caesar, who perhaps confounds it with Man (*B.G.* 5. 13, 3), and by Pliny, who says it was 200 miles from Camulodunum (*N.H.* 2, 187). In *A.* 14. 29, 3, Tacitus calls it 'incolis validam et receptaculum perfugarum', and describes graphically the attack on this Druid stronghold (c. 30). It is doubtful whether Paulinus' motive in attacking it was because it was a centre of Druidism or because it was a natural rendezvous for dissident refugees. See note on c. 12, 2. Agricola finally subdued it (c. 18).

vires rebellibus ministrantem: the expression is striking; for 'vires ministrantem' cf. Virg. *Aen.* 9, 764; Stat. *Silv.* 1. 4, 22; *Theb.* 8, 498; *rebellis*, which is common in Tacitus and in Virgil and Ovid, is not used by earlier prose-writers except Curtius. The phrase recalls Livy 33. 29, 1, 'ad rebellandum neque vires . . . habebant'.

terga occasioni patefecit, 'exposed his rear to opportunity', i.e. 'to attack', a novel expression with a bold use of *occasio*. It is modelled on Livy 4. 31, 2, 'aperuerunt ad occasionem locum', 'they gave the enemy room to take them at a disadvantage'.

15–16, 3. *The Revolt of Boudicca*

The attention of Roman governors had been principally directed to the military question of establishing a secure frontier by the subjugation of Wales, and Paulinus' interests and tastes seem to have been in any case primarily military. He overestimated the extent to which the province had been pacified, even though the removal of the Twentieth Legion from Colchester to the Welsh border dangerously weakened the internal security forces, and left the civilian administration largely in the hands of the procurator, Catus Decianus. A number of particular factors conspired to spark off the revolt. Ostorius had disarmed the British tribes, which had given great offence to the Iceni, who claimed immunity as an independent client-kingdom. The veteran colonists at Colchester behaved in an arrogant and provocative fashion towards the natives. Catus Decianus enforced what may have been a new and increased taxation census in order to cover the high costs of the military operations in the province, which induced Roman money-lenders (including probably Seneca, as Dio alleges) as a precaution to call in the loans which they had made to leading Britons, and he attempted to recover money from the Iceni which they had considered a gift but he regarded as a loan. The Druids in Anglesey may also have felt such an atmosphere ripe for encouraging a revolt that would distract the Roman army from attacking their island.

The revolt broke out in the late summer of A.D. 60. It was well timed and well co-ordinated. While Boudicca captured and sacked Colchester, a second army of Britons ambushed a relief force of the Ninth Legion under Petilius Cerialis which had hastened south from Lincolnshire[1] on the news of the revolt. Paulinus hurried from Anglesey with his cavalry, instructing the Fourteenth and Twentieth Legions to follow him and ordering the Second Legion, which was in the south-west, perhaps at Dorchester (Dorset) or Exeter, to meet him so that he could organize the defence of London. The acting commander of that legion, Poenius Postumus, perhaps because he was too hard pressed or because he did not have the confidence to act in such a confused situation, decided no' to move. Paulinus was forced to abandon London and Verulamium to their fate and fell back on his own legions advancing down Watling Street. For Boudicca ultimate success could come only from fighting and defeating Paulinus but in the eventual battle the strength of the Roman legions was decisive. For an account of the whole affair see Dudley and Webster, *The Revolt of Boudicca* (1962).

15, 1. Namque, etc. Tacitus makes no mention here of any special grievances, other than the ordinary 'mala servitutis', the 'cuncta magnis imperiis obiectari solita' (*H.* 4. 68, 5) such as Civilis unfolds in *H.* 4. 14: whereas in the *Annals* (14. 31–32) he specifies the exactions levied on the Iceni after the death of their king Prasutagus, the outrages on Boudicca and her daughters, the oppression of the Trinovantes by the veteran colonists of Camulodunum, the temple of Claudius as a standing monument of subjection, and the greed of the procurator Catus Decianus.

A similarly rhetorical speech is given to Boudicca by Dio (62, 3–5) in which she dwells upon the peculiarly Roman theme of the contrast between *libertas* and *servitium*. Tacitus has chosen to give here a general indictment of Roman Imperialism as it might seem to a critic of Rome rather than a particularized list of historical grievances in order to build up a drab background against which the merits and exploits of Agricola could

[1] The recently discovered 30-acre forts at Newton-upon-Trent and Longthorpe may be connected with operations of the Ninth Legion in this area: see *J.R.S.* 55 (1965), 75–76.

stand out in bold relief. He has drawn upon a stock of commonplaces many of which go back to Thucydides. For 'singulos . . . nunc binos imponi' cf. Livy 3. 9, 3; for 'si et se Britanni numerent' cf. Thuc. 7. 67, 3; for 'sibi . . . illis . . . causas belli esse' cf. Thuc. 7. 68, 2–3; for 'neve . . . eventu pavescerent' cf. Thuc. 7. 61, 2; for 'deos misereri' cf. Thuc. 7. 77, 4; for 'ipsos . . . deliberare' cf. Thuc. 6. 17, 4; 79, 3. The concluding epigram 'periculosius esse deprehendi quam audere' is proverbial; cf. Plut. *Lyc.* 18; Xen. *Resp. Laced.* 2, 8.

agitare . . . conferre. Cf. Livy 3. 34, 4 'agitarent . . . sermonibus, atque in medium . . . conferrent'. The phrase means 'discuss and compare'.

interpretando accendere, 'inflaming their wrongs by putting a (worse) construction on them', by suggestion of motives. *Accendere* is used of intensifying a feeling (hope, grief, etc.) or of aggravating the force of words (*A.* 1. 69, 5). Here *iniurias,* though it bears its ordinary sense with *conferre,* seems with *accendere* to take a pregnant meaning, 'sense of injury'. We have a still stronger figure, *delicta accendebat,* in *A.* 12. 54, 2. *Interpretando* (cf. *H.* 1. 14, 2 *deterius interpretantibus*) implies putting the worst construction on acts, tracing in them a set purpose to oppress and insult.

tamquam, 'as though', like ὡς, giving the ground, as it appeared to the rulers.

ex facili : a variation for the adverb. So *H.* 3. 49, 1 *cetera ex facili.*

15, 2. e quibus, etc., 'the governor to wreak his fury on our life-blood'—as having power to order levies and forced labour (c. 31, 1) and to put to death—'the procurator on our property'. The chief provincial taxes levied by the procurator were the land-tax (*tributum soli*), property-tax (*tributum capitis*), and customs-dues (*portoria*). For the corn-requisition, probably levied by the governor, which with a Roman garrison of four legions will have been the most onerous tax, see note on c. 19, 4.

alterius manus, etc., 'the tools of the one (the *legatus*), his centurions, those of the other (the procurator), his slaves'. *Manus* 'instruments' is used, e.g., by Cic. *Verr.* ii. 2, 27; Sen. *Const. Sap.* 8, 3 'homines potentes fortunae manus esse'; cf. Catullus 47, 1 'Porci et Socration, duae sinistrae Pisonis'. On the position of the procurator see note on c. 9, 4. Very similar language is used in *A.* 14. 31, 1 'ut regnum per centuriones, domus per servos velut capta vastarentur'. Cf. also *H.* 4. 14, 3 'tradi se praefectis centurionibusque, quos ubi spoliis et sanguine expleverint', etc.

miscere: 'combine': cf. *H.* 3. 74, 2 'minas adulationesque miscet'. *Vis* and *contumelia* are similarly paired by Caes. *B.G.* 3. 13, 4.

exceptum, 'exempted from': the dat. is not unusual: cf. Virg. *Aen.* 9, 271; Sen. *Contr.* 10. 5, 1; Sen. *Const. Sap.* 9, 4, etc.

15, 3. in proelio, etc., 'in battle the spoiler is at least the stronger', and the indignity therefore less.

nunc, 'as things are': cf. c. 1, 4.

ignavis . . . imbellibus: similarly *G.* 12, 1; 31, 1; Livy 26. 2, 11.

eripi domos. Cicero uses the same bold figure, 'domo per scelus erepta' (*de Dom.* 147). Here it is adapted to balance *abstrahi liberos.* The veteran citizens of Camulodunum are referred to, who 'pellebant domibus, exturbabant agris, captivos, servos appellando' (*A.* 14. 31, 3).

tantum, taken closely with *pro patria*: 'as though it were only for their country that they knew not how to die'. Alternatively it might be taken with *mori*—they had learned to tolerate every indignity: the only thing they had not learned to endure for their country was dying for it.

quantulum enim, i.e. 'but we will show that they are wrong in counting on our cowardice, for what a handful are our invaders in proportion to our own numbers!'

si et se. The Roman army, being organized by legions, was easily calculable. If the Britons were to count themselves as well, they would see how much more numerous they were than the Romans. *Et se,* the reading of *E,* has more point than the correction *sese.*

sic introduces an historical parallel to encourage the Britons to revolt. 'Thus, as we will, the Germanies threw off their yoke.' Cf. *H.* 4. 57, 2 'sic olim Sacrovirum . . . concidisse'; *A.* 4. 38, 5; 12. 25, 1, etc. The plural *Germaniae* is often used (like *Galliae*) of the two Roman military governments or provinces, and sometimes, as here, of portions of Germany that were subject at the time spoken of (cf. c. 28, 1; *A.* 1. 57, 2). The allusion is to the defeat of Varus in A.D. 9.

et, 'and yet': cf. c. 3, 1 (and note); 9, 3. *Flumine* is the Rhine.

15, 4. causas, 'motives': cf. c. 30, 1.

divus Iulius: possibly used sarcastically, like 'ille inter numina dicatus' (*A.* 1. 59, 5), but probably only as a distinctive title, as a Roman would use it: so in *A.* 12. 34 'Caratacus vocabat nomina maiorum qui dictatorem Caesarem pepulissent'.

modo, 'if only': cf. *A.* 2. 14, 4, etc. *Aemularentur* is the equivalent of *aemulemur* in *oratio recta.* For *aemulari virtutem* cf. *Bell. Afr.* 81, 2; Livy 7. 7, 3. The exhortation is carried on in *neve,* etc.

unius aut alterius, 'one or perhaps two': cf. c. 40, 4.

plus impetus felicibus. *Impetus* is here opposed to *constantia,* as it is opposed to *perseverantia* in Livy 5. 6, 8; 27. 16, 1 and Justin 41. 2, 8; and *felicibus* applies to the Romans who have the advantage for the time.

15, 5. fuerit, equivalent to *fuit* in *oratio recta*; 'we ourselves (contrast to *deos*), in meeting to deliberate, have already taken what has been in the past the most difficult step'. To have dared this is to have overcome the difficulty of disunion (c. 12, 2), and to dare all (*H.* 2. 77, 3 'nam qui deliberant desciverunt').

porro develops the argument, 'and indeed' or 'but' (so in *A.* 3. 34, 5; 58, 1, etc.). Elsewhere it adds a new argument ('further'), as in c. 31, 3; *G.* 2, 2, etc.

deprehendi, i.e. to hesitate until you are detected. The same sentiment is expressed in *H.* 1. 21, 2.

16, 1. in vicem, 'in turn', i.e. inspiring one another by such arguments. For *instincti* cf. Ovid, *Fasti* 6, 597; Livy 6. 14, 9; 9. 40, 7.

Boudicca. She was the widow of Prasutagus, king of the Iceni, who had died in A.D. 59. Tacitus' phrase 'a woman of royal rank' suggests that she did not become Queen regnant on Prasutagus' death as of right. Even if she had, the Romans did not automatically recognize hereditary successors as client-kings: each grant of the status of a client-king was personal to the holder and on his death the Romans reserved the right to deal with the kingdom as they thought fit. Paulinus may have had other plans. He may, for example, have intended to incorporate the kingdom into the Roman province of Britain. But the Iceni clearly regarded her as their natural leader.

The true form of the name is 'Boudica', from *bouda, 'victory' (cf. Welsh *buddug*), with the adj. termination *-ico*. *Boudicca* is read in the Medicean MS. of *A*. 14. 37, 2 and *E*'s *vo adicca* (*bovid icta* E^{2m}) indicates that this is what Tacitus wrote here too.

neque enim sexum. Tacitus' comment is exaggerated and at variance with his own remarks about the Germans (*G*. 45, 6) and with what is known of Celtic practice. Boudicca similarly exaggerates in *A*. 14. 35, 1 'solitum quidem Britannis feminarum ductu bellare'. In fact only one other queen is known, Cartimandua of the Brigantes (*A*. 12. 36, 1; *H*. 3. 45, 1), and her people are said to have rebelled 'ne feminae imperio subderentur' (*A*. 12. 40, 3). Tacitus is, however, referring to Boudicca's military leadership of an armed confederation of tribes (*imperiis*) rather than to her supposed position as queen of the Iceni.

sumpsere . . . bellum. This phrase, first used by Plautus (*Cist*. 300) is common in Sallust (*Jug*. 20, 5; 62, 9; 83, 1, etc.), and in later authors.

universi. The chief tribes involved in the rebellion were the Iceni and Trinovantes, but it is probable that some of the Coritani and the Cornovii in the Midlands and the Durotriges in Dorset also joined, and there will have been detachments from other restive areas, including the Brigantes and the Catuvellanni.

sparsos, etc. In *A*. 14. 33, 4 the Britons are described as only attacking defenceless places abounding in plunder, 'omissis castellis praesidiisque militarium'.

castella . . . praesidiis. The Romans used two distinct classes of permanent fortified posts, *castra* and *castella*, corresponding to the division of the army into legions and *auxilia*. (1) *Castra*, each between 40 and 60 acres in area, were garrisoned from Domitian's time by one legion each, i.e. nearly 6,000 men, almost all infantry. (2) *Castella* were far smaller posts of varying area, 2 to 7 acres, each garrisoned by one 'cohort' (infantry) or one 'ala' (cavalry), either 500 or 1,000 men strong. There is also evidence for a series of 30-acre forts (e.g. at Longthorpe) which probably housed half a legion. The *castra* were the larger central posts—in Britain: York, Chester, Caerleon, and for a time Lincoln, Wroxeter, and Gloucester. The *castella* were dotted along roads at strategic points (river-crossings, etc.) or along frontiers; for a distribution map of such forts at the time of Agricola see Fig. 2. In English *castra* should be rendered by 'fortress' (not 'camp', which implies canvas and temporary occupation) and *castellum* by 'fort'.

Praesidia is a general term, often equivalent to *castella* (c. 20, 3, and so sometimes in inscriptions, *RIB* 1583).

coloniam, Camulodunum: cf. c. 14, 1. The colony was unfortified (A. 14. 31, 4)—the still existing Roman walls of Colchester were built in the late second century—but the inhabitants managed to send word to the procurator in London, who was able to send only 200 men, and the news of the rebellion also reached Cerialis in Lincolnshire, whose attempt to relieve them ended in a disastrous ambush. The garrison occupied the precinct of the temple of Claudius but the stronghold was stormed in two days and the colony sacked. Archaeological evidence for the hastily improvised defence of the colony has been found on the site of the old British city where the discovery of large quantities of new and repaired military equipment (helmets, shields, breast-plates, weapons, etc.) points to the feverish manufacture of arms to counter the sudden emergency. Traces of the destruction of the colony are widespread, and include glass and pottery wares from shops melted and blackened by fire, and burnt coins. The temple of Claudius, which had been a particular object of grievance (see next note) and whose foundations were later used for the Norman Castle, was gutted. Some of the marble veneer, stripped from its walls, has been found re-used to make a culvert, but the head of Claudius, found in the River Alde near Saxmundham, which must have formed part of the loot of the colony (see Plate II), was probably torn not from the cult-statue but from an equestrian statue of the emperor set up in or near some other public building. See M. R. Hull, *Roman Colchester*.

ut sedem, 'looking upon it as the headquarters'. In *A.* 14. 31, 4, the temple is mentioned as especially regarded 'quasi arx aeternae dominationis'.

ira et victoria personified, 'fury and (the elation of) victory'. A similar, but softened, expression is used in *A.* 14. 38, 3, where it is said that a successor to Paulinus would be without 'hostili ira et superbia victoris'.

16, 2. quod nisi : the negative of *quod si*, used by Tacitus only in the *Agricola* (c. 26, 2; and *quod ni*, c. 37, 4).

subvenisset. An account of his march is given in *A.* 14. 33.

He marched ahead of his main force, along Watling Street, to London intending to link up with the Second Legion from the south-west, but was unable to save either that town or Verulamium, because of the failure of the Second Legion to move. Extensive traces of the sack of both cities have been discovered. The layers of burnt material of this period, including blackened Samian ware and burnt Claudian coins, have been recognized at many points within the City of London (R. Merrifield, *The Roman City of London*, p. 90 and Fig. 9). At Verulamium a clear level of burnt material from the first century reveals the extent of the conflagration (S. S. Frere, *Antiquaries Journal* 39 (1959), 3). In his great battle, fought probably in the Midlands (there is no wholly satisfactory conjecture about the exact site), he had with him the Fourteenth Legion, a detachment of the Twentieth, and auxiliaries making up the total to 10,000 men. The battle is described in *A*. 14. 34–37.

fortuna : best taken as abl.

veteri patientiae, 'to its old submission' (submissiveness, c. 15, 1). This is true in so far as the Britons ventured no more battles; but the context shows, and the account in *A*. 14. 38 further describes, the continuance of a stubborn passive resistance, and the devastation of rebel districts by troops quartered upon them.

tenentibus. There was no real capitulation or reconciliation, no willingness on the part of the Britons to acquiesce in Roman rule: they remained defiant and in arms, as they had been before the rebellion, alert to resist the reprisals which they anticipated from Paulinus. *Patientia* is opposed to the spirit of co-operation which Agricola inspired. *Tenentibus arma plerisque* is not a concessive ablative but describes one of the features of the *vetus patientia*, namely that the Britons were in arms but not in open rebellion as they waited to see what would happen. There is therefore no need for emendation.

proprius, 'personal' (cf. *H*. 3. 45, 1, etc.) implies that they were specially afraid as having been ringleaders. The whole phrase is modelled on Livy 23. 15, 7 'conscientia temptatae defectionis ac metus a praetore Romano'.

ne quamquam. The Britons were apprehensive that however estimable Suetonius might be in other respects (for *egregius cetera* cf. Sallust, *Hist*. fr 4. 70; Livy 1. 32, 2; 35, 6), he would be merciless to the rebels. They did not fear that he was corrupt or biased, as other governors had been, but merely that he was harsh. This estimate of Paulinus' character may not have been wide of the mark. He was a professional soldier, accustomed to the strict values of the camp, and seems to have been content to leave the civil administration to the procurator. [The correction *ne quamquam*, made by a friend of Walch, was probably also intended by E^2. The same error occurs at *H.* 4. 68, 1.]

in deditos, 'against those who surrendered'.

ut suae cuiusque iniuriae ultor : *E* reads *eiusque* which cannot be right as there is nothing for *eius* to refer to. The simplest correction is Wex's *cuiusque*, which must mean 'as a punisher of every injury to himself'. That is, Paulinus never forgave a personal injury; and inasmuch as the rebellion was an aspersion on his governorship, he would be counted on as likely to show the rebels no mercy. [For a somewhat different use of *sui cuiusque* cf. *A.* 14. 27, 3; Caes. *B.C.* 1. 83, 2; Madvig's note in his third edition of Cic. *de Fin.*, p. 689. Nipperdey's *quisque* would give the sense 'as everyone is apt to do in avenging an injury done to himself', but the change is less convincing palaeographically and the resulting sense does not fit the context as well.]

16, 3. missus igitur, etc. *Igitur*, i.e. because Paulinus' harsh policy encouraged the Britons to maintain their resistance. The circumstances are given more fully and clearly in *A.* 14. 38–39, where it is stated that a new procurator (succeeding Decianus) held out hopes to the people that Paulinus would be replaced by a milder governor (*clementer deditis consulturum*), and also wrote against him to Nero, who sent out his freedman Polyclitus to make inquiry, and on his report recalled Paulinus. The grounds for recall may have been strengthened by a naval reverse which he suffered (*A.* 14. 39, 3), perhaps off the Norfolk coast where the Iceni gathered to make a last stand. Decianus' successor was C. Julius Alpinus Classicianus. He and his wife both came from the Treveri, and his background and his age (he was over 40 when he came to Britain) fitted him for placating the provincials and reorganizing the financial administration. Fragments of his tombstone survive and have been reconstructed (see Plate III; Dudley and Webster, p. 145).

Petronius Turpilianus : P. Petronius Turpilianus was consul with Caesennius Paetus in A.D. 61 (*A.* 14. 29, 1). His tenure is unknown, probably only four months, perhaps less if *CIL.* vi.

1. 597, which mentions Paetus and P. Calvisius Ruso as consuls on 1 Mar., belongs to this year rather than *c.* A.D. 79. He will have superseded Paulinus in the early summer. Despite Tacitus' terse judgement both here and in *A.* 14. 39, 3 ('non irritato hoste neque lacessitus honestum pacis nomen segni otio imposuit') he must have accomplished a large measure of consolidation. He stayed only two years in Britain, being appointed *curator aquarum* at Rome in A.D. 63, but was awarded the *insignia triumphalia* for his services in A.D. 65 (*A.* 15. 72, 1; it is less likely that he was distinguished, as Groag thinks, with the *triumphalia* for his part in suppressing the Pisonian conspiracy). He was murdered, as a friend of Nero, by Galba in A.D. 68 (*H.* 1. 6, 1).

tamquam, 'as supposed to be', giving the ground of the government's action (cf. c. 15, 1, etc.). Tacitus might mean that the real cause of change was the intrigue of the procurator and freedman; but *tamquam* does not always imply a fictitious reason (cf. *A.* 3. 72, 3).

novus, 'a stranger to': cf. Sil. It. 6, 254 *novusque dolori. Paenitentiae,* abstract for concrete.

compositis prioribus, 'having pacified the previous turbulence'. Cf. *compositis praesentibus* in *A.* 1. 45, 1. More could hardly have been expected in two years.

Trebellio Maximo : M. Trebellius Maximus, son of a legionary commander (*A.* 6. 41, 1), held the suffect consulship with Seneca in A.D. 56. Five years later he was one of three consulars responsible for the census of the Three Gauls. He was governor of Britain A.D. 63–69 and was still alive in A.D. 72, when he was *magister Arvalium* (*CIL.* vi. 2053).

Some evidence for the success of his policy can be seen in the appearance *c.* A.D. 70 of simple rectangular houses, timber-framed on flint-and-mortar foundations containing four or five rooms, which replaced Iron Age C huts (e.g. at Lockleys, near Welwyn; Park St., near St. Albans; see Fig. 14), and which must reflect the adoption of a Roman style of housing. The discovery of stamped Neronian tiles at Silchester may also point to his encouragement of urbanization. His policy was to curb the excesses committed by the army among the provincials and to encourage the process of Romanization. Both aims are satirically judged by Tacitus, who attributes the former to the governor's meanness (*H.* 1. 60 'per avaritiam ac sordes') and stigmatizes the latter as *blandientia vitia.* The policy was doubtless unpopular with the army and contributed to the atmosphere in which Roscius Coelius, the commander of the Twentieth Legion, could lead a revolt in A.D. 69, but it was an essential stage in the consolidation of the province (see note on c. 7, 3).

et nullis . . . experimentis, 'and of no military experience' (for the abstract *experientia*, as in c. 19, 1; *Dial.* 22, 2, etc.).

curandi, 'of administration'. Cf. *A.* 11. 22, 3 *qui curarent*; so often in Sallust. Here, as Andresen points out, the juxtaposition of *provinciam* softens the absolute use of the word. For *comitate curandi* cf. Livy 1. 34, 11 'comitate invitandi'.

didicere iam, etc., 'the natives too now learnt to condone seductive vices'. Cf. *H.* 5. 4, 2, *blandiente inertia*, and the similar ironical expression in c. 21 *delenimenta vitiorum*. *Ignoscere* implies toleration of what was previously disliked, and the 'attractive weaknesses' appear to be Roman ways of life, as critically viewed by Tacitus. Agricola was not the first to encourage Romanization (c. 21).

civilium armorum = *civilis belli* (that of A.D. 69).

discordia, 'mutiny'. So in *A.* 6. 3, 2, etc.

otio lasciviret: see note on c. 11, 4. The phrase is taken from Livy (1. 19, 4, etc.).

16, 4. indecorus, 'unhonoured', i.e. despised.

There are three stages in Trebellius' downfall. First he is humiliated by having to take refuge from the anger of the army (*vitata . . . ira* abl. abs.); then (*mox*) he governs on sufferance; finally he and the troops came to dishonourable terms which temporarily end their open disaffection. 'Fuga ac latebris' is rhetorical; cf. Seneca, *H.F.* 1012; *H.O.* 1408.

precario, 'on sufferance'. The later developments of the mutiny are described in *H.* 1. 60, where it is stated that the feeling against Trebellius was inflamed by one of the *legati legionum* (see on c. 7, 5), and Trebellius was at last obliged to flee to Vitellius. This, however, cannot be the *fuga* here spoken of, which occurred at an earlier stage and resulted in his retaining nominal control.

ac velut pacta, etc., 'and a bargain as it were having been struck giving the army licence and the general his life, the mutiny came to a standstill without bloodshed'. Cf. *H.* 2. 15, 2 'ac velut pactis indutiis . . . revertere'; Livy 34. 19, 8 'velut communi pacto commercio'. The reading of the first hand of *E* can be retained as it stands with the correction of *facta* to *pacta* and the retention of *et* before *seditio*: *ac velut pacta*, etc., being taken with *precario praefuit* and *et* meaning 'and so', but the long participial tail drags unpleasingly behind. The whole sentence runs smoothly, if *et* be omitted; and the conjunction is often dropped by copyists and sometimes inserted afterwards or wrongly inserted. The correction of *et* to *ea* is not acceptable, because in that case Tacitus could hardly have failed to mention the final development, which drove Trebellius out of Britain (see preceding note). The confusion of *pacta* and *facta* and generally of *p* and *f* is very common.

[The second hand of *E* preferred the variant, 'ac velut pacti, exercitus licentiam, dux salutem'. With this reading *sunt* can be supplied, but not easily, and the whole sentence becomes cumbrous; it is also a distinct objection that Tacitus does not use *pactus sum* transitively. It is not possible to understand *essent* nor to read *esset* for *et* (Halm, W. Heraeus), since *pacti . . . esset* would be unparalleled: *essent* for *et* (Ritter, Urlichs, St.-Denis) is palaeographically improbable and produces the objectionable transitive use of *paciscor*. See Till, *Handschriftliche Untersuchungen*, pp. 48–49.]

seditio sine sanguine stetit, 'the mutiny came to a standstill without bloodshed'. *Stetit* 'stopped'; cf. *H*. 4. 67, 2 'belli impetus stetit'; *A*. 12. 22, 3.

16, 5. Vettius Bolanus: governor A.D. 69–71. See on c. 8, 1.

agitavit Britanniam disciplina, 'harassed Britain by keeping his army in training', an ironical expression explained by the following words.

nisi quod, 'except that' (cf. c. 6, 1): i.e. the only difference being that he was not, like Trebellius, despised by his troops. [*Agitavit* has been doubted (*castigavit* or *fatigavit* Madvig) but cf. *Dial*. 41, 2 'quod municipium . . . nisi quod aut vicinus populus aut domestica discordia agitat?'; Quint. 4. 1, 33.]

17, 1. recuperavit: an exaggerated expression. Paulinus could rightly be said to have 'recovered' a virtually lost province (c. 5, 4), and Cerialis had done the same in Lower Germany; but here Vespasian could only be said to have re-established a fully authoritative government. Suet. says of Vespasian, 'incertum diu et quasi vagum imperium suscepit firmavitque' (*Vesp*. 1).

magni duces, sc. *fuerunt.*

minuta, sc. *est.* For the asyndeton cf., e.g., c. 30, 4; 5, 2.

Petilius Cerialis (see on c. 8, 2): governor probably from the spring of A.D. 71 to that of A.D. 74.

Brigantum. This name is a participial stem based on Celtic **brig*, 'height', as in Brit. **briga* → W. *bre*, 'hill'. The Brigantes are 'The High Ones', 'The Overlords', and Brigantia the goddess of the Brigantes (*ILS* 4718) is 'The High One'. The territory occupied by the tribe is known from places mentioned by Ptolemy (*Geogr.* 2. 3, 16) and from the location of dedications to the goddess to have extended from the Tyne–Solway isthmus (with a tract of Dumfriesshire to the north-west) to the Mersey and Sherwood Forest in the south. It includes the whole magnesian limestone belt that runs from County Durham to the Trent. To the east the Yorkshire Wolds formed a separate enclave which was inhabited by the Parisi. The territory of the Brigantes was divided into at least a dozen separate districts with their own centres (Isurium, Camulodunum, Rigodunum, etc.) which were no doubt ruled by independent princes but may have been loosely federated. Cartimandua eventually succeeded to the overlordship of the tribe and was recognized by the Romans as a client-queen. The recognition was probably accorded under Ostorius Scapula who, anxious to consolidate the frontier of the Roman province, saw that it was necessary to reduce Wales and to that end advanced to the Irish sea, establishing a corridor from the Midlands to Chester and Flint which effectively isolated the Welsh from potential support in N. England. To safeguard the northern flank of that corridor it was essential to secure the neutrality of the Brigantes, and Ostorius achieved this by recognizing Cartimandua. This arrangement with some disturbances which required Roman intervention (*A.* 12. 32, 1; 36, 1; 40, 2; see note on c. 31, 4) lasted until A.D. 69 when Cartimandua was ousted by her former consort Venutius (*H.* 3. 45; Statius, *Silv.* 5. 2, 140) but by then the Roman control of S. England was firm enough to allow Cerialis to undertake active operations in the north. [See Richmond, *J.R.S.* 44 (1954), 43 ff. The coins which used to be attributed to the Brigantes and were alleged to carry the name of Cartimandua and her consort have been shown to belong to the Belgic-influenced Coritani (D. F. Allen, *Sylloge of Coins of the British Isles: the Coins of the Coritani*, 1963, pp. 26 ff.).] For their earlier relations with Rome, see *A.* 12. 32, 2; 36, 1; 40, 2. They are spoken of in c. 31, 5 as having joined Boudicca, and they were certainly in arms under Venutius in A.D. 69 (*H.* 3. 45).

perhibetur : cf. *perhibent* (c. 10, 5). In Agricola's time they must have been perhaps the best known of all Britons; but their numerical superiority to all others might still be only matter of rumour till the extreme north was more fully explored.

adgressus, aoristic.

aut victoria amplexus est aut bello, 'embraced within the range of victory or war'. *Amplexus* (cf. c. 25, 1) appears to express the range of his operations, which resulted in the permanent conquest (*victoria*) of one portion of their territory and the over-running (*bello*) of another. It seems that Cerialis launched an attack into the Brigantian kingdom with *legio IX* from York to Catterick, where he stormed the nearby native stronghold of Stanwick (R. E. M. Wheeler, *Soc. of Ant. Research Report*, xvii, 1954), and over Stainmore, where three legionary marching camps antedating the Roman road and so presumably pre-Agricolan have been found. The goal was Carlisle, but there is no evidence that Cerialis envisaged a permanent occupation. The Flavian pottery from Carlisle is not certainly pre-Agricolan and no forts in the area can be assigned to Cerialis. His advance was an exploratory and punitive expedition which is unlikely to have penetrated beyond Carlisle.

17, 2. alterius, 'any other' (than such a man as Frontinus). Cf. note on c. 5, 3.

curam, 'the administration'. Cf. note on *curandi* (c. 16, 3).

obruisset, 'would have effaced': cf. c. 46, 4, and *Dial.* 38, 2 'splendore aliorum obruebantur'.

sustinuitque molem : cf. Livy 36. 7, 10.

Iulius Frontinus : Sex. Julius Frontinus was a versatile and talented man. Like Agricola he came from Narbonensian Gaul (cf. *CIL.* xii. 1859) and he may have been picked by Galba for the senate from an equestrian career. He was *praetor urbanus* in A.D. 70 (*H.* 4. 39, 1) and (suffect) consul probably in A.D. 73, the short interval between praetorship and consulship indicating that he embarked late on a magisterial career. Between the two offices he may have seen military service in Gaul (cf. *Strat.* 4. 3, 14). He was governor of Britain from A.D. 74 until the appointment of Agricola in A.D. 78. As Tacitus makes clear,

Frontinus' objective was the subjugation of Wales. He established *legio II Augusta* in new legionary headquarters at Caerleon and also operated in N. Wales (cf. c. 18, 2 n.), being perhaps responsible for the foundation of such forts as Caersws and Caerhun and for planning the legionary fortress of Chester, but the task was still incomplete at the end of his term of office (see Introd., p. 53, M. G. Jarrett, *Arch. Journal* 121 (1965), 34–35). He may have accompanied Domitian to Germany in A.D. 83 (*Strat.* 1. 1, 8): otherwise nothing is heard of him during the later years of Domitian's reign. With the rise of Trajan he re-emerges into prominence, *curator aquarum* and member of an economic commission (Pliny, *Paneg.* 62, 2) in A.D. 97, consul II (with Trajan) in A.D. 98 and consul III (with Trajan) in A.D. 100. He died A.D. 103–4. His connexions were powerful—his daughter married Q. Sosius Senecio and P. Calvisius Ruso Julius Frontinus (*suff. cos.* 79) must be a relation —and he possessed villas at Tarracina (Martial 10. 58, 1) and Formiae (Aelian, *Tactica, praef.*). His interests were reflected in two (extant) treatises—the *De aquaeductu urbis Romae* and the *Strategemata*.

quantum licebat, i.e. as far as a subject could become great under an emperor. So Memmius Regulus is called 'in quantum praeumbrante imperatoris fastigio datur, clarus' (*A*. 14. 47, 1).

Silurum: see c. 11, 2, and note. Their pugnacity is dwelt upon in *A*. 12. 33, 1; 39, 2–3; 40, 1. Frontinus was responsible for completing the conquest of the Silures. See Fig. 2 for the forts probably established by him.

super = *praeter*, 'besides'. So in *A*. 1. 59, 1, etc., and often in Livy.

eluctatus, 'surmounted'. So *H*. 3. 59, 2 'nives eluctantibus'. The whole is conventional praise for a good general; cf. Vell. Pat. 2. 115, 2 'cum difficultate locorum et cum vi hostium luctatus (Lepidus)'.

18–38. AGRICOLA'S GOVERNORSHIP OF BRITAIN

18–19. *The First Season*: A.D. 78

For an examination of Agricola's achievements in his first year of office see Introd., pp. 53 f. His main effort was directed towards controlling N. Wales and establishing his position in the province as a whole. The picture of the busy and energetic governor contains many conventional features (see notes on c. 18, 2; c. 18, 3; c. 18, 6; c. 19, 2).

18, 1. hunc . . . has. The pronoun *hic* is used here, as at 39, 1, to introduce a new section. Livy similarly uses it to make the connexion between books (cf. 4. 1, 1; 7. 1, 1; 9. 1, 1, etc.).

vices: alternations of success and disaster, energy and inactivity, by which the status had been brought about.

media iam aestate transgressus. The summer of A.D. 78; see Appendix I.

Aestas normally meant to a Roman (as 'summer' means to us) mid-May to mid-August, and it is probably so used here, though in the sequel Tacitus uses *aestas* of the campaigning season, which included early autumn. *Iam* implies that *Agricola* arrived late, and is further explained by *cum et milites,* etc. *Transgressus* (sc. *in Britanniam*), 'crossed' the Channel to Britain.

velut: like *tamquam*, giving their opinion, 'as though campaigning for this year were dropped'. Cf. *H.* 2. 8, 1 'exterritae, velut Nero adventaret'.

occasionem, 'their opportunity': cf. c. 14, 3. *Verterentur,* deponent, 'were turning their thoughts'. The winter was their favourable time (c. 22, 3).

Ordovicum, 'the hammer people' (cf. Ir. *órd*), occupied north-central Wales (Powys) and their name survives in the Snowdon district; cf., e.g., Din-orddwig. Ptolemy (2. 3, 18–19) mentions Mediolanum and Brannoganium as their towns but these cannot be identified. The Ordovices had been associated with the Silures under Caratacus (*A.* 12. 33). After their conquest they seem to have remained under direct military rule and never to have formed a civil community of their own.

alam, an auxiliary regiment of cavalry, presumably an *ala quingenaria* (500 men). Such regiments usually occupied forts of 5·5 acres. The only forts of suitable size so far located in what is probably to be defined as Ordovican territory are Caernarvon (5·6 acres), Caersws (7·7 acres), and Forden Gaer (7·58 acres, although this might be in the territory of the Cornovii). See Fig. 2.

agentem, 'operating', 'stationed'; cf. *H.* 1. 70, 2, etc. The word itself does not determine whether this was a temporary bivouac or part of a permanent occupation but the context suggests that Frontinus would only have quartered a force of this size in Ordovican territory as a stage in their planned annexation.

obtriverat, 'had annihilated' (so in *A.* 15. 11, 1: *H.* 4. 76, 1). It is properly used of those crushed by a mass.

erecta (*est*), 'was excited', as in *A.* 14. 57, 2, etc. Cf. c. 4, 3 *erectum ingenium.* The verb is so used in Cic. and Livy, but more commonly with *ad aliquid,* or *aliqua re.*

18, 2. quibus bellum volentibus erat, 'those who wished for war'. This attracted dative is used by Sallust (e.g. *Jug.* 100, 4 'uti labor volentibus esset', 84, 3) and Livy. It is modelled on Greek usage; cf. Thuc. 2. 3, 2 *τῷ πλήθει οὐ βουλομένῳ ἦν.* Cf. also *H.* 3. 43, 2.

probare . . . opperiri, historical infin. The sense is: the disaffected and rebellious provincials approved (*probare* for *comprobare*; cf. *A.* 4. 43, 5 'probatum P. Rutilii exemplum') the example (of the Ordovices in exterminating a Roman detachment) and were curious to see how the new governor would react (lit. 'waited to see his temper'); if he was weak, they would themselves break out in open revolt. Notice the highly complex syntax in which Tacitus describes the conflicting factors in the situation that faced Agricola and which leads up to the simple statement of Agricola's simple decision—'ire obviam discrimini statuit'.

transvecta : so *H.* 2. 76, 3 'transvectum est tempus', the only similar instance. The remainder of the summer seems to have been spent in taking over the command, forming his plans, and concentrating his forces.

numeri, 'detachments', not of legionaries, since it would have been unusual to station legionaries in garrison detachments, but of auxiliary troops: cf. 'modica auxiliorum manu' below. So in *H.* 1. 87, 1.

praesumpta, the fact that the troops had already anticipated the cessation of military operations for the year: they had already relaxed in anticipation of the winter lull. This interpretation is more in accord with Tacitus' use of *praesumo* (cf., e.g., *H.* 1. 62, 2) than Anderson's who, understanding *animo,* interpreted it to mean 'the soldiers took their rest for granted'.

tarda et contraria : sc. *erant*, 'all these were factors which delayed and hindered an attempt to begin a campaign'. Alternatively, with Anderson, the clause may be taken in apposition to the three preceding clauses, *est* being supplied with *transvecta*, *sunt* with *sparsi*, and *est* with *praesumpta*, but *et plerisque . . . videbatur* follows less naturally after such a parenthetic apposition. For the neut. plural picking up masc. and fem. subjects cf. *H.* 4. 24, 2.

suspecta, 'suspected districts'. So *neglecta* (*H.* 3. 69, 4), *praesentia* (*A.* 3. 38, 4). *Potius* is an adjective.

vexillis, 'detachments', serving under a *vexillum*, or special flag, instead of their proper *signa*; also called *vexillationes* (Inscr.), and the men *vexillarii* (*A.* 1. 38, 1, etc.). He may have had one such body, comprising one or more cohorts, from each of his legions.

modica = *parva*, which was common and colloquial in Tacitus' time. In his developed style *parvus* is dropped except in certain old phrases and antitheses, and it disappears from late Latin.

ante agmen : sc. *incedens*. Cf. c. 35, 3. Such a gesture was in conventional rhetoric commonly expected of a brave general (cf. with this passage Caesar, *B.G.* 1. 25, 1; Sall. *Cat.* 59, 1 'remotis omnium equis quo militibus exaequato periculo animus amplior esset, ipse pedes exercitum instruit'). In fact it is unlikely that many generals (or Agricola on this occasion) did lead their troops in person. The thought goes back to Nicias in Thuc. 7. 77, 2.

erexit aciem, 'marched his troops up-hill', cf. c. 36, 2 'erigere in colles aciem'; *H.* 3. 71, 1; 4. 71, 5). It is a military term used also by Livy. The phrase probably refers to a decision to engage the Ordovices in their mountainous homeland, that is near Capel Curig and Betws-y-coed. An undated temporary camp of 9½ acres has been located at Penygwryd (*J.R.S.* 45 (1955), 121).

18, 3. instandum famae, 'prestige must be followed up'. Cf. *A.* 13. 8, 3, etc. The desirability of exploiting the advantage caused by a first impression was a traditional theme; cf. Thuc. 6. 49, 2; 7. 42, 3.

prout prima cessissent, 'according to the issue of the first attack would be the terror inspired by his other operations'. *Ceteris* is neuter: *cessissent*, 'turned out', as in *bene cedere*, a use apparently not found before Celsus but common in Seneca's letters (e.g. *Ep.* 36, 6; 83, 5; 104, 6) and Curtius. We should expect 'so would be the prestige of his arms in the future', a conventional thought for which cf. *H.* 2. 20, 2; Cic. *ad Att.* 10. 18, 2; [Sallust], *Ep.* 1. 1, 10, etc. But Tacitus passes from the general thought to the particular case of Agricola, whose first operation had inspired terror.

Monam : see c. 14, 3.

cuius. The preposition, omitted by *E* and easily restored after *insula,* is perhaps not needed; cf. Löfstedt, *Syntactica,* i, p. 294 who adduces Virg. *G.* 4, 88 and Livy 25. 36, 2.

possessione, 'occupation'. Cf. *A.* 2. 5, 4.

intendit, usually without *animo* when followed by an infinitive.

18, 4. ut in subitis, 'as was natural (or, as usually) in hastily formed plans'. Cf. Caes. *B.G.* 3. 8, 3; Livy 35. 35, 5.

ratio et constantia, 'the methodical planning and determination'. For *ratio* cf. c. 6, 4; 20, 3. The words are paired by Cicero in a slightly different sense (*Parad.* 22; *de Div.* 2, 18).

auxiliarium : probably his Batavians (cf. c. 36, 1), who were famed for their skill in this style of swimming (*H.* 4. 12, 3; *A.* 2. 8, 3). For the German fondness for swimming see Anderson's note on *Germ.* 22, 1.

quibus nota vada, 'those who knew the fords'; cf. *H.* 5. 15, 1; Caesar, *B.G.* 3. 9, 6; 4. 26, 2. There will have been people in Agricola's army in A.D. 78 who had taken part in Suetonius Paulinus' attack on Anglesey in A.D. 60 and who were familiar with the fords that were used on that occasion (*A.* 14. 29, 3 'equites vada secuti aut altiores inter undas adnantes equis tramisere'). It is in fact possible to ford the straits on foot over Caernarvon Bar at certain conditions of low tide (see Admiralty Chart 1464: Menai Strait) but not at Bangor or elsewhere. If the Romans crossed over the sands, it would explain the element of surprise in the attack since the British were doubtless expecting a sea-borne attack from Bangor (see note on *qui mare* below). The Romans must have secured both the coastal approach to Bangor from Chester, where the Roman road later ran along the Dee estuary, and the inland approach by way of Corwen and the upper Conway valley to Betws-y-coed, Capel Curig and, by the Nant Ffrancon pass, to Bangor (see note on c. 18, 2: Richmond in Foster and Daniel, *Prehistoric and Early Wales*, pp. 158 ff.).

seque et. In this combination of conjunctions, used frequently by Tacitus after Sallust and Livy, *que* is almost always joined to *se*, *sibi*, or *ipsi*. It seems to have been affected by historians (see Ogilvie's note on Livy 1. 43, 2). *Arma, viri, equi* formed a stereotyped tricolon in Latin but different authors favoured different arrangements of the words: Livy preferred *arma viri equi* (22. 39, 11; 23. 24, 9, etc.), Cicero *arma equi viri* (*Phil.* 8, 21), Tacitus *viri arma equi* (*H.* 1. 51, 2).

qui mare, like *qui naves*, is rhetorical amplification, to express vividly the thoughts in the minds of the islanders. They were looking out for a fleet, for ships, for an attack by sea, not imagining that the sea could be crossed like a river. *Naves* or *classes* is often added to *mare* for closer definition or for emphasis: *H.* 2. 12, 1 'possessa per mare et naves maiore Italiae parte'; *H.* 3. 1, 2 'superesse . . . mare, classes'. So *mare ac naves* frequently in Livy (cf. Weissenborn on 22. 19, 7). Paulinus had used 'naves plano alveo' (*A.* 14. 29, 3). [Other interpretations ('high tide', Philippson; 'a maritime operation', Marín y Peña; 'navigation', Gudeman, cf. Tibullus 1. 3, 50) do violence to the Latin. Emendation (*cameras* Boucholt; *per mare* Baehrens) is not required.]

invictum, 'invincible', as often in Sallust, Livy, etc.

18, 5. clarus ac magnus, a frequent collocation: cf. Cicero, *de Orat.* 2, 19; *Phil.* 14, 33; Sallust, *Cat.* 53, 1 'clarus atque magnus habetur'; *Jug.* 92, 1; Livy 38. 50, 4.

quippe cui. Tacitus uses this form here only, *ut qui* frequently.

officiorum ambitum, 'soliciting attentions'. *Officia* are the attentions, services, or respect due to a superior (cf. *H.* 5. 1, 1 'comitate et adloquiis officia provocans'); *ambitus* is commonly used of soliciting votes in elections (cf. *A.* 13. 29, 1 'ambitu suffragiorum suspecto'). Governors liked to gratify their vanity by extorting such shows of deference. They are deprecated by Ulpian, *Dig.* 1. 16. 4, 3.

labor et periculum. A rhetorical formula: cf. Sallust, *Or Macri* 18; *Jug.* 44, 1, etc.

18, 6. nec applies both to *usus* and *vocabat* ('he did not—nor did he'), as in *A.* 14. 32, 2 'neque motis senibus et feminis iuuentus sola restitit'. [Strictly, *nec* qualifies the finite verb and then the negatived finite verb is further qualified by the participial phrase.]

victos continuisse, apparently his own modest expression, he had 'kept in hand tribes already conquered', and did not call that a campaign or a victory. Cf. Livy 29. 36, 10; *H.* 1. 49, 4.

laureatis : sc. *litteris*, expressed in Livy 5. 28, 13, etc. The custom of wreathing dispatches announcing victory with bay leaves is described by Pliny (*N.H.* 15, 133).

auxit. The thought is a commonplace; cf. Sall. *Cat.* 54, 6 '(Cato) quo minus petebat gloriam eo magis illam adsequebatur'. Tacitus attributes a similar modesty to Germanicus (*A.* 2. 22).

aestimantibus, 'when men considered'; probably a concise abl. abs., such as is often used by Tacitus (and sometimes by earlier writers) not only where a subject has been recently expressed, but also where it can be inferred from the context: cf. *H.* 1. 27, 2 *requirentibus*.

quanta . . . spe, etc., 'how great must be his hopes for the future, when', etc., the modal abl. containing the predicate of the sentence.

tam magna, somewhat stronger than *tanta*: cf. *G.* 37, 1; *A.* 11. 36, 2.

19, 1. animorum, 'the feelings'. *Prudens* is so used in *A.* 3. 69, 5; *H.* 2. 25, 1.

experimenta : cf. c. 16, 3; *Dial.* 34, 6 'eruditus alienis experimentis'. The reference is to Boudicca's revolt.

excidere, 'to extirpate'. Similarly *excidere mala* in Brut. *ad Cic.* 1. 4a, 3; Pliny, *Paneg.* 34, 2.

19, 2. a se suisque orsus. It was a common rhetorical claim that a good commander began by setting his own house in order; cf., e.g., Cic. *pro Leg. Man.* 13; Livy 34. 18, 4–5, etc. The establishment (*domus*) of a provincial governor amounted to over 60 persons. In addition to the senior advisers (cf. note on c. 5, 1) there were several grades of clerical staff, most of whom were seconded from the legions. At their head was the *princeps praetorii*, a centurion (*centurionem* below; cf., e.g., *IGR* iii. 1230) and under him there were usually three *cornicularii* and three *commentarienses* (secretaries), ten *speculatores* for each legion, and a large number of lesser officials called *beneficiarii*. This establishment, which had existed under the Republic (cf. Caes. *B.C.* 1. 75, 2), was concerned with the routine administration of the provincial army, and was distinct from the civil staff which the procurator would have had (tax-assessors, etc.). There was recognized promotion within it (for a typical career cf. *ILS* 2118). See A. H. M. Jones, *Studies in Roman Government and Law*, pp. 161 ff.

[**primum** : *primam* (*E*) would mean that he reformed his own establishment before reforming the establishments of others.]

nihil per libertos, etc. For the omission of *agere* cf., e.g., Cic. *Phil.* 1, 6 'nihil per senatum, multa per populum'. The pious conduct attributed to Agricola should be compared with the advice urged in Cic. *ad Q.F.* 1. 1, 20. The freedmen of the governor were, on a smaller scale, apt to resemble those of the emperor. A reform promised by Nero at his outset was 'discretam domum et rem publicam' (*A.* 13. 4, 2). Vitellius began by transferring the *ministeria principatus* from freedmen to equites (*H.* 1. 58, 1).

publicae rei : as in *G.* 13, 1, with *nihil*. The inversion is rare, usually, as in *G.* loc. cit., and Cic. *ad Att.* 1. 17, 6, emphasizing the distinction between public and private. Here the contrast with *privatis* is purely formal.

studiis privatis, 'from personal likings', as opposed to the recommendation and entreaties of others.

adscire, 'to take upon his staff', usually with some explanatory word added.

19, 3. exsequi, 'to punish' as in *A.* 12. 20, 2; Livy 3. 13, 3: not 'to investigate' (Peerlkamp, Gudeman) which would hardly reflect credit on Agricola's perseverance. Cf. Suet. *Jul.* 67 'neque observabat omnia neque pro modo exsequebatur'; Pliny, *Pan.* 70, 3; 80, 3.

commodare is here used in its wider sense as the equivalent of *praebere* (cf. c. 32, 1; Plautus, *Rud.* 435; Livy 23. 48, 11).

nec poena semper : abl., like *paenitentia*, with *contentus*; cf. Seneca, *H.O.* 295 'o nulla dolor contente poena'; Amm. Marc. 30. 8, 3. The sentence is an epigrammatical paradox. Agricola was often content not with punishment but with less than punishment, repentance.

officiis, 'functions'.

non peccaturos, 'men not likely to transgress'. For the thought cf. note on c. 7, 3 *moderatione*.

19, 4. **frumenti et tributorum exactionem**: for the taxes see note on c. 15, 2. The abuses involved in the requisition of grain are illuminated by Cic. *Verr.* ii. 3, 163–203. The Britons were required to provide two main categories of grain: (1) for the provisioning of the legions and (2) for the support of the governor's household and staff (*frumentum in cellam*). The regulations respecting the requisition of grain lay not with the procurator but with the governor and the amount of corn to be supplied under these two heads was probably determined at Rome. The responsibility for collecting and delivering it fell upon the individual tribes or cantons. The grain supplied was paid for at a fixed rate, but at less than its market value. Hence it was common practice in the provinces to commute the grain requisitioned for the governor's household for a sum of money: instead of supplying corn on which they received only a nominal return, the provincials prepared to discharge their obligation by paying cash (*frumentum aestimatum*). The provisioning of the legions raised different problems. In Britain, no doubt, the supply of corn was scarce or plentiful according to locality and the legionary garrisons were not always situated in corn-producing areas. Indeed corn imported from the Mediterranean has been found at the legionary fortress of Caerleon although this was being used for making beer (Helbaek, *New Phyt.* 63 (1964), 158–64). In consequence the provincials were sometimes faced with two dilemmas. (1) When they did not have enough corn of their own to meet their requisition, they were forced to buy it from the imperial granaries at whatever price was demanded and then to hand it over as their requisition and receive a nominal payment for it. Thus they might have to buy the corn at 12 sesterces a peck and sell it back at 4 sesterces a peck (cf. Cic. *Verr.* ii. 3, 178; below, 'emere ultro frumentum ac luere pretio'). In such transactions the corn would not actually leave the granaries at all. (2) Where they had corn, they might be ordered to deliver it at some great distance, and were thus induced to pay money to get excused from this vexatious and expensive transport (cf. Cic. *Verr.* ii. 3, 140). During the early period of the Roman occupation depots were established at Richborough (Rutupiae) and Fishbourne and probably at Fingringhoe, Hamworthy, and Topsham, but the legionary garrisons also had their own granaries and Agricola constructed a large depot for his Scottish campaigns on the Tay estuary (Horrea) at or near the site of Carpow where only Severan occupation has as yet been proved. For an account of the provincial tax-system see G. H. Stevenson, *Roman Provincial Administration*, pp. 89, 133 ff.

[*exactionem* is clearly the right reading. Some have thought that *auctionem* might refer to the increase of tribute generally under Vespasian (Suet. *Vesp.* 16); but it is hardly possible to suppose that Tacitus would so use the word, and the confusion of *auctio* with *actio* occurs elsewhere (e.g. *H.* 1. 20, 2; cf. *actum* for *auctum* in c. 39, 1 below).]

aequalitate munerum, 'by equalizing the burdens' (viz. the *frumentum* and *tributa*), but in what precise way Tacitus does not explain: he limits himself to describing the abuses abolished. Perhaps no corn was demanded from districts poor in grain, but fixed sums were made payable by way of indemnity.

circumcisis, etc., 'cutting off the devices for profit'. Cf., e.g., Livy 32. 27, 4.

reperta. Tacitus uses *reperire*, the literary equivalent of the popular *invenire*, sparingly in the minor works but with increasing frequency as his style develops. (The figures are: *Agr. invenire* 8: *reperire* 1; *Germ.* 2: 1; *Dial.* 7: 2; *Hist.* 9: 7; *Ann.* 10: 80; see Löfstedt, *Aetheria*, pp. 232 ff.)

per ludibrium, 'in mockery'. The mockery consisted in the fiction of purchase and redelivery (see note above), and in their being kept waiting, like beggars praying for the concession, at the doors of granaries (cf. Sen. *Ep.* 4, 10 'superbis adsidere liminibus'), which were not in fact opened at all. The device of making people buy corn from his own granaries, to meet their obligations, was practised by Verres. For an instance from the later Empire see Amm. Marc. 28. 1, 18.

horreis. Such imperial granaries are found in many provinces, and no doubt existed in all. They were under the control of the governor. In provinces which did not send corn to Rome they served the needs of the troops and could be used to meet the needs of the province itself or other provinces (Pliny, *Pan.* 31–32).

ultro, 'what is more', cf. c. 31, 2 'semel veneunt atque ultro a dominis aluntur'. It would be expected that they came to the granaries to deliver and sell their corn: instead, they have to buy it.

luere pretio. *Luere* is elsewhere used by Tacitus in the sense of *expiare* and of *solvere*, but in the latter case only with *poenas* or *supplicium*. Here it is used absolutely, with the meaning 'to make amends' by a money payment, as in Livy 30. 37, 12 'cetera quae abessent aestimanda Scipioni permitti atque ita pecunia luere Carthaginienses (sc. placuit)'. Cf. also 38. 37, 5 'luendam pecunia noxam'. *Ludere* (*AB*) is a mere error and emendation is not required.

divortia, etc., 'by-ways and distant districts', at which the corn was to be delivered. The device here described was also one of those practised by Verres and other governors (Cic. *Verr.* ii. 3, 190), and it continued in the fourth century (*Cod. Theod.* 11. 1, 9; 7. 4, 15, etc.). *Divortium itinerum*, lit. 'a bifurcation of roads', clearly means a side-road rather than a junction of roads in Amm. Marc. 28. 5, 6, but in Livy 44. 2, 7 it means a junction. The parallel from Amm. justifies the interpretation 'side-roads' which is required here.

ut, etc. This clause is best taken as depending on *indicebatur* and explanatory of the nominatives. For *indico ut* cf. Livy 1. 50, 1; 52, 5. Alternatively *ut* may be taken as final.

proximis hibernis, usually taken as a concessive abl. absolute, 'even though there were permanent quarters for troops close by', but such a use is hard to parallel in Tacitus and it is easier to regard it as a plain abl. of separation with *deferrent* (cf. Sil. 3, 631; 9, 396, etc.; Puteolanus wished to insert *a*). They had to deliver it not to near-by camps but to remote and inaccessible places.

donec, etc. 'Till a service easy for all (i.e. in which there need have been no difficulty for all the Britons) should become profitable to a few' (Romans), by bribes received to escape this needless transport. The subjunctive expresses the purpose in the minds of the officials. *In promptu* usually means 'obvious' or 'ready to hand' (cf. *H.* 5. 5, 2) but it can also mean 'easy' (cf. [Sallust], *Epist.* 2. 7, 1, contrasted with *difficile est*; Ovid, *M.* 2, 86; 13, 161) and that is the sense required here.

20–21. *The Second Season*: A.D. 79

The campaigns during this year secured N. England and completed the work which had been begun by Cerialis. For details see Introd., p. 57. Agricola will have operated from York and Chester. He is throughout depicted as a model general (see notes on c. 20, 2; 21, 1) but this description is probably based on what was expected of a good commander rather than on actual evidence.

20, 1. famam . . . circumdedit: cf. *Dial.* 37, 6; *H.* 4. 11, 2. περιβάλλειν and περιτιθέναι are similarly used in Greek.

vel incuria vel intolerantia. The combination of the two nouns is interesting and typical of Tacitus' choice of words. *Incuria* 'negligence', first found in the older Cato, was clearly felt to be archaic since it is used only twice by Cicero but is liked by the archaizing writer of [Sallust], *Epist.* 1. 1, 3 and by later archaizers. *Intolerantia* 'arrogance', on the other hand, which is used by Tacitus only here, doubtless for the balance of sound with *incuria*, is found earlier in Cicero only twice (*Clu.* 112; *de Agr.* 2, 33). Both words are, therefore, unusual and their combination is forceful.

20, 2. multus in agmine, 'present everywhere on the march', imitated from Sallust's description of Sulla (*Jug.* 96, 3), 'in agmine . . . multus adesse'. Cf. c. 37, 4 *frequens ubique*.

modestiam, 'discipline', abstract for concrete, answering chiastically to *disiectos*, 'stragglers'.

loca, etc. Cf. Statius' description of Vettius Bolanus' operations in the East (*Silv.* 5. 2, 41 ff.):

> Bolanus iter praenosse timendum,
> Bolanus tutis iuga quaerere commoda castris,
> metiri Bolanus agros, aperire malignas
> torrentum nemorumque moras.

Cf. also c. 33, 5. Livy 35. 28, 1 ff. (Philopoemen). The personal supervision of camp-sites, the exploration of rivers, etc., was evidently part of the stock description of a good general.

aestuaria. On the west coast north of Wales, now out of the picture, there are the estuaries of the Mersey, Ribble, Lune, Kent, Duddon, Esk, and Solway. On the east coast only the Tees can properly be described as an estuary until the Forth is reached.

praetemptare, 'explored'; so in poets and the elder Pliny.

nihil . . . quietum pati, from Sall. *Jug.* 66, 1 'nihil intactum neque quietum pati, cuncta agitare'.

interim : while keeping his own troops in discipline.

quo minus : with the force of *quin* as an epexegetic adversative conjunction. So often in Tacitus: cf. c. 27, 2.

excursibus, 'sorties'; cf. *Bell. Alex.* 19, 2; Caes. *B.C.* 3. 92, 2.

rursus, 'on the other hand': cf. c. 29, 1; *A.* 1. 80, 2, etc.

invitamenta pacis. *E* read *inritamenta*, corrected by *E*[2] to *irritamenta*, but *irritamentum* is always used of provocation to vice (*A.* 14. 15, 2; *H.* 1. 88, 3; *gulae, libidinis*, etc.); cf. especially Justin 31. 7, 9 'belli irritamenta non pacis blandimenta'; note also *A.* 13. 46, 2. Editors who retain *E*'s reading assume that Tacitus is cynically equating peace with subjection and moral decline, or deny that *irritamentum* must be pejorative (cf. Ovid, *Met.* 9, 133). Neither is probable. The same objections apply to *A*'s conjecture *incitamenta* (cf. *A.* 12. 34, 1 'aliis belli incitamentis'; Amm. Marc. 31. 5, 7). Both *incitamenta* and *inritamenta* are too brusque. The appropriate word to make the antithesis with the preceding *terruerat* is *invitamenta*; cf. Livy 2. 42, 6 'largitiones temeritatisque invitamenta'; Vell. Pat. 2. 67, 3; Curtius 4. 10, 24.

20, 3. ex aequo egerant, 'had lived on a footing of equality with others (i.e. independent)'. *Ex aequo* is so used in *H.* 4. 64, 3, and the elder Pliny (e.g. *N.H.* 6, 112 'pertinent Parthi ad Scythas cum quibus ex aequo degunt'). Elsewhere it has rather the adverbial force of 'equally'.

praesidiis castellisque. On these terms see note on c. 16, 1.

tanta ratione curaque. Agricola's special skill in planting forts is praised in c. 22, 2.

inlacessita. The meaning is that Agricola's policy of building a ring of forts (for which see Introd., p. 57) ensured that the northern tribes were incorporated into the province without having to sustain attacks from independent tribes on their frontier who regarded them as traitors for coming over to the Roman side (for *transire* cf. *H.* 3. 61, 1; Livy 26. 12, 5) and that this was the first part of Britain of which this was true. 'Nulla ante nova pars' is the equivalent of 'haec prima pars'. There seems no reason to assume that an adverb of comparison has dropped out. *Inlacessitus* is used only here and in *G.* 36, 1 in Latin.

On Agricola's policy of encouraging the Romanization of Britain cf. Introd., p. 5. He was following the lead already set by Trebellius Maximus (cf. note on c. 16, 3).

21, 1. absumpta, cf. *A.* 2. 8, 2; Livy 27. 17, 1 'cum hiemem totam reconciliandis barbarorum animis absumpsisset'.

dispersi: living separately, like the Germans (*G.* 16, 1; cf. Strabo 4, 200). Tacitus, and before him Caesar (*B.G.* 5. 21, 3), exaggerate when they imply that there were no urban centres in Britain. The extent of urbanization varied greatly from region to region and from culture to culture but the Belgae and those under their influence (Iron Age C) developed large, if architecturally squalid, settlements. Examples are known e.g., at Colchester, Prae Wood (St. Albans), Wheathampstead, Braughing, Dorchester (Oxon.), Minchinhampton, Bagendon. They were unplanned conglomerations of huts, surrounded by dykes. Elsewhere Iron Age A and B favoured hill-forts, which served, like medieval castles, less as towns than as strong-points to which the neighbouring population could rally. Notable instances are Maiden Castle, Spettisbury, Hod Hill, and Ham Hill, and they are found in Wales and northern England as well (e.g. Stanwick). Roman policy was, first of all, to found urban capitals of the cantons (*civitates*) to serve as administrative centres. Many of these were on or near the site of existing Belgic settlements (e.g. Ratae Coritanorum (Leicester), Calleva Atrebatum (Silchester), Camulodunum (Colchester), Verulamium (St. Albans), Corinium Dobunnorum (Cirencester). Where no such native towns existed the Romans sometimes founded the capital near a large hill fort (e.g. Durnovaria (Dorchester), capital of the Durotriges, near Maiden Castle, or else created an entirely new centre, as Isca Dumnoniorum (Exeter). The network of roads encouraged proliferation of smaller towns which grew up round the posting-stations, and the legionary fortresses also attracted urban development round them (e.g. at Lincoln and Gloucester). Similarly many towns are known to have grown up on the sites of forts which were dismantled early in the Roman occupation after the pacification of the country, e.g. at Dorchester (Oxon.), Water Newton, Great Casterton. For a fuller account see A. L. F. Rivet, *Town and Country in Roman Britain*² (1964), pp. 72 ff.

eoque = *ideoque*, as often in Tacitus and in Sallust and Livy.

faciles = *proni*, like *facili ad gaudia* (*A.* 14. 4, 1) and *faciles*

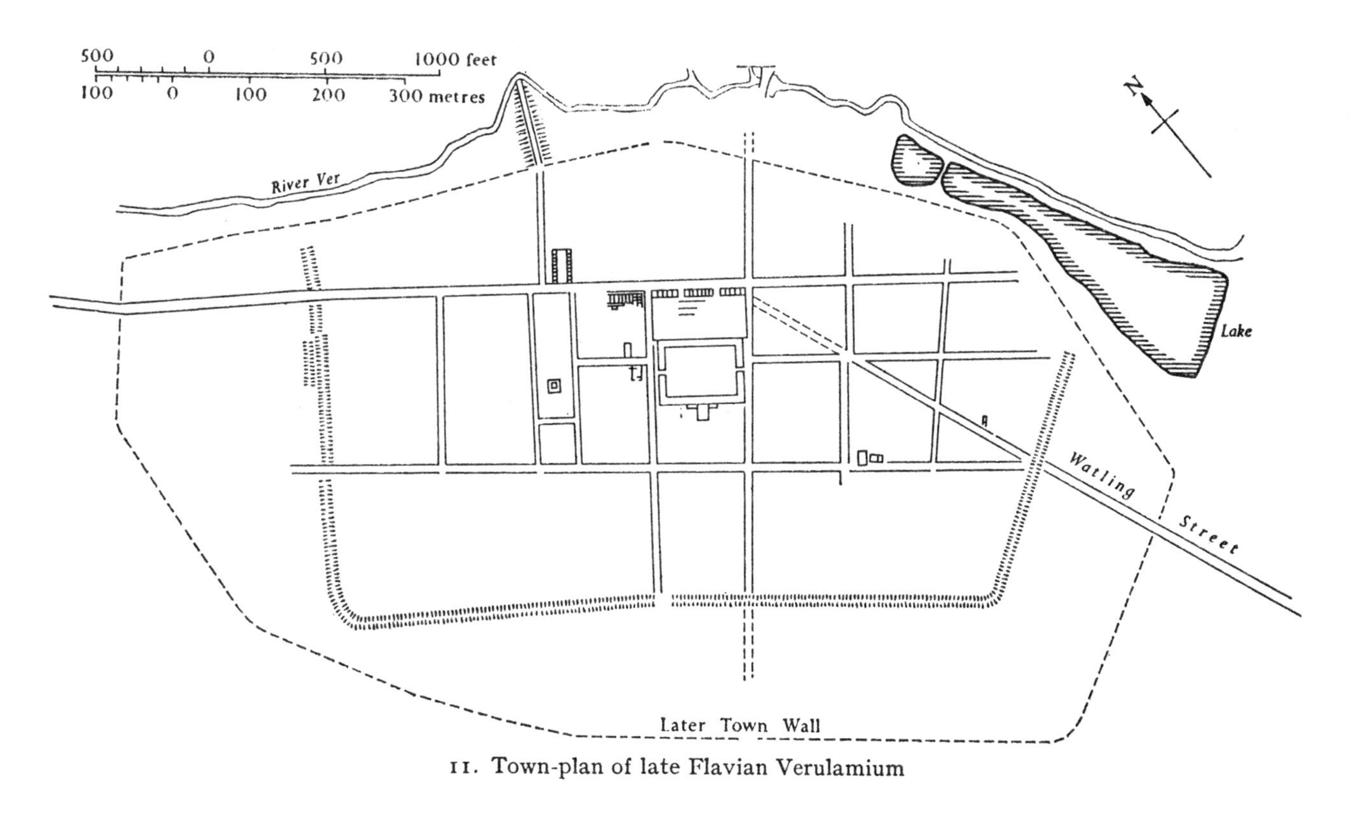

11. Town-plan of late Flavian Verulamium

ad fera bella manus (Ovid, *A.A.* 1, 592). Cf. Livy 2. 15, 2 'in perniciem suam faciles'.

quieti et otio, as in c. 6, 3.

privatim . . . publice : best taken, with Andresen, as referring to the object, 'he encouraged individuals and helped communities' (cf. *G.* 10, 2; *A.* 11. 17, 2, etc.), since if they are taken as referring to the subject ('by personal (i.e. unofficial) encouragement and official assistance (i.e. government grants and technical aid)'), they would imply a distinction between Agricola's official and unofficial activities which is pointless in a sentence describing his policy as governor. The same two approaches are recommended for provincial governors by Ulpian, *Dig.* 1. 18, 7.

templa, etc. Tacitus summarizes the characteristic elements of a Roman town which sharply distinguished it from its Iron Age predecessor. It was planned round a rectangular gravelled space, the *forum*, which served as a market-place and a civic centre, and which was bordered on three sides by a colonnade (*porticus*) giving access to shops and offices and on the fourth side by the municipal offices (*basilica*). The rest of the town was laid out on a grid-system of streets. The other distinctive features of a Roman town were the official hotel (*mansio*), the theatre, the public baths (see below) which served as a public meeting-place and were among the earliest buildings to be constructed in a town, and the temples. For the plan of Verulamium in the Flavian period see Fig. 11. The most striking confirmation of Tacitus' statement comes from an inscription erected between July and December A.D. 79 over the entrance to the new forum at Verulamium which mentions Agricola as governor. See Plate Ia and R. P. Wright, *Ant. Journ.* 36 (1956), 8–12. There is evidence of town planning and the erection of public buildings in the last decades of the first century at a number of other places, e.g. Wroxeter, Caerwent, Exeter, and Cirencester.

Temples were a regular feature of Roman towns and, although a few are dedicated to classical Roman deities, it was normal to assimilate local cults into Roman cults (cf. *RIB* 309, 310, 949 *Mars Ocelus*). The typical Romano-Celtic temple, a square-roofed shrine surrounded by a square portico (cf. Fig. 12 for an example at Verulamium dating from the Flavian period) was evolved on the Continent in the first century B.C. and reflects Roman architectural techniques applied to the primitive open shrine of the Celts. Although at

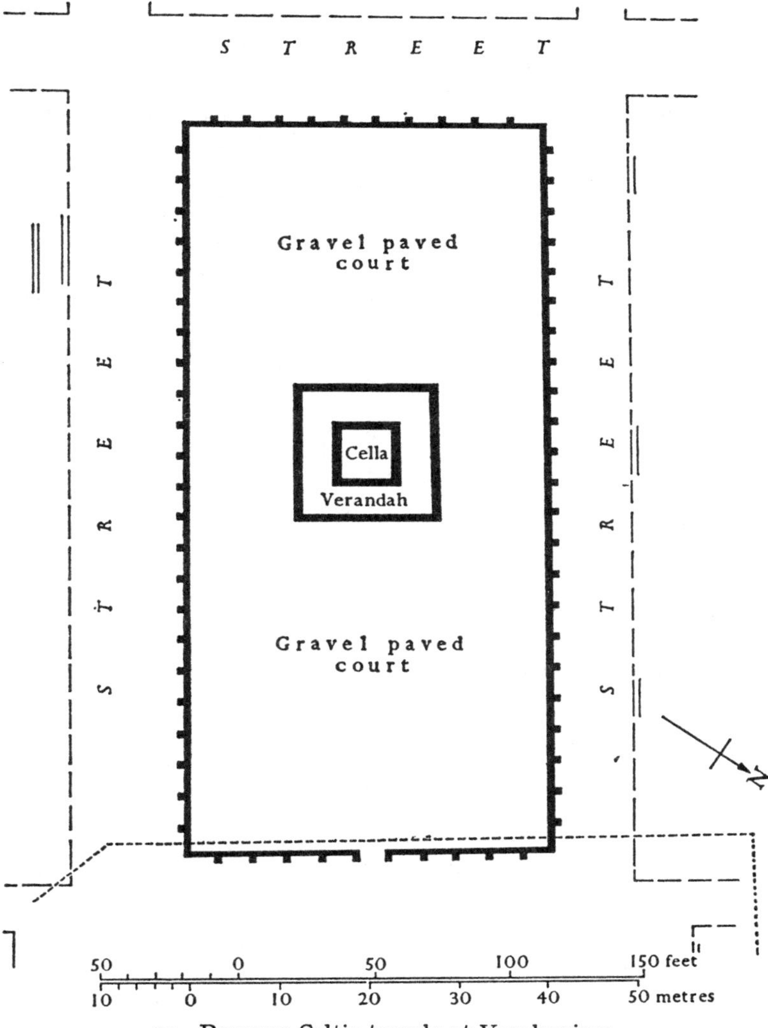

12. Romano-Celtic temple at Verulamium

a number of sites in Britain (e.g. Frilford, Woodeaton, Worth) there is evidence for the continuity of cult from the Iron Age to the Roman period, there is no certain example of a Romano-Celtic temple as such before the occupation; the Heathrow temple remains a unique building, perhaps to be explained in some other way. Thereafter they are numerous. Pre-Agricolan temples in Britain include those in towns at Colchester (*A.* 14. 31, 4: imperial rather than Romano-Celtic cult) and Silchester, and those in the country at Woodeaton and Farley. It is not possible to attribute any temples specifically to Agricola's governorship but it is likely that Verulamium (no. 1), Springhead (no. 1), and Frilford were built during those years and there must be many more that have not been discovered. There were in addition other shrines of oriental and purely local cults. See M. J. T. Lewis, *Temples in Roman Britain* (Cambridge, 1966), esp. pp. 49 ff.

domos: Roman town and country houses, both of which Tacitus probably includes in this term, are quite different in plan from Iron Age dwellings (see Fig. 13). For the country houses see note on c. 16, 3 and Fig. 14; for town houses of late-first-century date see Fig. 15. They may be compared with the huts of Belgic type found in the earliest phase of Roman Canterbury and at Colchester.

castigando, with words. *Laudando . . . castigando* is a cliché for the good general's reactions: cf. Livy 3. 63, 3–4.

honoris, etc., 'competition for honour (that of being praised) took the place of compulsion'. For *aemulatio* with the gen. cf., e.g., *A.* 2. 44, 2.

21, 2. iam vero, 'further' (c. 9, 3). **principum**: cf. c. 12, 1. One of the people employed by Agricola in his educational programme was probably Demetrius of Tarsus, a schoolmaster (γραμματικός) who gives an account of some of his experiences in Britain in Plutarch's *De Defectu Oraculorum* (cf. especially 412 c for his scholastic character: for his voyage see Introd., p. 32). He is to be identified with the Demetrius Scribonius who made two dedications at York, one to 'the gods of the governor's residence' where he will have lived and taught the sons of the local princes (*RIB* 662–3). For the spread of grammar schools in the western provinces see Sherwin-White on Pliny, *Ep.* 4. 13, 16.

ingenia . . . anteferre: often taken to mean that he 'expressed a preference for British abilities over Gallic industry', i.e.

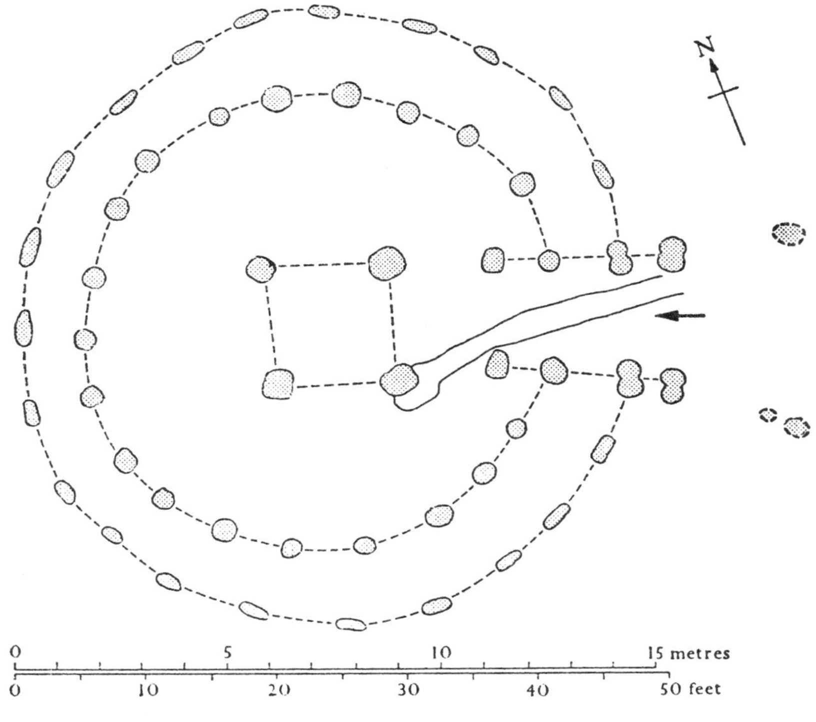

13. Iron-Age house at Little Woodbury

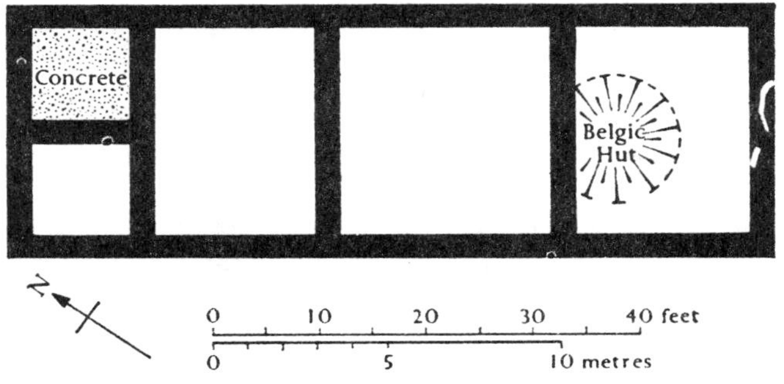

14. Roman house at Lockleys, Welwyn

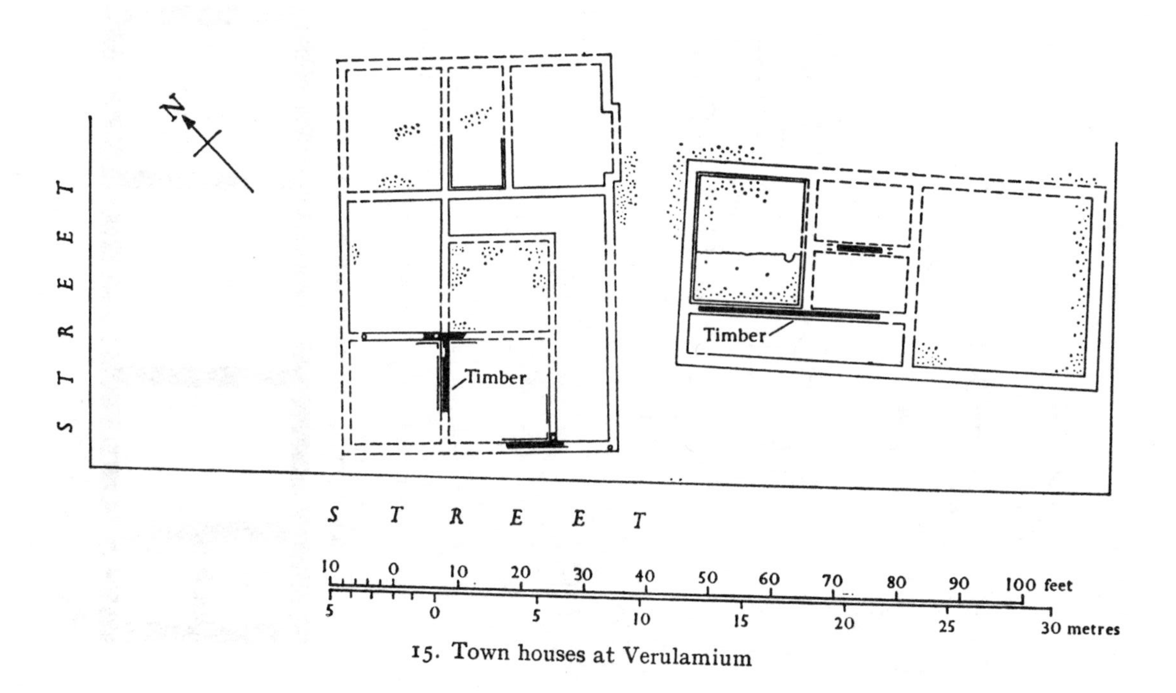

15. Town houses at Verulamium

flattered them by saying that their native wit would do more for them than diligent culture did for the Gauls. But no such antithesis seems to be intended, any more than in the similar passage in *Dial.* 1, 4 'qui nostrorum temporum eloquentiam antiquorum ingeniis anteferret', a contrast is intended between *eloquentia* and *ingenium* (cf. Gudeman, ad loc.). Agricola 'expressed a preference for British abilities (as brought out by training) over the trained abilities of the Gauls', implying that when they were properly trained they would be better orators than the Gauls. From the *studia* in which the *ingenia* of both were exercised it could be seen that the British were the superior (so Andresen). As elsewhere, lucidity of expression is impaired by straining after conciseness. The premium set on eloquence in Gaul is noted in *H.* 4. 73, 1: cf. also Juv. 7, 148. The comment may have added significance if, as is likely, both Agricola and Tacitus himself were Gallic in origin.

linguam Romanam : Latin was the language of the governing classes in Roman Britain, of civil administration, of the army, and of trade. The rural upper classes were probably bilingual. Graffiti on tiles and potsherds show that some of the urban lower classes could write Latin but since Latin was the only language of writing they do not prove that they spoke Latin rather than British. The peasantry in southern Britain, who formed the majority of the population, spoke British and probably knew little Latin, while the language of northern Britain was almost exclusively British. The failure of Latin to survive as the native language of Britain shows that it was not deeply rooted in the country. The evidence is carefully reviewed by K. H. Jackson, *Language and History in Early Britain*, pp. 94–105. Cf. also Pliny, *Ep.* 2. 10, 2.

eloquentiam concupiscerent. In A.D. 96 Martial says 'dicitur et nostros cantare Britannia versus' (11. 3, 5). About A.D. 128 Juvenal casually speaks of British pleaders trained by Gallic eloquence (15, 111). The appearance of two distinguished lawyers, Javolenus Priscus and Salvius Liberalis, *legati iuridici* in Britain soon after A.D. 80 indicates an extension of Roman law courts in the island (*ILS* 1011, 1015) and implies that the task of reconciling Roman and Celtic law was being taken in hand.

habitus nostri honor, sc. *apud eos erat*, 'our dress came to be esteemed'. *Habitus*, here explained by *frequens toga*, often means 'dress' (c. 39, 1), but is frequently used in a wider sense (c. 11, 1; 44, 2, etc.).

discessum, 'they went astray' from the right path (simplicity of life); so 'discedere ab officio, a fide' (Cic. *Off.* 1, 32, etc., cf. Varro, *de Ling. Lat.* 8, 31 'cum discessum est ab utilitate ad voluptatem'; Amm. Marc. 28. 6, 25). *Descensum* would be more usual, but no alteration seems needed.

delenimenta vitiorum, 'the allurements of evil ways', demoralizing luxuries: cf. c. 16, 3 *vitiis blandientibus*; Arnob. *Nat.* 1, 54 *voluptatum delenimenta.* Elsewhere in Tacitus *delenimentum* denotes 'a means of soothing' (*curarum, vitae,* etc.).

balineas. Dio (62. 6, 4) makes Boudicca deride warm baths as a Roman effeminacy. Cf. also Livy's account of the degeneration of Hannibal's army at Capua (23. 18, 12 'somnus . . . vinum et epulae et scorta balineaeque et otium consuetudine in dies blandius enervaverunt corpora animosque'). Public baths from an early period of the Roman occupation have been found at Wroxeter, Wall (Letocetum), and Caerwent. At Silchester the baths and the *forum* were built in the last decades of the first century and antedate the grid system of streets as can be seen from the differences of alinement. For a plan of the baths at Silchester see Fig. 16. The Roman baths at Bath (Aquae Sulis), which were also being built in the Flavian period (*RIB* 172), were part of the cult centre of a deity presiding over thermal springs and a healing establishment and thus fall in a different category.

[*E* reads *balinea*, *E*[2m] *balnea*, but the neut. plur. of *bal*(*i*)*neum* is only commonly found in the dactylic poets: instead a fem. form *bal*(*i*)*neae* was used (cf. Varro, *de Ling. Lat.* 8, 48). Tacitus has *balineae* at *H.* 3. 32, 3; 83, 2: *balneae* at *H.* 3. 11, 3; *A.* 15. 52, 1. We should read either *balineas* (Rittèr) or *balneas* here too.]

idque: referring to all these innovations. An attraction would be usual in classical Latin (like *is . . . honos* in c. 46, 2): cf. c. 43, 2 *illud.*

humanitas, 'civilization'. Cf. Caes. *B.G.* 1. 1, 3 'a cultu atque humanitate provinciae longissime absunt (Belgae)'.

pars, 'a part', as at *A.* 6. 27, 1. What the innocent called 'humanitas' was just as much part of the enslavement of Britain as the military control. This Roman method of 'enervating' subjects is alluded to in *G.* 23, 2, and *H.* 4. 64, 3 'voluptatibus quibus Romani plus adversus subiectos quam armis valent'. The attitude of Tacitus, himself an administrator, towards the policy of conquest by assimilation is remarkable.

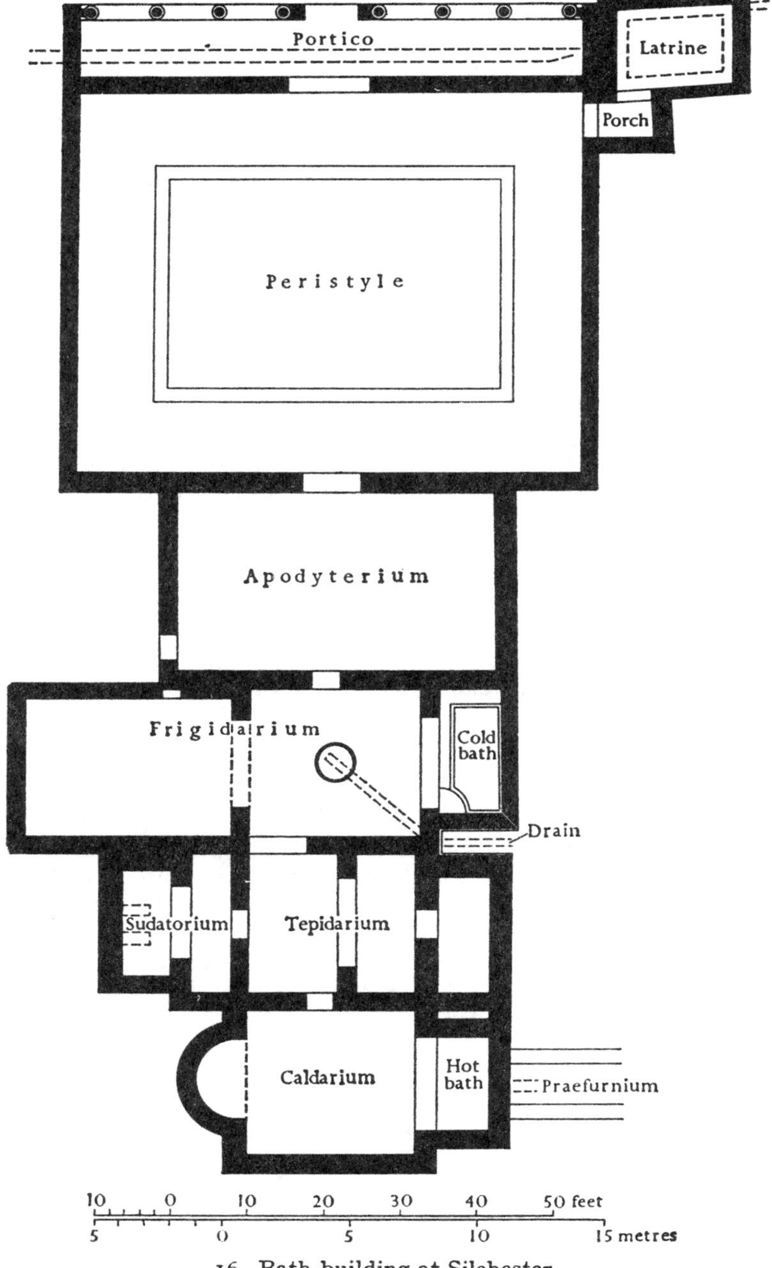

16. Bath-building at Silchester

22. *The Third Season*: A.D. 80

For an account of the campaign see Introd., p. 57. Agricola overran the whole area of the Scottish Lowlands between the Tyne–Solway and Forth–Clyde isthmuses and even probed as far as the Tay. Tacitus continues to stamp Agricola with the hall-marks of a good general (cf. notes on c. 22, 2; 22, 4).

22, 1. novas gentes: cf. c. 34, 1; 38, 3. They are distinguished from the *ignotae gentes* of c. 24. 1. Among them will have been the Votadini and the Selgovae. Agricola probably advanced by a pincer-movement from both Carlisle and Corbridge.

aperuit, 'opened up'. So *quos bellum aperuit* (*G.* 1, 1): cf. *H.* 2. 17, 1; 4. 64, 3; *A.* 2. 70, 2.

Taum. See Introd., p. 57, n. 2. The river Tay is meant, called by Ptolemy *Ταούα εἴσχυσις* (2. 3, 4).

aestuario nomen est: for the dat., instead of the more usual gen., cf. *A.* 11. 30, 1; 12. 51, 4.

conflictatum, 'harassed'; so *hieme conflictatus* (*H.* 3. 59, 2).

ponendis . . . castellis: for a list of these forts see pp. 338–9 and cf. Fig. 4. Outstandingly good situations picked at this time are the Tyne bridge-head at Corbridge, the Tweed bridge-head at Newstead, Tassieholm in upper Annandale, and Inveresk on the Forth.

spatium, 'time to spare': cf. *A.* 1. 35, 5.

22, 2. adnotabant periti: a phrase repeated in *A.* 12. 25, 2 and *H.* 3. 37, 2 with reference to antiquarians, as here to military experts. The imperf. is analogous to an epistolary imperf. and implies that Tacitus consulted written sources before composing the Agricola. It does not mean either 'they used to observe' or 'they observed at the time'.

non alium: cf. c. 5, 2 *non alias*. Agricola was not the first general to be praised for his capacity for selecting camp-sites. Livy similarly praises Hannibal (35. 14, 9 'nemo elegantius loca cepisse, praesidia disposuisse') and Philopoemen (35. 28, 1–9; cf. *H.* 2. 5, 1). For an interesting account of the principles involved cf. Hyginus, *de mun. cas.* 56 and for an illustration of the construction of a camp see Trajan's Column (C. Cichorius, *Die Reliefs der Traiansäule* (Berlin, 1896–1900), i. Taf. XXX).

opportunitates locorum, for *opportuna loca*, 'suitable sites'. Cf. c. 19, 4 *longinquitas regionum*.

pactione ac fuga desertum, 'abandoned by arrangement (with the enemy, capitulation) or by retirement (without negotiations)'. We should have expected *vel*, as in *Dial.* c. 28, 1; but *ac* may be explained by the fact that the two nouns form a pair of ideas which are almost two parts of one idea, opposed to *vi expugnatum* (evacuation after and without negotiations): cf. *H.* 2. 37, 2 'aut exercitus . . . aut legatos ac duces'. To translate 'by capitulation and (consequent) evacuation' would be to attribute to Tacitus an improbable tautology.

desertum, sc. *est*, not *esse*. The comment is by Tacitus not by the experts.

crebrae eruptiones. The sites were so well chosen that they could not be carried by force or induced to capitulate: on the contrary, they allowed the garrisons, even when invested, to make frequent sallies to harass and disorganize the enemy; for (*nam*) they were well provisioned to face a siege which would otherwise have forced them to suspend any aggressive countermeasures and to concentrate on mere survival. In this obscure and compressed sentence Tacitus seems to be saying that the ample provisions which were carried in reserve enabled the garrisons to take advantage of their superior situations and make counter-attacks on a besieging army; a poorly victualled garrison could not have afforded such a risk and would have been reduced to a military standstill (*moras obsidionis*).

[The connexion of thought is abrupt and has encouraged editors to transpose *crebrae eruptiones* after *hiems*. The transposition is not necessary and is not attractive because it would make *nam . . . firmabantur* explain why no forts were captured, whereas Tacitus is clearly attributing the resistance of the forts not to their reserve supplies but to their strategic siting by Agricola. Perret's transposition, based on the hypothesis that the archetype of *E* had 13 letters to the line, is impossible because of the position of *ibi*.]

Crebrae eruptiones is a military technical term: cf. Caes. *B.G.* 7. 22, 4; *B.C.* 2. 2, 6; 8, 1; 22, 1; Livy 37. 5, 6, etc.

annuis copiis, 'supplies to last a year', i.e. to last a year from the beginning of a siege, practically therefore a year's reserve supplies. This interpretation accords with the archaeological facts, which show that the average military granary had storage space for a two years' ration from the time of any given harvest. For an example of such granaries see Fendoch (*Proc. Soc. Ant. Scot.* 73 (1938/9), 129–32). See also Fig. 8 above. For the use of *annuus*, cf. *A.* 3. 71, 2, etc.; for that of *copiae*, *G.* 30, 3, etc.

22, 3. intrepida ibi hiems, 'winter there (i.e. in the forts) was free from anxiety'. *Intrepidus* is rare in this sense before Tacitus (cf. Sen. *Brev. Vit.* 10, 4).

sibi quisque praesidio, i.e. no commandant of a fort required any help from outside.

inritis, 'baffled'. So used of persons in poets and post-Augustan prose. Cf., e.g., Vell. Pat. 2. 63, 2.

soliti plerumque : cf., e.g., Caes. *B.C.* 3. 8, 2; Sall. *Jug.* 7, 5.

eventibus, 'successes': cf. c. 8, 2; 27, 1; *A.* 2. 26, 2.

pensare, 'to counterbalance', a post-Augustan use for *compensare*.

iuxta = *pariter*, 'alike', an adverbial use common in Sallust and Livy; cf. also, e.g., Pliny *N.H.* 2, 136 'iuxta hieme et aestate'.

22, 4. intercepit, 'took credit to himself for'. Cf. Cic. *Leg. Agr.* 2, 3 'honos interceptus'; Curtius 8. 1, 29. For the converse of this commonplace see note on c. 8, 3. Agricola would not be present at these operations, which took place far from head-quarters.

praefectus : an officer of equestrian rank commanding an auxiliary infantry (or cavalry) regiment of 500 men, which would form the garrison of a smaller *castellum*.

habebat, 'had in him'. The use in *H.* 3. 65, 2 is different.

ut erat. Tacitus frequently concludes a character sketch with a generalization introduced by asyndeton (cf. below *honestius . . . odisse*) so that there is no need to insert *et* (*et ut erat* Peerlkamp; *et erat ut* Henrichsen). The assertions about Agricola are all commonplace and should be compared with Pliny, *Paneg.* 66. So too Germanicus displayed 'comitas in socios' (*A.* 2. 72, 2).

comis bonis, etc. The dative is varied to the accus. with *adversus*, as in *H.* 1. 35, 2. *Iniucundus* is not found elsewhere in Tacitus, and is generally used of things, but cf. Quint. *Inst.* 10. 1, 124.

ceterum, etc., 'but none of his resentment remained hidden away in his mind, so that (cf. c. 12, 3) you need not fear his silence', i.e. that he was silently brooding over his grievance, with a view to future vengeance. For the potential *timeres* in a consecutive clause cf. *H.* 3. 83, 2; Livy 31. 7, 11. A contrast to Domitian is evidently suggested, who 'secreto suo satiatus, optimum . . . statuit reponere odium' (c. 39, 3): cf. 'quo obscurior, eo inrevocabilior' (c. 42, 3). It was a characteristic trait of tyrants (*A.* 1. 69, 5; 16. 5, 3). This punctuation gives an excellent sense, and no emendation is wanted.

offendere, 'to give offence', by open rebuke: cf. 'dum offendimus' (*A.* 15. 21, 3). It is thus contrasted with *odisse*, 'to harbour dislike'.

23. *The Fourth Season*: A.D. 81

The season was spent in constructing a chain of forts across the Forth–Clyde isthmus (for which see App. 3, p. 323) and in consolidating the area behind it. See Introd., p. 57. The number of forts that had to be constructed can be judged from the distribution map (Fig. 4).

23, 1. obtinendis, 'securing' by military occupation. Cf. c. 24, 3, and 'percursando quae obtineri nequibant' (*A.* 15. 8, 2), etc.

percucurrerat: Tacitus seems elsewhere to use the reduplicated form only in the perfect, to distinguish it from the present. He has the simple form at *H.* 3. 80, 1 (*occurrerant*) and that form should perhaps be read here with *B* and at c. 37, 1.

exercitus: E^2 corrected to *exercituum* but the singular is required because Agricola, although operating with several legions, had only one army.

pateretur. The imperf. is used because the words were still applicable when he wrote (as Andresen notes). Half-conquests were not the Roman policy of the time.

inventus: sc. *erat*.

in ipsa Britannia, 'within Britain itself' = *citra finem Britanniae*. The line of the forts is that separating Britannia proper from Caledonia (cf. c. 10, 3), but the former term is generally used for the whole. With the advance beyond this line *terminus Britanniae patet* (c. 30, 3), *finem Britanniae tenemus* (c. 33, 3).

Clota et Bodotria. The Clyde and Forth. On this line and on the forts planted along it see App. 3. The estuary of the Clyde has the same name in Ptol. 2. 3, 1, while that of the Forth is called *Βοδερία εἴσχυσις* (2. 3, 4 there is a variant reading *βογδερία*). The Ravenna geographer calls it Bdora (438, 5). The name Clota comes from the root **clou* 'wash' (cf. Lat. *cloaca*; O. Ir. *Cluad*). The etymology of Bodotria is obscure. Watson (*CPNS*, 51–52) would read *Bodortia* and connect it with, e.g., O. Ir. *bodar* 'deaf'—'the Silent One'. But the etymology is not certain enough to justify the change. The name does not appear to have survived the classical period: Forth is derived from a different root.

aestibus, etc., 'carried far back (inland) by the tides of opposite seas'. For the sense of *diversus* cf. c. 11, 2, and *diversa maria* in Livy 21. 30, 2; 40. 22, 5. The idea seems to be that the river-water is driven back by the tide. *Per inmensum*, as in Ovid, *M.* 4, 261. *Revectae* appears to be used only here of water being driven back but cf. *invehi* in Plin. *N.H.* 2, 224; Solin. 37, 5.

firmabatur, 'was being securely held'. Cf. note on c. 14, 3.

omnis propior sinus, 'the whole sweep of country nearer' (i.e. southward)—a rhetorical exaggeration or a misconception, for only the eastern half of southern Scotland was held by Agricola. See Introd., p. 57. *Sinus* is so used of a projecting stretch of land in *G.* 29, 3; 37, 1; *A.* 4. 5, 2, and in Livy and the elder Pliny, and has often no reference to sea-coast.

summotis . . . hostibus : the phrase is echoed by *S.H.A.*, *Ant. Pius* 5, 4 'Britannos per Lollium Urbicum vicit legatum alio muro caespiticio summotis barbaris ducto', referring to the Antonine Wall constructed in A.D. 140–2. In that passage the author appears to be saying that the barbarians were pushed back behind the wall, that is, virtually deported. Here Tacitus does not mean that there was any forcible displacement of the whole civilian population, but only that those who were not prepared to acquiesce in Roman rule but wished to continue armed militant resistance (*hostibus*) were ejected. *Summovere* is similarly used in Florus 2, 28.

in aliam insulam : the tract of Caledonia, wholly cut off by the occupation of the isthmus.

24. *The Fifth Season*: A.D. 82

For Agricola's fifth campaign see Introd., p. 59. The area of his operations is determined by the mention that his troops arrived at a point facing Ireland. This must mean that he was campaigning in Ayrshire and Galloway where substantial traces of Agricolan forts have been found (Dalswinton, Glenlochar, Loudoun Hill, Gatehouse-of-Fleet). See also the discussion by S. N. Miller, *J.R.S.* 38 (1948), 15 ff.

24, 1. nave prima transgressus. *Nave prima* means 'in the leading ship', and emphasizes Agricola's personal initiative (cf. Livy 37. 23, 8; 26. 39, 14; 37. 29, 5; 25. 27, 11; Caes. *B.G.* 4. 23, 2; 4. 25, 6; *Bell. Alex.* 45, 3; 15, 3: for the word-order cf. *H.* 2. 49, 2 *luce prima*: see Lacey, *C.Q.* 7 (1957), 118 ff., who gives a summary of earlier interpretations and emendations). *Transgressus*, however, remains difficult since the context does not make clear what Agricola crossed with his fleet *en route* for Galloway and Ayrshire. The stretch of water most easily understood from the context is the Clyde (cf. c. 23; 25, 1), which would imply that Agricola sailed down the Clyde, perhaps even as far as Irvine Bay, on the first stage of his campaign. Richmond suggested that *Anavam* (R. Annan) should be inserted before *transgressus*, which is palaeographically easy, but the river is hardly large enough to justify such an allusion to a sea-borne operation. P. E. Postgate (*Proc. Cam. Phil. Soc.*, 1930, p. 8) read *Itunam* (Solway Firth) for *nave prima*. Although the deletion of *nave prima* is wrong, the insertion of *Itunam* deserves consideration if Agricola wintered at York rather than on the Forth–Clyde line: in that case Maryport would be the obvious port to sail from. I am inclined to believe that Tacitus deliberately left the water unspecified (cf. c. 18, 1 where *transgressus* is again used without a supplement expressed), either because he was unable to verify the exact names and details (for a similar omission in Thucydides cf. Rehm, *Philologus* 89 (1934), 133 ff.) or because he felt that the addition of the name would not enlighten his readers. Other suggestions that have been made so far have nothing to recommend them.

ignotas ad id tempus: for the expression, cf. c. 10, 4. These tribes would have included the Novantae of Galloway.

quae Hiberniam aspicit, 'which faces Ireland': cf. 'mari quod Hiberniam insulam aspectat' (*A.* 12. 32, 1), 'qua . . . Pannoniam aspicit' (*G.* 5, 1). *Aspicere* in this sense, for which *spectare* is more common, is not earlier than the Augustan period (cf. Colum. 8. 8, 2, etc.; Sil. 1, 198; Stat. *Silv.* 2. 2, 46). The locality occupied was probably the Rhinns of Galloway from which Ireland can be clearly seen across the North Channel.

in spem, etc., 'in hope (of future action) rather than out of fear (of the Irish)'. It should be noted that, despite Juv. 2, 159–60, this passage effectively disposes of any supposition that Agricola invaded Ireland. It would be unthinkable that an expedition would not have been specifically mentioned here if it had ever taken place. For the contrast of *in* and *ob* cf. c. 5, 1; and for *in spem, A.* 14. 15, 5.

si quidem . . . miscuerit. *Si quidem,* used by Tacitus only here and in *G.* 30, 1, provides the justification for a conclusion already stated: 'if, as is assuredly the case, . . .', 'since'. In *G.* 30, 1 and in Cicero it is used, as would be expected, with the indicative. *Miscuerit* should, therefore, be the fut. perf. indic., not the perfect subj. The clause expresses Tacitus' belief that Ireland, if conquered, would serve to unite the strongest sections of the empire. It does not express Agricola's opinion, for then a subj. in historic sequence would be required (cf. c. 24, 3 *si tolleretur*).

medio : often so used by Tacitus, who is fond of such local ablatives. On the geographical conception see note on c. 10, 2.

opportuna, 'conveniently situated for', 'within easy reach of'. Cf. *A.* 3. 38, 2. 'insula . . . Thraeciae opportuna'. Tacitus regarded the south coast of Ireland as much nearer to Gaul than it is.

valentissimam imperii partem, i.e. Gaul, Spain, and Britain. In *H.* 3. 53, 3 Gaul and Spain are called 'validissimam terrarum partem', but here Britain seems clearly to be included. Gaul and Spain were great military recruiting districts, and in the pre-Flavian period specially heavy demands were made on them. Their economic resources were also very great: see the estimates in Tenney Frank (ed.), *An Economic Survey of Ancient Rome,* 3, pp. 126 ff.; 385 ff. In both respects the small province of Britain was of much less importance, though its military contribution was not inconsiderable.

magnis in vicem usibus, 'with great mutual advantages'. Persson would translate 'with much reciprocal trade' but *usus* is only used in this sense in the singular (cf. Cic. *ad Fam.* 7. 32, 1; *G.* 5, 2). The adjectival use of *in vicem* = *mutuis* is found commonly in Livy (2. 44, 12, etc.). For a somewhat different use cf. c. 16, 1.

24, 2. spatium: its extent. The area of Ireland is 32,337 sq. miles, that of Sicily approx. 10,000 sq. miles. Caesar was accurate when he claimed (*B.G.* 5. 13, 2) that 'Hibernia dimidio minor ut aestimatur quam Britannia', for the area of Britain is 83,355 sq. miles.

nostri maris: the Mediterranean. So *nostri orbis* (c. 12, 3).

differunt. The singular cannot satisfactorily be defended where *ingenia cultusque* are so closely coupled by the genitive *hominum* as the nearest subject: cf. Mela 2. 1, 9 'ingenia cultusque gentium differunt'. The following clause is corrupt. *In melius* cannot be taken either with *differunt*, since it contradicts *haud multum*, nor with *cogniti*, since it cannot be imagined that Tacitus was so ignorant of the truth as to suppose that Ireland was better known than England. Nor does emendation help: *melius* (Rhenanus) can hardly mean 'fairly well (known)' since this would be a most ambiguous way of stating it. It is easier to assume that the words are interpolated than to suppose that something has dropped out, because there is nothing obviously missing from the sense. A patriotic motive for such an interpolation would be ready to hand if Irish monks had been concerned in the transmission of the text (Norden, *German. Urgesch.*[2] p. 457 n. 1) but this cannot be definitely established (see Introd., p. 85). A similar interpolation has been suspected in Mela 3. 6, 53.

per commercia. Cf. c. 28, 3. It is not clear whether the merchants were British or Roman (or Gallo-Roman). The similarities of the Iron Age cultures in Britain and Ireland would make the former very difficult to detect, but there is increasing, although still slight, evidence for Roman trade with Ireland in the first century A.D. The details are collected by S. P. O'Riordain, *Proc. Roy. Irish Acad.* 51 C (1947), pp. 35 ff. They include Arretine pottery and Flavian coins. The principal sites are Lagore, Ballinderry, and Lambay Island (off Dublin) but no systematic search of Roman contacts has yet been undertaken. An *olla* of early second-century date, dredged off Porcupine Bank, is likely to have come from a sunken merchantman. For Ptolemy's information about Ireland, also derived from merchants, see Introd., p. 46.

24, 3. expulsum seditione domestica. So British princes had been received by Augustus (*Mon. Ancyr.* 5. 54; 6. 2), Gaius (Suet. *Cal.* 44, cf. Introd., p. 50), and Claudius (Dio 60. 19, 1). This *seditio* may be connected with the Revolt of the Vassals, mentioned in Irish legends. The Cruitnin revolted against their overlords, the men of Mil, and murdered them all at a banquet except for three women who managed to escape. The government set up by the Vassals was so incompetent that the three sons of the three fugitives were later recalled and became respectively the kings of Munster, Ulster, and Tara. The revolt can only be approximately dated but Feradach, the King of Tara, was probably born about A.D. 60. The *regulus* could, therefore, either be one of the three sons or one of the Vassals who had fled to Britain on the overthrow of their government. Tacitus' picture does, however, fit the disturbed conditions prevailing in Ireland at this time. See C. E. Stevens, *Rev. Ét. Anc.* 42 (1940), 371 ff. [The historicity of the Irish legends is rejected by T. F. O'Rahilly, *Early Irish History and Mythology* (1946), pp. 155 ff.] For the phrase cf. *H.* 4. 12, 2: *seditio domestica* is a Livian expression (cf. 2. 42, 3; 45. 19, 13).

in occasionem, i.e. to make use of him, if he should invade the island.

ex eo. For other references to Agricola's own testimony cf. c. 4, 3; 44, 5. Some wrongly take *eo* here of the Irish prince. To excuse Agricola from having so grossly underestimated the difficulty of conquering Ireland, the Bipontine editors wished to read *audivit,* also taking *eo* of the prince: 'Agricola was often told by him . . .'. But the prince can hardly be credited with the unpatriotic sentiments *idque . . . tolleretur.* In fact the question of the number of troops required to conquer Britain was evidently a traditional subject for discussion: cf. Strabo 4, 200 τοὐλάχιστον μὲν γὰρ ἑνὸς τάγματος χρῄζοι ἂν καὶ ἱππικοῦ τινος (referring to an invasion of Britain mooted by Augustus); Quint. 7. 4, 2 'haec in suasoriis aliquando tractari solet . . . quo numero militum (Britannia) adgredienda'. Agricola may be transferring the discussion from Britain to Ireland.

debellari, word first found in Virgil and Livy, who uses it more than 50 times.

adversus, i.e. in the Roman relations towards: cf. c. 12, 2.

arma, sc. *essent,* often omitted when a co-ordinate subjunctive clause follows.

25–27. *The Sixth Season*: A.D. 83

For an account of the campaign see Introd., pp. 60 f. Agricola first secured the people of Fife, Angus, and the Mearns ('amplexus civitates trans Bodotriam sitas') and then turned his attention to the tribes of the Highland plateau, Caledonia proper.

25, 1. Ceterum: marking the return from a digression, as in c. 11, 1; *G.* 3, 2, etc.

amplexus: cf. c. 17, 1. Agricola enveloped and overran the area immediately to the north of the Forth–Clyde line. [Anderson wrongly took *amplexus* metaphorically, sc. *animo* 'embracing in his plans'.] Traces of Agricola's policy can be seen in the line of forts which seal off the glens leading from the Highland plateau and which isolate and enclose Fife, Angus, and the adjoining regions.

quia, etc., explains 'portus classe exploravit' (see below).

ultra: used as an adjective denoting what lies beyond a boundary, here the Forth–Clyde line; cf. *G.* 2, 1 'immensus ultra Oceanus'.

infesta hostili exercitu itinera: 'routes that were dangerous because (of the threat) of a hostile army'; for the use of *infestus* with an abl. in Tacitus cf. *A.* 2. 23, 3 'insulas saxis abruptis infestas'. Such perils on the road were a common fear; cf. Cic. *Prov. Cons.* 4 'via illa nostra . . . excursionibus barbarorum infesta'; Livy 39. 1, 6 'itinera infesta insidiis'; Apul. *Met.* 9, 10; *Paneg.* 4. 36, 5. For *exercitus* of a non-Roman army cf. *G.* 30, 2, etc.

exploravit: see Introd., p. 32. So far no archaeological traces of the operations of the fleet on the N.E. coast of Scotland have been found, but Tacitus must mean harbours north of the Tay, that is, such as Arbroath, Montrose, and Stonehaven.

adsumpta in partem virium, 'taken up to form part of his force' (cf. c. 13, 3 'adsumpto in partem rerum'). The *classis Britannica* is mentioned in A.D. 70 (*H.* 4. 79, 3), and doubtless existed in some form from the first invasion; but it would appear to have been previously used rather as a means of transport and supply (cf. c. 24, 1), and by Agricola first as an essential branch of the attack. It follows from c. 28 that Agricola had also ships on the west coast. Normally the British fleet had its main station at Boulogne (Gesoriacum, later

Bononia), although this was outside the control of the legate of Britain. In the first century the English base-port seems to have been Richborough (Rutupiae; cf. c. 38, 4 n.) but in the second century brick stamps indicate that there was a spaced series of naval bases along the south coast, viz. at Dover, Folkestone, Lympne (cf. *RIB* 66), Richborough, and Bodiam). The fleet seems to have patrolled only the narrow sea dividing Kent and the Thames from Gaul and the Rhine. See C. G. Starr, *The Roman Imperial Navy*, pp. 152–6; B. Cunliffe, *Richborough*, V (Soc. Ant. Res. Rep.), 255 ff.

impelleretur, 'was being pushed forward'. *Bellum impellere* is elsewhere found only in Lucan 5, 330. *Egregia specie* is a descriptive abl. qualifying *quae* (= *classis*) and describes the imposing spectacle made by the fleet (cf. c. 35, 3; *A.* 2. 6, 2: for the abl. cf. *A.* 15. 9, 1). *Cum . . . impelleretur . . . adtollerent . . . compararentur* elaborates *egregia specie*, 'as the war was being pushed forward', etc. [There is no need to put a full stop after *sequebatur* and read *egregia species* (nom.: so Mützell). It is however necessary to emend *E*'s *impellitur* to *impelleretur* (Rhenanus). Although changes of mood and, occasionally, alternations between indic. and subj. are found in a *cum*-clause (cf. Livy 29. 37, 8; see Kühner–Stegmann 2, p. 345 n. 2; Madvig on Cic. *Fin.* 2, 61), Tacitus uses the vivid present ind. in historic sequence only in a temporal *cum*-clause and then only when the clause is the logical equivalent of a main clause: cf. § 3 below.]

pedes equesque : coupled closely as the land force. *Iisdem castris* is local abl.

mixti copiis et laetitia, 'sharing their rations and exultation'. Cf. *H.* 1. 9, 3, 'nec vitiis nec viribus miscebantur'; the ablatives expressing that in respect of which they were *mixti inter se*. For *copiis* cf. c. 22, 2; for the coupling of different ideas, *nox et satietas* (c. 37, 5).

adtollerent = *extollerent,* as in several places in *Hist.* and earlier in Vell. Pat. 2. 65, 1; Seneca, *Epist.* 94, 72.

profunda, 'the ravines', where danger would lurk. The substantival use of neuter plural adjectives, often (as here) with a partitive genitive following, is very common in Tacitus.

hinc . . . hinc : for *hinc . . . illinc.*

victus also goes with *terra et hostis.*

iactantia (cf. c. 39, 1, etc.): a word not apparently found earlier than Quint. and the younger Pliny. The classical *iactatio* is used by Tacitus only in his minor works (c. 5, 1; 42, 3; *G.* 6, 1). The telling of adventures by the troops is a conventional touch; cf. Lucan 4, 196 ff.; *A.* 2. 24, 3–4.

25, 2. Britannos quoque . . . obstupefaciebat, i.e. 'the sight of the fleet affected them also, but with stupefaction (not with joy)'.

tamquam: expressing their thought, 'as though, by the opening up of the recesses (c. 31, 3, etc.) of their sea, their last refuge was closed against them'. The ingenious verbal contrast of *aperto* and *clauderetur* is an intentional oxymoron.

25, 3. ad manus et arma (as in c. 33, 5): virtual synonyms, the latter word defining the former. The collocation is common; cf., e.g., Cic. *Verr.* ii. 1, 82; Vell. Pat. 2. 115, 4; Stat. *Theb.* 9, 183; *H.* 3. 20, 2.

Caledoniam incolentes populi. By Caledonia Tacitus means the whole Highland region (c. 10, 3; 31, 5), part of which, from the Great Glen to the upper Tay valley, was occupied by the tribe of Caledonii (Ptol. 2. 3, 8–12; Dio 76, 12: see Introd., p. 60).

paratu: used for *apparatus*, as, e.g., in the *Hist.* and *Ann.* and in Sallust, *H.* fr. 1. 88 M.

uti mos, etc., applying only to *maiore fama*: report usually exaggerates. A similar thought is expressed in c. 30, 3 'omne ignotum pro magnifico': it is a commonplace, cf. *A.* 3. 44, 1; Thucydides 6. 34, 7.

oppugnare ultro castella adorti. The phrase is from Livy (35. 51, 8; 37. 5, 5; 43. 21, 4 'castellum adortus oppugnare'). *Ultro* implies an offensive movement (cf. c. 19, 4; 42, 1, etc.). *Castella* (E^{2m}) should be preferred to *castellum* (*E*) because sporadic attacks on several forts are more likely to have inspired panic (*metum*) than a single attack. It is not certain whether the attacks were launched on the forts of the Forth-Clyde isthmus, behind Agricola's line, or on the more recent forts with which he garrisoned the territory acquired north of the isthmus (cf. note on § 1 above). In view of the reaction 'regrediendum citra Bodotriam' the latter is more probable. See Introd., p. 64.

metum, etc., 'had created the greater panic, as taking the offensive'.

quam = *quam ut,* used by Sallust and by Livy.

cognoscit. The omission of the subject (not expressed till the next sentence) is thought harsh, but *Agricolam* is naturally supplied as the object of the preceding *admonebant*.

pluribus = *compluribus*, as in c. 29, 2, and often. The modal ablative is much used in describing military formations.

25, 4. ne . . . circumiretur. From the threefold division of his army, and from the isolation of his weakest legion, the Ninth, it has been inferred that Agricola had only three legions, each of which formed the nucleus of a division. It is conceivable that a legion had to be left behind somewhere in garrison; but we cannot assume that the three divisions were all of equal strength. Confirmation of this account of Agricola's disposition of his troops has been found in the discovery of a series of temporary camps of approximately 25 acres (which would hold a legion and a few auxiliaries) in the area, notably at Dealgin Ross and Stracathro on the North Esk. These are of Agricolan date (see Introd., p. 64).

et ipse, 'himself also': cf. *G.* 37, 3; *H.* 3. 82, 2, where it comes, as here, in the middle of an abl. abs. So also in Livy 4. 44, 10; 29. 2, 2; 45. 10, 2.

26, 1. repente is always used by Tacitus instead of the popular *subito*.

nonam. This legion, part of the original invading army, had been almost cut to pieces in A.D. 60 (*A.* 14. 32, 3), after which it had been reinforced (14. 38, 1). The explanation of its present weakness in A.D. 83 is that, like the other legions in Britain, it had sent a vexillation to Germany for Domitian's Chattan war of A.D. 83. An inscription to L. Roscius Aelianus (consul A.D. 100), found at Tivoli and set up about A.D. 118 (*ILS* 1025), records him as *trib. mil. leg. ix Hispan(ae) vexillarior(um) eiusdem in expeditione Germanica, donato ab imp. Aug. militarib(us) donis. Ab imp. Aug.* is the formula for Domitian after the *damnatio memoriae*. Cf. Appendix, p. 320.

inter, 'in the midst of', 'during': cf. *H.* 4. 1, 3 'inter turbas et discordias'.

edoctus: so with accus. several times in Tacitus, after Sall. *Cat.* 45, 1; *Jug.* 112, 2.

vestigiis insecutus, 'following close on the track'; Livy has *vestigiis sequi* (6. 32, 10, etc.); Justin has *vestigia insequi* (32. 3, 14).

adsultare appears first in the poetry of Germanicus (*Arat.* 299). The military sense is common in Tacitus and also in Silius Italicus and Statius.

signa : of his legionary force.

26, 2. ancipiti malo, as in Sall. *Jug.* 67, 2; *Cat.* 29, 1; Livy 3. 28, 9.

securi pro salute, 'without fear as to deliverance'. So 'pro me securior' (*H.* 4. 58, 1), 'pro . . . Catone securum' (Sen. *Const. Sap.* 2, 1).

de gloria, i.e. disputing the honours of victory with the relieving force ('utroque exercitu certante', etc.). The antithesis between *salus* and *gloria* is conventional; cf. Dem. *Olyn.* 1, 5; Sall. *Jug.* 114, 2; Livy 21. 41, 13.

ultro quin etiam. *Ultro* as in c. 25, 3. *Quin etiam* is in anastrophe here alone in this treatise, but five times in *Ger.*, once each in *Hist.* and *Ann.* Cf. c. 3, 1 *quippe*.

portarum angustiis. Some of Agricola's marching camps had a highly unusual gateway plan wherein the ditch forms on one side an external *clavicula* and on the other an oblique external spur. This feature, which shows up clearly on air-photographs (see Introd., p. 62 and Fig. 6), will have made it more difficult for an attacker to force his way into the camp. The detail of the battle given by Tacitus here is, therefore, likely to be authentic.

tulisse opem . . . eguisse auxilio, echoing Livy 7. 35, 4 'digni estis qui pauci pluribus opem tuleritis ipsi nullius auxilio egueritis'.

texissent. Cf. Livy 3. 22, 9 'deletusque exercitus foret ni fugientes silvae texissent'.

debellatum. Although Agricola seems to have been fortunate to have avoided a serious set-back, his victory was probably more substantial than editors allow who attribute Tacitus' comment to rhetorical exaggeration. Ptolemy (2. 3, 7) mentions a fort named Victoria in the territory of the Damnonii and, although it cannot be identified for certain because of the distortion of measurements taken from Inchtuthil (see Introd., p. 40), it must be one of the forts constructed in this campaign (c. 25, 1 n.), perhaps Strageath or, as Roy suggested, Dealgin Ross. The name must be inspired by the rescue of the Ninth Legion, in the vicinity of whose camp the fort must lie, and not by the battle of Mons Graupius which took place far to the north. Nor could such a name have been officially bestowed if the action had not in fact been creditable.

27, 1. conscientia ac fama, 'the consciousness and report', the former applying to those who had taken part in the battle, the latter to the rest of the army. The same terms contrast personal feeling and report of others in *A*. 6. 26, 1. The terms are often paired (cf., e.g., Pliny, *Ep*. 1. 8, 14; 5. 1, 11, etc.).

invium : a variation of the proverbial 'fit via vi': cf. Ovid, *M*. 14, 113 'invia virtuti nulla est via'; Curt. 9. 2, 9.

penetrandam. Cf. 'longius penetrata Germania' (*A*. 4. 44, 2). This transitive use is first found in prose in Vell. Pat. 2. 40, 1. Being *trans Bodotriam*, c. 25, 1, they were already within Caledonia.

Britanniae terminus, 'the furthest bounds of Br.': cf. note on c. 23, 1.

fremebant, 'were clamorously demanding'. Collectives, whether nouns or pronouns (*quisque* and *uterque*), frequently take a plural verb in Tacitus.

illi : the 'ignavi specie prudentium' of c. 25, 3. For the thought cf. *H*. 1. 35, 1. *Cautus* and *sapiens* are similarly linked by Cicero, *ad Att*. 14. 14, 2. *Magniloquus* occurs first in Ovid, here alone in Tacitus, and apparently in no earlier prose. Cf. also Stat. *Theb*. 3, 192; Mart. 2. 43, 2.

prospera, etc. The same sentiment is found in *A*. 3. 53, 3; 14. 38, 3 (cf. *H*. 4. 52, 1). It may have been suggested by Sallust's 'in victoria vel ignavis gloriari licet, advorsae res etiam bonos detractant' (*Jug*. 53, 8) but it is common; cf. Aesch. *Sept*. 4–8; Dem. *De Cor*. 212; Nep. *Alc*. 8, 4.

27, 2. occasione et arte ducis, 'the opportune action and skill of the general', i.e. the discovery of their design by Agricola and his prompt action on it.

quo minus : cf. c. 20, 2.

conspirationem . . . sancirent, 'ratify the confederacy'. Such sworn confederacies, united by common oaths and sacrifices, were commonly encountered by Rome in her early expansion in Italy (cf. Livy 10. 38, 2 (Samnites)).

atque ita, etc. Cf. 'atque ita infensis utrimque animis discessum' (*A*. 13. 56, 1), where a colloquy had taken place. *Inritatis animis* is a Livian phrase (4. 54, 8; 1. 17, 4; 8. 32, 16; 23. 44, 5, etc.). Here the words, in dramatic fashion, prepare the reader for the denouement (cc. 29 ff.), before which comes the interlude of c. 28.

28. *The Mutiny of the Usipi*

For details of the voyage of the Usipi see Appendix 2. Tacitus follows the practice of Sallust and Livy in marking off and introducing major sections of his work by digressions. Just as he had prefaced the account of Agricola's governorship with the digression on Britain (cc. 10–13), so here he pauses before the climax of Mons Graupius to relate the strange adventure of the Usipi. The story serves also to high-light the difficulties against which Agricola had to contend (cf. c. 32, 3) and to remind readers of Agricola's achievement in ordering the circumnavigation of Britain (c. 10, 4; 38, 4).

28, 1. Usiporum: the Usipi of Mart. 6. 60, 3; Usipetes (with Celtic termination) of *A*. 1. 51, 2 and Caes. *B.G.* 4. 1, 1. The name perhaps means 'good horsemen' (*us-*+*ip*(*i*)*o*; cf. Gall. *epos*, Gk. εὔιπποι). In A.D. 14 they lived across the Rhine between the rivers Lippe and Yssel. In A.D. 69 they were associated with the Chatti and Mattiaci (*H*. 4. 37, 3) in attacking Mainz, and apparently occupied the region of the river Lahn (Nassau) north of the Taunus mountains, where they still were in A.D. 98 next to the Chatti of Hessen (*G*. 32, 1). Mommsen supposed that they were annexed by Domitian early in his campaign of A.D. 83, and that the recruits here mentioned were at once enrolled in the auxiliary forces and immediately sent off to Britain, whence they made their escape very soon after their arrival (*Prov*. i. 150, note). But the Usipi lay close to the border and, like the Frisii, they may have been subject to conscription even before they were formally annexed to the Roman province.

Germanias, the two military districts on the Rhine, see note on c. 15, 3. The geographical vagueness is characteristic of Tacitus.

magnum ac memorabile, emphasizing the remarkable character of their adventure; cf. Ter. *Heaut*. 314; Virg. *Aen*. 4, 94; Livy 39. 51, 10, etc.

militibus, sc. *legionariis*. Dio mentions a tribune (χιλίαρχος) and centurions, which would indicate that a milliary cohort was involved. But 1,000 men could not have been accommodated on three Liburnians (see below) and Dio has probably made a false inference, unless only a part of them mutinied. It is more likely to have been a *quingenaria* cohort. Such drill instructors of recruits were usually centurions and other veterans of distinction (cf. Pliny, *Pan.* 13).

habebantur, 'were attached': cf. *A.* 1. 73, 2.

liburnicas. The term was used loosely for the small warships, commonly two-banked open galleys (pentekontors), carrying a small sail, which were in service with the Roman navy. These had a crew of approximately 60 rowers and others and normally carried 30 troops, although in emergencies they could carry up to 60 (Hdt. 7. 184, 2–3). The Usipi presumably acted as their own rowers but even so the three vessels could hardly have carried more than 400–450 men in all. See also Kromayer, *Philologus* 56 (1897), 489.

remigante. We retain, with misgiving, *E*'s reading *remigante* 'rowing' and assume that one of the helmsmen was made to take his place among the rowers, and that the other two were killed. It is clear from 'per inscitiam regendi' (§ 3) that none of the professional helmsmen was in charge at the end of the voyage. But although the Usipi may not have trusted the actual steering to any of the former crew for fear that they would direct the ships into Roman hands, they may well have been anxious to keep at least one of the helmsmen for advice in case of emergency. Some editors translate *remigante* 'steering' but *inscitia regendi* shows that there was no professional at the helm and, even although the rudder was in effect a large oar, *remigare* cannot mean 'to steer' (at Claud. *Rapt. Pros.* 2, 178 it means, as normally, 'to row').

[The word has been more emended than any other in the *Agricola* and may be corrupt, but the emendations do not convince. They fall into three classes: (1) assuming that one helmsman stayed at his post, Bitschofsky proposed *rem agente* 'performing his task' (so also Iliffe), Wex *morigerante*, Doederlein *regente*; but it has been seen that the assumption must be wrong and *rem agente*—the most satisfying palaeographically—is linguistically impossible because *rem agere* is only used with an adverb, e.g. *rem bene, satis, agere.* (2) Assuming that one tried to escape or to reach the shore, Puteolanus conjectured *remigrante*, Peerlkamp *remeante*, Baehrens *renatante*, Currie *in oram agente*: as a variation G. Ammon proposed *se mergente* 'drowning himself'. (3) Assuming that one refused the duty and so cast suspicion on the others, Walter suggested *rem repigrante*, Porter *renuente*, Lynch *se negante*. *Alii alia.*]

suspectis duobus: presumably of some intended treachery, such as steering into a Roman port.

nondum vulgato rumore. So in Dio's account they are said to have escaped detection when they landed at points on the east coast. Since news of the mutiny had not spread, they were able to pass themselves off as troops on a legitimate mission.

ut miraculum, not only because people could not account for their appearance, but because of their erratic course.

praevehebantur = *praetervehebantur*; so in *H.* 5. 16, 3, etc. Cf. *praelegere*, c. 38, 4.

28, 2. mox ubi aquam, etc. The text in *E* is badly corrupt (see App. Crit.) but the sense is clear: the Usipi made periodic landings to pick up water and supplies. It can be compared with Livy 29. 28, 5–6 'escensiones in agros maritimos factae erant: raptisque quae obvia fors fecerat, recursum semper ad naves . . . fuerat'. Syntactically a temporal clause is required, followed by the two participles *congressi . . . pulsi*, leading up to the main verb *venere*.

[A clue to the proper reconstruction of this passage may be provided by Perret's argument that the exemplar of *E* had approximately thirteen letters to the line (*Recherches sur le texte de la 'Germanie'*, pp. 76 ff.). The widespread corruption is explained if the exemplar had read

praevehe
bantur mox ubi a
quam adq. utilia
raptum exissent

Damage to the end of the lines would account for *E*[2] *praebebantur* (from *praevebantur*), for *utilla*, and for the loss of *nt* in *exissent*, and if similarly *ia* had been lost at the end of the second line, *ubquam adq.* would easily be emended or misread as *adquam adq.* The further telescoping of *raptumexisse* into *raptisse* is accounted for by the appearance of *bantur mox* in the corresponding position of the line but one above.

In arriving at a reconstruction, W. Heraeus rightly pointed out that *ubi* is the temporal conjunction to be expected: *mox ubi* is common in Tacitus (cf. *A.* 12. 47, 2, etc.). *Utilia* 'supplies' may be safely restored by comparison with Sall. *Jug.* 86, 1; *Or. Cottae* 7, etc., so that there is no need for Selling's *utensilia*. *Exire* is the standard word for 'to make a landing' (cf. Livy 1. 1, 7; Cic. *Verr.* ii. 5, 91, etc.) and *exissent* is palaeographically more attractive than Anderson's *adpulissent* and more explicit than the bare *issent* (Eussner). Editors have, however, largely accepted the correction *ad aquam* made by E^2, comparing Caes. *B.C.* 1. 81, 5 'universas ad aquam copias educunt' and assuming a stylistic variation between *ad aquam* (= *ad aquandum*) and *utilia raptum* (supine). But the passage from Caesar is not a true parallel since *ad aquam* is evidently a technical expression for a formal military operation whereas the Usipi were making a sudden raid. E^2's reading is likely to be no more than a conjecture, influenced by the following *adq.* for *atq.*]

defensantium : the frequentative form of *defendere* is used once, for effect, by Plautus (*Bacch.* 443) and is thereafter favoured as the stronger word by historians (Sall. *Jug.* 26, 1, etc.) and poets (Ovid, *M.* 11, 374, etc.). In the same way Livy prefers *imperitare* to *imperare* (cf. 1. 24, 3, etc.).

eo ad extremum inopiae, 'at last came to such need'. *Eo* with genit. is often used by Tacitus (cf. c. 42, 4), also by Sallust and Livy. Cf. Livy 23. 19, 13 'ad id ventum inopiae' and for *ad extremum* 'finally' cf. Livy 22. 23, 5, etc.

vescerentur. The accus. with this verb (used here alone in Tacitus) seems an archaism, like that with *fungi* and *potiri*, but is found in Sallust, the elder Pliny, etc.

28, 3. circumvecti Britanniam. These words are the sole indication afforded by Tacitus of the direction which the mutineers took. Since they had circumnavigated Britain and arrived on the north coast of Germany, the inference would be that they rounded Cape Wrath and had started on the west coast, as Dio in fact affirms. His account runs: *ἐς πλοῖα κατέφυγον καὶ ἐξαναχθέντες περιέπλευσαν τὰ πρὸς ἑσπέραν αὐτῆς (τῆς Βρεττανίας), ὥς που τό τε κῦμα καὶ ὁ ἄνεμος αὐτοὺς ἔφερε, καὶ ἔλαθον ἐκ τοῦ ἐπὶ θάτερα πρὸς τὰ στρατόπεδα τὰ ταύτῃ ὄντα προσσχόντες. κἀκ τούτου καὶ ἄλλους ὁ Ἀγρικόλας πειράσοντας τὸν περίπλουν πέμψας ἔμαθε καὶ παρ' ἐκείνων ὅτι νῆσός ἐστιν.* 'They fled to some vessels and putting out to sea sailed round the west of Britain as the wind and the sea carried them. They touched at the forts on the other side of Britain but escaped detection. As a result both from them and from other expeditions which he sent to attempt the circumnavigation Agricola learnt that Britain was an island.' This means that they sailed round the western coast and put in unawares at at least one of the Roman forts on the other side, i.e. they sailed round the north and called on the east coast. The troops will, therefore, have been stationed in the vicinity of the Clyde and have sailed from a nearby port such as Ardrossan or Irvine. Independent evidence for the operations of a fleet in these waters is provided by the account of Demetrius of Tarsus, for which see p. 32.

amissis . . . navibus : on the north European coast.

habiti, 'taken for', 'treated as'.

primum . . . mox, i.e. some by the former, the remnant by the latter.

Suebis. In *G.* 38, 1 this is a generic name for a very large number of German tribes, living mostly east of the Elbe. Their original home was the Mark of Brandenburg round Berlin and they later formed the core of the great empire of the Marcomanic king Maroboduus (*A.* 12. 29, 1, etc.), but the Suebi meant here must have dwelt on or close to the coast, presumably east of the Frisii, in Schleswig, where Tacitus locates the Reudigni (*G.* 40, 1) or between the Ems and the Elbe, where he puts the Chauci (*G.* 35, 1). They cannot have been the detachment of the tribe that was settled by Augustus between the Scheldt and the Rhine (Suet. *Aug.* 21), since it is clearly implied that they were outside the empire (cf. *nostram* below). Tacitus may be using the name vaguely.

Frisiis. The Frisii occupied the north of Holland from the Old Rhine to the lower Ems; their name still survives in that of Friesland. On their history see *G.* 34, 1, with Anderson's note.

per commercia : cf. c. 39, 1.

nostram : the west or Gallic bank of the Rhine.

indicium . . . inlustravit, 'the story of such great adventures made famous'. The indic. after *sunt qui*, common in poetry but very rare in prose, is used here because only a definite few are meant (= *nonnullos*).

29–38. *The Campaign of Mons Graupius*: A.D. 84

The climax of Agricola's operations was to bring the Highland tribes to battle. He achieved this by threatening the thickly populated plain bordering the Moray Firth, the fertile heart of northern Scotland, which the natives could be calculated to defend. The Battle of Mons Graupius has thus close analogies with the Battle of Culloden—and was no less decisive.

Tacitus narrates it as a miniature exercise in historical writing with two carefully balanced speeches by the opposing generals (cc. 30–32; 33–34) prefacing the actual engagement. To intensify the drama he makes use of the technique of *peripeteia*, exploited by Livy, by which the final Roman triumph is preceded by a threat of defeat (c. 35, 4). The turning-point is Agricola's personal intervention. The details of the battle are authentic (cf. notes on c. 35, 2; 37, 2). It is in no sense an imaginary battle modelled on a famous earlier engagement such as Pharsalus (Nutting, *Class. Week.* 23 (1929), 65). See Introd., p. 29, for the heightened language which Tacitus uses to delineate the pathetic aftermath.

29, 1. Initio aestatis. Clearly this is not the summer of c. 28, 1, but the following one (cf. c. 34, 1). As it is difficult to understand *insequentis* from the context, Brotier suggested that *vii* (*septimae*) has been lost after the last syllable (*vit*) of the preceding chapter, while Koestermann supplied *insequentis* after *initio*. But a new year is decisively indicated by *proximo anno* (c. 34, 1) and since this is Agricola's last year of command Tacitus may have dispensed with the formal designation of the year.

ictus . . . amisit. Agricola was apparently accompanied by his wife in Britain, as in Asia (c. 6, 2). This was the second son; on the loss of the first see c. 6, 2. The participle can be taken aoristically with *amisit*; cf. c. 5, 1.

fortium, 'brave'. Notice the contrast between *virorum* and *muliebriter* (for which cf. c. 46, 1; Livy 25. 37, 10). Tacitus does not disparage the fortitude with which they met such losses, but regrets that they flaunted it (*ambitiose* 'ostentatiously': cf. c. 42, 4 *ambitiosa morte*, and note). The reference is quite general and should not be confined to professional Stoics.

per : taken nearly as in c. 4, 2, the sense of a modal (as well as instrumental or causal) abl. being often given by the accus. with this preposition. Cf. c. 37, 4; 38, 1; 40, 4; 44, 5.

inter remedia. Tacitus so describes the practice (*A.* 4. 13, 1) and sentiment (*A.* 4. 8, 2) of Tiberius. It is a commonplace (cf. Plut. *Alex.* 72).

29, 2. incertum, 'vague', expressing the uncertainty of defenders as to the quarter next to be attacked.

expedito: 'without heavy baggage'. Cf. Caes. *B.G.* 5. 2, 4 'ipse cum legionibus expeditis quattuor'.

ex Britannis fortissimos. The additional words *longa pace* show that they were enlisted from southern Britain. *Longa pax* had not yet had the effect of weakening them, as Tacitus observes in c. 11, 4. Conscription is also alluded to in c. 13, 1; 15, 3; 31, 1; 32, 1 and 3. British troops had been recruited since the creation of the province and although exceptionally employed in Britain itself (as here), were usually, as was the Roman practice for security reasons, stationed in other provinces. The first British cohorts to be mentioned are those from Germany which fought at Bedriacum in A.D. 69 (*H.* 1. 70, 2) and it is likely that, as the province extended, Agricola raised several new units. Present knowledge allows the following list of units, probably created before the death of Domitian (A.D. 96), to be drawn up.

Units	*Location*	*First Attested*	*Evidence*
Alae			
I (Flavia) Augusta Britannica M.C.R.	Germania Sup.	A.D. 69	H. 3. 41, 1
Brittonum V.	Mauretania Caes.	*c.* A.D. 110	CIL. viii. 9764
Cohorts			
I Brittanica M.C.R.	Pannonia	A.D. 80	Dipl. 26
I Brittonum M.E.	Pannonia	A.D. 85	Dipl. 31
II Brittonum M.C.R.P.F.	Moesia Sup.	A.D. 100	Dipl. 46
II Brittonum	Mauretania Caes.	A.D. 107	Dipl. 56
III Brittannorum	Raetia	A.D. 107	Dipl. 55
III Brittonum	Moesia Sup.	A.D. 100	Dipl. 46
(IV and V Brittonum	Inferred)		
VI Brittonum	Unknown	Unknown	CIL. ii. 2424
I Flavia Brittonum	Dalmatia		CIL. iii. 2024
II Flavia Brittonum E.	Moesia Inf.	A.D. 99	Dipl. 45

Graupium: this name, of which the version Grampius, printed by Puteolanus, is the fictitious basis for the name of the Grampian mountains, was regarded by W. J. Watson (*Celtic Place Names of Scotland*, p. 55) as more properly being spelled Craupius and derived from Celtic **crup* ('the hill of the hump'; cf. Welsh *crwb*). But the implied vocalism is difficult and K. H. Jackson (in F. T. Wainwright, *The Problem of the Picts*, p. 135) considers the Celtic nature of the name unproven. We have retained the traditional spelling. It is likely to be Knock Hill, for the nature of the campaign and the series of marching camps suggests that it may have been in the vicinity of the Pass of Grange. For recent discussions see O. G. S. Crawford, *The Topography of Roman Scotland*, pp. 130 ff. (Raedykes); Sir David Henderson-Stewart, *Trans. Anc. Mon. Soc.* 8 (1960), pp. 85 ff.

pervenit. Aerial photography has revealed a number of marching camps stretching from Inchtuthil to the Moray Firth, probably on the approximate line of Agricola's march. Three of them (Stracathro, Ythan Wells, Auchinhove) have gateways of the distinctive type attributed to Agricola; most of them, however, are not datable at present, and the 120-acre series is likely to belong to Severus' campaigns in the early third century (see Introd., p. 62, and Fig. 5).

29, 3. pugnae prioris: the battle described in c. 26.

expectantes. The sense of 'awaiting' suits both substantives sufficiently to make it unnecessary to suppose a zeugma.

tandemque docti: cf. c. 12, 2. The thought is conventional (cf. e.g. Thuc. 4. 61, 2).

legationibus et foederibus: probably not a hendiadys; some came in answer to embassies, others after concluding pacts. The tribes will have included the Taezali of Buchan, the Vacomagi of Inverness-shire, the Decantae of Cromarty, the Smertae of Ross, and others listed by Ptolemy (2. 3, 10–12).

29, 4. triginta milia. Ancient writers often overrated the numbers of a barbarian enemy. But there is no need to suppose an error here (though errors are always possible in figures), especially as the context speaks of subsequent additions, and as the victorious auxiliaries numbered only about 13,000 to 14,000 (c. 35, 2; 37, 1; Introd., p. 78). Cumberland had 6,400 foot and 2,400 horse at Culloden, Prince Charles Edward Stuart not much more than 5,000 men. See note on the casualties, estimated at 10,000, in c. 37, 6.

adhuc, 'still further', cf. c. 33, 1; *G.* 10, 1, etc.

cruda ac viridis, 'fresh and green' (not sapless and withered), taken from Virg. *Aen.* 6, 304 'cruda . . . viridisque senectus', which itself expresses the Homeric ὠμογέρων (*Il.* 23, 791). [Guarnieri's palaeographically elegant correction of *E*'s *viris* is only an emendation and, if the marginal *virens* could be proved to have manuscript authority, it should perhaps be preferred (cf. Hor. *Odes* 1. 9, 17). *Viridis* occurs only once elsewhere in Tacitus (*Dial.* 29, 1: met.), *virere* in *H.* 2. 78, 2 (lit.). The Virgilian phrase became something of a tag but the later allusions only comprise *cruda senectus* (Sil. 16, 331; Apul. *Apol.* 53; Symm. *Ep.* 8, 69) and disregard *viridis*. It was conventional to adapt slightly a borrowing from an earlier author (cf. the analogous procedure of Persius: see D. Henss, *Philologus* 99 (1955), 277 f.), which might favour *virens* here but adaptation is already present in the substitution of *-que* for *ac*.]

decora, 'military decorations'. The Romans were perhaps impressed by the fine torques worn by Celtic warriors.

praestans nomine C., 'one excelling . . ., called C.' For such a concise use, answering to Greek uses of *τις*, cf. *A.* 2. 74, 2; 13. 15, 3; 55, 1; *H.* 4. 82, 1, in all of which *nomine* is thus used to introduce foreign names. The coupling of *virtute* and *genere* is a cliché; cf. Hor. *Sat.* 2. 5, 8; Prop. 3. 18, 11; Cic. *Verr.* ii. 5, 180, etc.

Calgacus: otherwise wholly unknown. The suffix has a long first vowel. The word, connected with Ir. *calgach*, means 'swordsman' (K. H. Jackson, in F. T. Wainwright, *The Problem of the Picts*, p. 135).

in hunc modum locutus fertur. The speech is obviously a composition by Tacitus.

30–33. *The Speech of Calgacus*

The speech is a *declamatio*, a rhetorical exercise, and may be compared with similar speeches in Sallust (*Jug.* 10 Micipsa; 14 Adherbal), in Caesar (*B.G.* 7. 77 Critognatus) and in the *Histories* (4. 32 Civilis). In particular it has many points of contact with the letter of Mithridates preserved in Sallust's *Histories*. Tacitus does not attempt to give arguments which Calgacus might actually have used, although he does convey something of a barbarian's boastfulness (cf. note on c. 31, 4 *expugnare*) but contents himself with the traditional Roman criticisms of imperialism such as were voiced in the schools. Calgacus could not have known and would not have denounced the world-wide character of Roman power (c. 30, 4 n.) nor would he have been familiar with the cosmopolitan organization of the Roman army (c. 31, 1). The attack on *avaritia*

(c. 30, 4) and the contrast between freedom and slavery are typically Roman attitudes. Above all, the example drawn from the position of slaves in a household (c. 31, 2 n.) presupposes Roman society. Hence it is not surprising that many of the ideas can be closely paralleled from earlier sources (cf. notes on c. 30, 1 'nullae ultra terrae'; 30, 3 'omne ignotum'; 30, 4 'raptores', 'scrutantur' and 'satiaverit'; 32, 1 'contractum'; 32, 2 'nulla patria'; 32, 3 'auri fulgor').

In composing it Tacitus has conformed to strict rhetorical principles. After a formal opening calling attention to their present plight (c. 30, 1–3: *principium a rebus ipsis*) Calgacus advances arguments drawn from self-respect (c. 30, 4–31, 4: *dignum*) and superiority (c. 32, 1–2: *tutum*) and ends with an emotional exhortation (c. 32, 3–4: *προτροπή*). He employs the techniques of rhetoric (anaphora, e.g. c. 30, 4 'non Oriens, non Occidens'; 31, 2 'cotidie emit, cotidie pascit': chiasmus, e.g. c. 30, 4 'defuere terrae, mare scrutantur': balanced clauses, e.g. 31, 2 and note: etc.) and the clausulae recommended by Cicero for oratory (esp. the cretic trochee, e.g. c. 30, 1 clāssĕ Rōmānā, and the double trochee, e.g. c. 30, 4 cōncŭpīscūnt). His language is rich in echo and reminiscence (see the notes on c. 30, 1; 30, 4; 31, 1; 31, 4).

[The speech is discussed by R. Ullmann, *La Technique des discours dans . . . Tacite*, p. 199; H. Fuchs, *Der geistige Widerstand gegen Rom*, p. 47; G. Walser, *Caesar, das Reich und die fremden Völker*, p. 155. See also R. Reitzenstein, *Nachr. d. Gesellsch. der Wissensch. zu Göttingen*, 1914, pp. 262 ff.; H. Volkmann, *Gymnasium Beihefte*, Heft 4 (1964), pp. 17 ff.]

30, 1. Quotiens, etc. The opening words are inspired by the beginning of Cicero's *pro Marcello* 'diuturni silenti . . . finem hodiernus dies attulit; idemque initium . . . dicendi' and by Sall. *Cat.* 58, 18 'quom vos considero, milites, . . . magna me spes victoriae tenet. Animus aetas virtus vestra me hortantur, praeterea necessitudo'.

causas belli. In c. 15, 5 the Britons make the motives to be 'sibi patriam coniuges parentes, illis (the Romans) avaritiam et luxuriam', and the thought is the same here: their determination to escape oppression will secure them victory.

necessitatem, 'peril': cf. *A.* 1. 67, 1 *necessitatis monet.*

animus est, 'I have confidence', here constructed with accus. and inf. on the analogy of *spes est,* or *confido.*

hodiernum diem consensumque vestrum, forming one idea in thought, 'your union as this day witnessed'. *Hodiernus dies*, only here in Tacitus, is taken from Cicero (see above).

initium libertatis. Similar language is put into the mouth of Caratacus in *A*. 12. 34: 'illam aciem . . . aut reciperandae libertatis aut servitutis aeternae initium fore'. Here the alternative is deferred till later in the speech.

universi coistis. Although now obscured by a smudge, *E* undoubtedly read *coistis* 'you are all united', picking up *consensum vestrum*. [*T*'s *colitis*, preferred by some scholars including Marouzeau (cf. *G*. 16, 1), is a mere misreading by Crullus of *E* whose illegibility may also have been responsible for the omission of the word in *AB*.]

nullae ultra terrae, sc. *sunt*. The idea is repeated in 'terrarum extremos' and 'sed nulla iam ultra gens' (§ 3). The thought is very old; cf. Homer, *Iliad* 15, 734 ff.

securum, 'free from danger': cf. *Dial*. 3, 2; *H*. 1. 1, 4. So used of things, for *tutus*, in Livy (39. 1, 6) and afterwards, but rarely.

classe. Cf. c. 29, 2.

30, 2. pugnae . . . habebant. By a bold personification the battle is put for the combatants: and the thing hoped for (*subsidium*) is coupled with the hope.

eoque in ipsis penetralibus siti. By a flight of rhetoric the speaker is made to say that because they were the noblest race in Britain, Fortune had located them in the innermost sanctuary of the island, the better to preserve them undefiled. Similarly the Semnones claimed to be the *nobilissimi* of the Suebi (*G*. 39, 1). *Penetralia*, the sanctuary of a temple, is also used metaphorically in *Dial*. 12, 1. *Situs* is used of persons (cf. *A*. 12. 10, 1) after Sallust.

servientium : substantival (cf. c. 4, 1), 'of slaves', i.e. of the Gauls, who were within sight of south Britain (c. 10, 2).

oculos quoque a contactu. A similar bold figure is found in *A*. 3. 12, 4 'corpus contrectandum vulgi oculis'. The prep. with abl. seems adapted to the personification.

30, 3. recessus ipse et sinus famae, etc. 'The remote position itself and a land of rumour' (that is, by hendiadys, 'the very remoteness of this land of rumour') 'have protected us to this day, the last people of earth or of freedom'. This difficult passage is to be interpreted in the light of *G.* 29, 3 'sinus imperii et pars provinciae habentur' where *sinus*, often a gulf of the sea, is an enclave or tract of Roman-occupied land bulging into Germany (cf. *A.* 4. 5, 2). *Sinus famae* is a tract or pocket of rumour, an outlying land about which there is no certain knowledge, only rumour. The whole phrase corresponds to 'longinquitas ac secretum ipsum' (c. 31, 3). [Lipsius paraphrased it by 'vix fama noti, in sinu famae conditi', that is, 'the protecting folds of fame' but *sinus* is more easily understood in a territorial sense. De Witt, interpreting *sinus* as a nook or cranny, translated 'the meagreness of information about us' which gives a strained meaning both to *sinus* and to *fama*. Peerlkamp, Ernout, and others take *famae* as dat. with *defendit* (cf. Virg. *Ecl.* 7, 47), 'our remote corner preserves us from being known', but that would require *famam nobis defendit*. The same sense is got by Constans' emendation but no change is required.] *Nos* is emphatic, opposed to *priores pugnae*.

terminus Britanniae : the furthest bounds, i.e. the remotest tract forming the limit, of the island. The remoteness is constantly harped upon by both sides, see below *nulla iam ultra gens* and cf. c. 27, 1; 33, 3 and 6. Interest in the ends of the world was strong at this time (cf. Sen. *Med.* 964 ff.).

atque omne ignotum pro magnifico est ; sed. The sequence of thought is 'hitherto our isolation and obscurity have defended us: now our land lies exposed, and our very obscurity, so far from guarding us, is a positive disadvantage as the unknown is always thought grand (and therefore invites expropriation; the lure of the unknown is irresistible to the insatiable Romans, *raptores orbis*, etc.); for us battle is the only course: we have no longer any refuge (as others had, *iam*), the sea and the Roman fleet are behind us, and in submission there is no hope of mercy'. 'Nulla iam ultra gens, nihil nisi', etc., repeats § 1 'et nullae ultra terrae ac ne mare quidem', etc.; *iam* is opposed to *priores pugnae* (§ 2). For *pro magnifico* cf. *A.* 6. 8, 5 (to be known to Sejanus' freedmen) 'pro magnifico accipiebatur' and *G.* 34, 2. The sentiment is a commonplace (cf., e.g., Caes. *B.C.* 3. 36, 1 'plerumque in novitate fama antecedit'; Livy 28. 44, 3; also c. 25, 3), as is the theme that there is no refuge left for the enemies of Rome (cf. Joseph. *B.J.* 2. 16, 4). [Brueys' transposition, adopted recently by St.-Denis, whereby the clause is made to amplify 'recessus ipse et sinus famae', is in no sense necessary and produces an awkward asyndeton between *patet* and *nulla . . . gens*. Perret transposes the words to follow *ambitiosi* (c. 30, 4) which would obscure the antecedent of *quos* (*Recherches sur le texte de la 'Germanie'*, pp. 95–96).]

infestiores, i.e. *quam haec.*

obsequium. Similarly Mithridates complains that their good behaviour has not resulted in amelioration of their treatment by the Romans (Justin 38. 5, 5).

effugias : potential subj.

30, 4. raptores orbis. So Mithridates is made to call them 'latrones gentium' (Sall. *Epist. Mithr.* 22), and Telesinus 'raptores Italicae libertatis lupos' (Vell. Pat 2. 27, 2, the earliest prose use of *raptor*). Similar is the Scythian jibe to Alexander (Curtius 7. 8, 34 'omnium gentium latro').

mare scrutantur : cf. *G.* 45, 4. The sentiment is rhetorical; cf. Sen. *de Clem.* 1. 3, 5 'sive avarus dominus est, mare lucri causa scrutamur sive ambitiosus . . .'; *de Ben.* 7. 2, 5.

ambitiosi, 'eager for glory'.

satiaverit : best taken as fut. perf.: even when they have conquered the whole world they will not have been satisfied. The commonplace that Roman imperialism wished to embrace East and West is frequent in such diatribes; cf. Joseph. *B.J.* 2. 16, 4; Sall. *Epist. Mithr.* 17 'an ignoras Romanos postquam ad Occidentem pergentibus finem Oceanus fecit, arma huc (sc. in Orientem) convortisse'. It was inherited from the panegyrics of Alexander's world-empire (cf. Virg. *Aen.* 6, 794 ff.).

omnium : best taken with *soli*: cf. *G.* 45, 4.

opes atque inopiam, 'wealth and want', i.e. every acquisition, great and small. Cf. Sall. *Cat.* 11, 3 'avaritia . . . neque copia neque inopia minuitur' and Dio 60, 33, 3.

30, 5. auferre trucidare rapere : used as substantives, 'plunder, murder, rapine'; the first relates to things, the second to men, the third to both. Similar collocations are common in Cicero (e.g. *Phil.* 10, 12 'omnia vastaret diriperet auferret').

falsis nominibus : cf. *H.* 1. 37, 4, to be taken with both *imperium* and *pacem*.

pacem. For the thought cf. 'servitutem falso pacem

vocarent' (*H.* 4. 17, 2), and *A.* 12. 33. The blessings of peace as the accompaniment of empire figured much in Roman literary propaganda (cf. Virg. *Aen.* 6, 852–3; Vell. Pat. 2. 131, 1; Sen. *de Prov.* 4, 14). The Ara Pacis Augustae was voted in 13 B.C. and Pax is regularly represented on Roman coins (as lately as A.D. 80; *B.M. Coins R. Emp.* II. pp. xlvi ff. 98–103), To the Romans stable government, even if achieved by force of arms, was the necessary precondition of peace (H. Fuchs. *Augustin und der antike Friedensgedanke* (Neue Philologische Untersuchungen, 1926), pp. 126 ff.). The epigram itself is an old jibe against the Romans; cf. Curtius 9. 2, 24 'postquam solitudinem in Asia vincendo fecistis'; Pliny, *N.H.* 6, 182.

31, 1. voluit: viewing nature as a lawgiver. The thought is an old philosophical commonplace (cf. Cic. *Off.* 1, 11–13) going back at least to Plato (*Rep.* 9. 574 c).

alibi servituri: used bitterly of the conscription.

coniuges, etc. Cf. *A.* 12. 34, 1 and 14. 31, 1.

nomine amicorum, etc., i.e. by persons professing to be friends. Clearness of construction is sacrificed to conciseness.

ager atque annus, 'the land and yearly produce'. For the use of *annus* (for *annona*) cf. 'expectare annum' in *G.* 14, 3: it occurs also in Lucan, Statius, etc. On the requisitions of corn cf. c. 19, 4, and note. Notice the rhetorically balanced structure of the sentence: there are three clauses of which the last is the longest (a *tricolon auctum*) and the subject in each clause is a pair of nouns (*bona fortunaeque, ager atque annus, corpora ac manus*).

emuniendis (here alone in Tacitus), 'making roads through'. The usual sense of the word is to fortify, but *munire* is similarly used in *ILS* 39 *Tempe munivit* and Livy 21. 37, 2. The force of the compound *e-* is to convey the method by which Roman roads were built on an embankment (*agger*). This distinctive feature was designed to improve the drainage and to give a solid foundation. Agricola's advance will have been accompanied by an enormous programme of road-building utilizing native labour (Dere Street probably belongs to this period) which will have left a lasting impression as much on the Britons as on Tacitus' father-in-law. The leading grievances of subjects are all brought together, conscription, tribute, corn requisition, forced labour.

contumelias, probably verbal abuse rather than physical assault; cf. c. 15, 2.

31, 2. nata servituti: in indignant contrast to the free-born Britons. So Livy speaks of the Asiatic Greeks as 'vilissima genera hominum et servituti nata' (36. 17, 5); Cic. *Prov. Cons.* 10.

semel veneunt, etc., 'are sold once for all and, what is more, are fed by their masters; whereas Britain daily pays for her own slavery (by tribute) and daily feeds it (by corn supply), i.e. feeds its enslavers'. Cf. Dio 62. 3, 3. The logic is sacrificed to rhetorical point.

sicut. Notice the close balance between 'sicut in familia' and 'sic in hoc . . . famulatu'.

recentissimus quisque, 'the last newcomer'. Calgacus is made here to speak as if he knew a Roman household.

novi nos et viles in excidium. A further point is introduced: not only are we, like all new slaves, objects of derision, but so worthless and contemptible in our masters' eyes that they wish only to extirpate us. *Novi* and *viles* are so coupled by Hor. *Ep.* 2. 1, 38.

neque . . . arva. Caledonia had only mountain wastes and pastures. In Caesar's time this was believed to be the general condition of the remoter parts (*B.G.* 5. 14, 2 'interiores plerique frumenta non serunt').

metalla. On the working of mines under the Romans see Appendix 4. The labour was supplied by slaves, hired freemen, prisoners of war, and condemned criminals.

exercendis. On the dat. of purpose cf. c. 23, 1. The verb is used with *agri* and *metalla* and of other kinds of trade or industry. So with *portus* it means working the harbours, i.e. acting as dockers, etc.

31, 3. porro: giving another reason why they should expect annihilation.

secretum, 'our seclusion': cf. c. 25, 2, and for the thought, c. 30, 3. *Tutius,* while they are free; *suspectius,* if they are conquered (as Andresen explains).

sublata spe veniae repeats c. 30, 3.

sumite animum, 'take courage', as in Livy 6. 23, 3 and in Ovid and other poets. In *A.* 14. 44, 1 it means 'to form a plan'; in *H.* 1. 27, 1 'to adapt one's attitude'.

31, 4. Brigantes. In fact the Brigantes took no part in the revolt of Boudicca and, if the text is sound, they must be mentioned here by error. Since they were perhaps the most notorious and largest of the British tribes, the mistake would be natural, and is more likely to be Tacitus' own than one deliberately foisted by Tacitus on Calgacus to characterize barbarian boastfulness and inaccuracy (cf. note on *expugnare* below and c. 32, 3 *municipia*: see G. W. Clarke, *Historia* 14 (1965), 336). Camden proposed to read *Trinobantes,* i.e. Trinovantes (cf. *A*. 14. 31, 2).

coloniam, Camulodunum: cf. c. 16, 1.

castra. Presumably that of the Ninth Legion is meant, but the narrative in *A*. 14. 32, 3 says that the remnant were saved by flying to it. No doubt the speaker is here made to exaggerate.

nisi felicitas . . . vertisset, 'had not success turned to inactivity'; so Livy 3. 64, 1 'victoria in luxuriam vertit'. It seems to be meant that only gross negligence prevented them from annihilating the army of Paulinus; and this, though not stated in the narrative in the *Annals*, is certainly borne out by it.

exuere iugum, as in Livy 35. 17, 8; *H*. 4. 17, 2.

potuere : used as an ordinary indicative ('were able') with *exurere* and *expugnare*, but with *exuere* it means 'had it in their power to', although as events turned out they did not throw off the yoke. In the apodosis of unreal conditional sentences, the indic. of *possum* and other modal verbs is regular, but the perfect is rare.

laturi (*E*) is certainly corrupt. It has been defended either in the intransitive sense ('will lead towards') which is only found of roads and the like (cf. c. 10, 6 *ferre*; but cf. Nep. *Dat*. 4, 8), or by assuming the ellipse of *nos* ('who are about to betake ourselves to . . .': so Spilman, Lenchantin, Büchner, and others). Such an ellipse of the reflexive pronoun can be paralleled in a few formulaic phrases (cf. Livy 28. 22, 13, etc.) and perhaps by *Dial*. 10, 5 where Acidalius read *te ferat*. But *laturi* is impossibly colourless to describe the probable consequences of a desperate battle. The corruption is probably limited to *laturi*. Calgacus is favourably contrasting the prospects of the Caledonians who are *integri* and *indomiti* with the achievements of the Brigantes despite all their disadvantages (having a woman as general, etc.). *Paenitentia* is found as a euphemism for submission (*H*. 4. 37, 2) but it naturally means a change of mind, 'second thoughts'. Pursuing the theme of the difference between the Caledonians (*nos*) and the Brigantes, Calgacus is made to contrast the *libertas* of those who had never submitted to Rome (*integri*) with the *paenitentia* of those who first accepted Roman rule and then, regretting their decision (cf. c. 30, 3 n. *obsequium*), attempted to throw it off (i.e. in the Revolt of Boudicca). All emendations that alter 'in libertatem non in paenitentiam' must be disregarded. Two main lines of approach have been followed: (1) to supply an object for *laturi*, e.g. *nos* (R. Meister), *arma* (Wex; cf. *A.* 4. 48, 3), *rem* (Tucker). Such supplements are either weak (see above) or produce a false sense (the Caledonians were already *arma ferentes*). (2) To alter *laturi* to another future participle, e.g. *ituri* (Breithaupt; cf. c. 32, 4), *laboraturi* (Wex), *vindicaturi* (Halm), *aemulaturi* (Borleffs), *certaturi* (Brotier). The best of these are Gronovius' *periclitaturi* (cf. *G.* 40, 3) or Koch's *bellaturi*, but the force of *in* is difficult; although it often means 'with a view to' in Tacitus (cf. c. 8, 3; Livy 24. 2, 4 'in libertatem pugnaretur'), it would naturally be understood as 'against'. As Muretus and Baehrens saw, we should expect *in libertatem*, etc., to describe a distinctive present quality of the Caledonians rather than some future situation, that is, a present or past rather than a future participle. Muretus' *nati* is perhaps rendered less likely by the proximity of *nata servituti* in c. 31, 2. We have considered *educati*. The Caledonians have been trained for freedom not for revolt. (The *ductus litterarum* has over-influenced restoration of this passage.)

primo . . . congressu, as in Caes. *B.C.* 1. 46, 4; 47, 2; Sall. *Jug.* 74, 3; Livy 4. 33, 1; 10. 1, 9; 38. 17, 6.

seposuerit, 'has kept in reserve': cf. 'in usum proeliorum sepositi' (*G.* 29, 1).

32, 1. An, etc., 'Or do you suppose . . .', i.e. you should take courage, unless you think, etc. Such expostulations are common in speeches of this kind (cf. e.g. Curt. 4. 14, 23).

dissensionibus and **discordiis** are synonyms, as in *Dial.* 40, 4; Cic. *Har. Resp.* 40, etc. On the fact cf. c. 12, 2.

contractum. The dangers of a heterogeneous army were proverbial; cf. Polyb. 1. 67, 4–6; Veget. *de Re Mil.* 3, 4.

nisi si here puts ironically a supposition dismissed as impossible, as in Cic. *Cat.* 2. 6.

pudet dictu is here alone used for *pudendum dictu* (*H*. 2. 61, 1, etc.) or *pudet dicere*. On the foreign auxiliaries in this Roman army see c. 29, 2; Introd., p. 78.

commodent, 'lend', an emendation supported by 'nomen . . . commodavisse' (*A*. 15. 53, 4), 'vires . . . commodando' (Livy 34. 12, 5), etc. Cf. c. 19, 3, note.

adfectu, 'attachment': cf. 'militia sine adfectu' (*H*. 4. 31, 1), a use not found before Ovid. A similar state of feeling among auxiliaries is referred to in *H*. 4. 76, 4.

32, 2. metus ac terror, the two words are synonymous, balancing, and contrasted with *fide et adfectu*. Cf. *Dial*. 5, 5; Cic. *Sest*. 34, etc. *Est* is retained by some with the sense 'exists between them'; but with *est* we should expect *vinclum*, and the corruption of *sunt* to *est* is easy.

infirma vincla caritatis: a bitter understatement, as they are not really bonds of affection at all. The thought is at least as old as Thucydides (3. 12, 1).

timere . . . odisse, a paradoxical variation of the proverbial 'oderint dum metuant'.

victoriae incitamenta, 'incentives to victory'. So the wives and children present are called 'hortamenta victoriae' in *H*. 4. 18, 2. The British women were present in the battle against Paulinus (*A*. 14. 34, 2); the German custom is described in *G*. 7, 2; 8, 1; and that of the Thracians in *A*. 4. 51, 2. The same enumeration of 'coniuges, parentes, patria' is made in the appeal of Civilis (*H*. 5. 17, 2).

aut nulla plerisque patria, 'most of them have no home or an alien home', i.e. are a *colluvies* with no homeland feeling (having forgotten their *patria*) or have a different home from the Romans for whom they fight, like Gauls, Germans, etc., who gave their name to cohorts and *alae*. The chequered origins of the Romans were often used as matter for political abuse (Livy 1. 8, 5). Cf. Sall. *Epist. Mithr*. 17 'convenas olim (Romanos) sine patria, parentibus'; Justin 38. 7, 1.

ignorantia: explained by 'caelum . . . circumspectantis', which perhaps contains a reminiscence of Sall. *Jug*. 72, 2 'circumspectare omnia et omni strepitu pavescere'. A similar argument is advanced by Boudicca in Dio 62, 5.

vinctos, 'spellbound'. So used of panic-stricken or hampered soldiers *A*. 1. 65, 4; *H*. 1. 79, 2.

32, 3. auri fulgor atque argenti. Cf. *A*. 15. 29, 2 'fulgentibus aquilis signisque', and c. 26, 1 'fulsere signa'. The reference is to the gold *aquila* and the silver decorations of the standards (*signa*), not to the soldiers' decorated shields, etc., which were covered in leather cases during active service (Joseph. *B.J.* 3. 5, 5). The disconcerting appearance of an enemy is regularly belittled on such occasions; cf. Livy 10. 39, 11; Justin 11, 13; Curtius 3. 2, 12.

nostras manus, troops who will be on our side. The emphatic position of the verbs in this and the next sentence is noteworthy.

adgnoscent . . . suam causam, 'will see that our cause is their own'.

tam . . . quam : as in c. 2, 3; *H*. 1. 83, 3, etc., with the force of *non minus quam*. *Tamquam* (E^{2m}) is not found in this correlative use in Tacitus.

ultra : beyond the army facing us. *Formido* here of that which can cause fear, as in Sall. *Jug*. 23, 1; 66, 1. Cf. *metus*, *A*. 1. 40, 1.

vacua castella, 'forts drained of their garrisons'. This may well refer to the evacuation of many southern garrisons in order to provide the army of occupation of the North.

coloniae : a rhetorical plural since Camulodunum (Colchester), established in A.D. 49, was the only *colonia* yet founded. Lincoln came later in Domitian's reign (*Arch. Journ.* 103 (1946), 16 ff.), Gloucester under Nerva (*ILS* 2365).

inter, often used with the force of an abl. abs. or causal sentence, 'where subjects are disobedient, and masters tyrannical', 'cum alteri male pareant, alteri iniuste imperent'. So *A*. 1. 50, 4 *inter temulentos* = *cum temulenti essent*.

aegra, 'feeble', 'sickly', opposed to *validus* in *H*. 1. 4, 1; cf. *aegram Italiam* (*A*. 11. 23, 2), etc.

municipia : strictly, non-Roman towns which by natural development and the growth of Roman civilization were considered fit to receive the Roman citizenship and a constitution of the Italian type. The plural is presumably rhetorical. At least Verulamium alone is known as a *municipium* (*A*. 14. 33, 2). Londinium had no such status (*A*. 14. 33, 1) *Discordantia*, 'quarrelling': cf. c. 16, 3.

32, 4. hic dux, hic exercitus : referring to themselves, 'on this side you have a leader and a national army, on that side bondage and all belonging to it'. Calgacus is called *dux* implicitly in c. 29, 4. *Hic* and *ibi* are opposed in *A.* 15. 50, 4, *hic* and *illic* in *A.* 1. 61, 4, *hinc* and *inde* very often, *hinc, hinc* in c. 25, 1.

metalla : used concisely for mine labour, and as a type of servile labour (cf. c. 31, 2).

statim ulcisci, 'here and now', for, though the penalties are not yet inflicted, they are certain to be imposed in the event of defeat.

in hoc campo est, 'depends on this field'. *Est* = *positum est.* Cf. c. 33, 5 *in his omnia.* The same idea is elsewhere put in other words, e.g. 'illos esse campos in quibus' (*H.* 3. 24, 1).

proinde, hortatory, 'accordingly'.

maiores vestros, etc., 'think both of your ancestors and of your descendants'. *Cogitare,* 'to think of', with the acc., as always in Tacitus, is first found in Cic. *Fin.* 5, 2; *Rep.* 3, 47. The concluding thought is a commonplace (Curt. 4. 14, 25).

33, 1. alacres : *alacritas* is the emotion conventionally inspired by such speeches (cf. c. 35, 1; Caes. *B.G.* 1. 41, 1 'hac oratione habita . . . summa alacritas innata est', etc.).

ut . . . moris qualifies the words which follow: cf. c. 11, 1. For *moris* cf. c. 39, 1; 42, 4, etc. The Romans often remarked upon the alarming sound of the Celtic battle-cries (cf. Polyb. 2. 29, 5; Caes. *B.G.* 5. 37, 3; Livy 5. 37, 8, etc.).

dissonis, 'confused', to Roman ears inarticulate. So Sen. *Vita Beat.* 1, 2 'fremitum et clamorem dissonum'.

agmina, etc., 'there were bodies of troops in movement, and flashes of arms as the boldest darted before the ranks'. The omission of a verb like *aspiciebantur* is in Tacitus' manner. The ablative is that of attendant circumstances (= 'audentissimo quoque procurrente'). The rare plural *fulgores* is used of separate flashes of lightning in Cicero and Seneca; cf. Hor. *Odes* 2. 1, 19–20 'iam fulgor armorum fugaces terret equos'. The adjectival *audens* is found in Quint. 12. 10, 23; the superlative in Aul. Gell. 6. 2, 10.

adhuc, 'still further': cf. c. 29, 4.

ita disseruit. Cf. the words used in c. 29, 4. This verbal pattern for introducing a speech is common in Livy; cf., e.g., 25. 38, 1 'adhortandos sibi milites ratus, contione advocata ita disseruit'.

33, 2–34. *The Speech of Agricola*

Agricola may well have made a speech before the battle but we cannot tell whether Tacitus preserves anything of it. The arguments used suit the Roman situation (cf. notes on c. 33, 4 *extrusi*; 33, 5 *commeatuum*) but the plan is artificially rhetorical. After a conventional opening (c. 33, 2 with note: *principium ab auditoribus*), Agricola develops two lines of argument—(1) a decisive battle is their only hope of safety (c. 33, 3–6: *tutum*); (2) victory is assured (c. 34, 1–3: *facile*)—and ends with an exhortation. Moreover much of the substance of the speech is derived from the two speeches made by Scipio and Hannibal before Ticinum (Livy 21. 40–44; cf. notes on c. 33, 2 *septimus*; 33, 5 *superasse* and *commeatuum*; 34, 1 *ignota* and *decora*; 34, 2 *ignavorum*; 34, 3 *non restiterunt*) and it is unlikely that Agricola would have indulged in literary reminiscence on such an occasion. The language itself has a few touches appropriate to a true soldier (cf. notes on c. 33, 2 *commilitones*; 33, 6 *decretum*) but is throughout heavily influenced by the phraseology of Sallust (cf. notes on c. 33, 2 *naturam*; 33, 4 *omnia prona*; 33, 5 *in his omnia*). Like Calgacus, Agricola uses the sophisticated techniques of rhetoric, e.g. anaphora (c. 33, 4 *quando . . . quando*; 34, 1 *vestra decora recensete, vestros oculos interrogate*), and double-trochee clausulae (c. 33, 3 *Britannia et subacta*, etc.).

33, 2. septimus annus est. The manuscript copyists may easily have confused *vii* and *viii* in their exemplar, and the correction is required by the chronology (cf. the parallel confusion in c. 44, 1). Against the supposition that a year has been lost must be set the fact that the sixth year (c. 25, 1) is referred to below as *proximus* (c. 34, 1); and the previous years are accounted for.

This type of opening is conventional; cf. *H.* 1. 29, 2 'sextus dies agitur, commilitones, ex quo . . .'; *A.* 14. 53, 2; Lucan 1, 299. The sentiment is the same as that expressed by Hannibal in Livy 21. 43, 13 'ut viginti annorum militiam . . . taceam, . . . vincentes huc pervenistis'.

commilitones, a familiar and inspiring mode of address; cf. Suet. *Jul.* 67 'nec "milites" sed blandiore nomine "commilitones" appellabat'; Livy 2. 55, 6, with Ogilvie's note.

virtute et auspiciis imperii Romani, fide atque opera nostra. The difficulty of this phrase lies in the word *virtute.* The phrase as a whole is an expansion of the technical formula which stated the legal nature of a command—e.g. 'auspiciis Imp. Domitiani, ductu Cn. Iulii Agricolae' (cf., e.g., *Inscriptions of Roman Tripolitania* (Reynolds and Ward-Perkins), no. 301). In the Republic a general would usually have had the *auspicia* and, therefore, the formula would have been, e.g., 'T. Quinctii ductu et auspicio' (Livy 3. 1, 4) but under the Empire only the *princeps* had the *auspicia* (cf. c. 40, 1 n.). The two parts of the phrase can be considered separately. (1) As Gronovius saw, Tacitus uses the periphrasis *imperii Romani* instead of naming Domitian directly in view of his *damnatio memoriae* (cf. *G.* 29, 2(3) with Anderson's note). The addition of *virtute*, which can be paralleled by the mock-formulaic expression in Plaut. *Epid.* 381, is illuminated by the personification of Virtus Augusti which is represented on coins from Vespasian onwards, particularly those of Domitian after A.D. 83 (*B.M. Coins R. Emp.* II. pp. xci ff.). It signifies the power of the state as symbolized in the emperor. (2) In the second phrase Tacitus combines two ideas, Agricola's leadership and his troops' loyalty (cf. below 'neque me militum neque vos ducis paenituit'; Xen. *Hell.* I. 1, 27, etc.). Hence there is no need to alter *nostra* (= *mea et vestra* 'our combined') to *vestra*. *Opera*, although commonly used of the services of troops (Livy 23. 46, 6, etc.) is also found of a general's responsibility (e.g. Livy 4. 40, 4; Cic. *Cat.* 3, 14, etc.). *Fides* is the regular term for denoting the obedience of troops and Fides Militum, Exercituum, etc., is a common legend on coins (*B.M. Coins R. Emp.* I. pp. ccxxiii ff.). After *nostra*, *vicimus* rather than *vicistis* might be expected but Agricola is made to emphasize the achievement of the troops.

expeditionibus . . . proeliis: to be taken as quasi-local ablatives with *paenituit*.

adversus . . . naturam, 'against Nature herself', i.e. storms (c. 22, 1), marshes, mountains, rivers (§ 4), forests, etc. (§ 5); not merely 'the elements'. Cf. Sall. *Jug.* 75, 2 'omnes asperitates supervadere ac naturam etiam vincere aggreditur'.

neque . . . paenituit. For the shape of the sentence cf. Livy 5. 27, 15 'nec vos fidei nostrae nec nos imperii vestri paenitebit'.

33, 3. non fama . . . tenemus: a rhetorical expression, which might mean 'our hold is not a matter of report or rumour, but of armed occupation'. But the paraphrase which follows shows that the meaning is: 'we are no longer dependent on report or rumour for our knowledge of the remotest tract of Britain, we have discovered it and subdued it'. There seems to be a play on the double meaning of *tenere* (1) 'know' as in *Dial.* 32, 3 (*non teneant* = *ignorent*) and (2) 'hold'. *Fama* is virtually synonymous with *rumor* (*fama* emphasizes the content, *rumor* the sources of the rumour) and the words are paired to balance *castris et armis*.

inventa . . . subacta : rhetorical exaggeration. Cf. c. 10, 4 'incognitas . . . invenit domuitque'. *Inventa*, discovered what was only vaguely known from *fama*. *Subacta* is an optimistic anticipation, assuming both that the issue of the battle will be favourable and that it will complete the conquest; cf. c. 10, 1 *perdomita*. Introd., p. 76.

33, 4. dabitur . . . animus For *dabitur* cf. *daretur pugna* (*A*. 2. 13, 3). *E*'s *animus* is meaningless. The sense is clearly 'when will we come to grips with the enemy?' but it is difficult to restore the text. It is possible that *animus* conceals another noun to balance *hostis*, e.g. *acies* (Rhenanus), but the emphatic position of the plural *veniunt* at the beginning of the following sentence suggests that it repeats a plural form of *venire* at the end of the preceding sentence as Walter proposed, presumably in the future tense to balance *dabitur*. The loss of such a word by haplography is common (cf. Livy 5. 5, 7; 5. 54, 6) and would account for the subsequent corruption to *animus*. We could then restore either *cominus* (Anderson) or *in manus* (Walter) or *ad manus* (Till) *venient*. *In manus venire* means 'to come to close quarters' or, more generally, 'to come to grips with' (*H*. 4. 76, 3 'venturos in manus . . . Civilis et Classici'); *cominus venire* means 'to come to close quarters' (Livy 38. 21, 12; Amm. Marc. 16. 12, 43; 23. 13, 13); *ad manus venire* should mean, more generally, 'to come to blows' but there are no precise parallels for the expression (cf. Cic. *Verr.* ii. 5, 28; *Cluent.* 136; *H*. 2. 88, 3). *In manus venient* is perhaps the best emendation so far proposed.

e latebris suis extrusi, cf. Curt. 4. 14, 4 'ex latebris erutos'. This refers to the work of the fleet in provoking the British to action. Culloden was precipitated in the same way.

vota virtusque in aperto, 'your prayers and prowess have a free field', a concise way of saying 'your wishes are realized and it is open to you to show your valour'. For *in aperto* cf. c. 1, 2; and for the sentiment, cf. Livy 34. 13, 5, 'tempus quod saepe optastis venit quo vobis potestas fieret virtutem vestram ostendendi'; Virg. *Aen.* 10, 279; Lucan 7, 251, etc.

omniaque prona victoribus: repeated in *H.* 3. 64, 1, and taken from Sallust's 'omnia virtuti suae prona esse' (*Jug.* 114, 2). The passage shows also a general reminiscence of *Cat.* 58, 9. For *pronum* cf. c. 1, 2, and for *victoribus* ('if you conquer') *A.* 13. 57, 2.

33, 5. superasse, cf. Livy 21. 43, 9 'tantum itineris per tot montes fluminaque et tot armatas gentes emensos'.

in frontem, to be taken adverbially with *superasse, evasisse,* and *transisse,* 'frontwards': 'It is a fine and glorious thing to have . . . crossed the estuaries frontwards'. The clause as a whole chiastically balances *fugientibus . . . periculosissima* and illustrates Tacitus' tendency to use different grammatical constructions to express parallel ideas.

ita. Peter defended the MSS. *item* from Cic. *Off.* 2, 51; *Tusc.* 5, 9, but the antithesis *ut . . . ita* is constant in Tacitus; and compendia of the two words could very easily be confused, or the preceding syllable *-tem* was repeated by a copyist.

manus et arma: cf. c. 25, 3. *In his omnia,* more fully *in armis omnia sita* (Sall. *Jug.* 51, 4).

commeatuum. The Romans had advanced *expedito exercitu* (c. 29, 2), the enemy had settled down (*insederat*) to await their arrival (ibid.). The same argument is advanced by Hannibal (Livy 21. 44, 7 'nihil usquam nobis relictum est nisi quod armis vindicarimus').

33, 6. decretum est = *statui.* This passive use seems to be colloquial (cf. Plaut. *Aul.* 573; *Bacch.* 516, etc.).

terga tuta, a familiar sentiment, for which cf., e.g., Sall. *Cat.* 58, 16; Livy 27. 13, 7.

proinde, here = *igitur*: cf. *H.* 1. 21, 2, etc. *Et . . . et* should be understood as 'not only . . . but also' rather than 'both . . . and' here (cf. Plaut. *Trin.* 340). The thought, which is commonplace (cf. Nep. *Chab.* 4, 3 'praestare honestam mortem turpi vitae'; Sen. *de Tranqu. An.* 11, 4, etc.), is: 'Fight bravely, for either you will die (which is better than a shameful life) or you will survive gloriously (for bravery gives you the best chance of survival).'

eodem loco sita sunt, i.e. go together.

fuerit : fut. perfect.

naturae, 'the world', 'creation', here synonymous with *terrae* and emphasizing the remoteness of the spot. Usually it includes the ocean: cf. *G.* 45, 1, 'illuc usque . . . natura', and Pliny, *N.H.* 30, 13, 'arte oceanum transgressa et ad naturae inane pervecta'.

34, 1. novae . . . ignota. It was normal for a general to encourage his men by stressing that they were facing a familiar, and therefore contemptible, enemy. So Scipio: 'ne . . . hostem ignoretis, cum iis est pugnandum quos terra marique priore bello vicistis' (Livy 21. 40, 5).

constitisset, 'had stood to face you': cf. c. 35, 3; Livy 1. 1, 7 etc.

decora, 'glorious deeds'. So 'tanti decoris testis' (*A.* 15. 50, 4), 'referre sua decora' (Livy 21, 43, 17), etc.

hi sunt, for the form of expression cf. Livy 27. 13, 3 'nempe iidem sunt hi hostes quos vincendo priorem aestatem absumpsistis'.

proximo anno, see c. 26, 1. This is decisive for the chronology of the campaigns and for the differentiation between *eadem aestate* of c. 28, 1 and *initio aestatis* of c. 29, 1.

furto noctis, 'by surprise at night'. So 'furtum noctis obstare non patiar' (Curt. 4. 13, 9), 'furto unius diei' (Livy 26. 51, 12).

clamore, 'by a mere shout'.

ceterorum . . . fugacissimi. This idiom, common in Greek, is the equivalent of the superlative with *omnium* ('most of all'; cf. Cic. *N.D.* 2, 130) or the comparative with *aliis, ceteris,* etc. ('more than others'). It recurs in Macrob. 7. 8, 9 '(Aegyptus) regionum aliarum calidissima'.

34, 2. quo modo : often used by Tacitus for *quemadmodum,* and thus followed by *sic.*

silvas saltusque, as often in Virgil (e.g. *G.* 3, 40, etc.).

penetrantibus : best taken as a dat., sc. *vobis.* Cf. c. 11, 3.

fortissimum quodque . . . ruere . . . pellebantur. The use of *agmen* and the past tense *pellebantur* show that the simile is not general, as, e.g., in Curt. 3. 8, 10, but intended to refer to the campaigns which they had gone through. *Ruere* is perfect (aorist) indicative, answered by *ceciderunt.* The variation between the aorist and the imperfect contrasts the occasional charges of the bolder animals with the continuous process of withdrawal by the timid ones (A. C. Moorhouse, *C.R.* 61 (1947), 12). Tacitus consistently throughout his writings preferred the form *-ere* (37 times in the *Agr.*) to the more popular form *-erunt* (11 times in the *Agr.*). R. H. Martin (*C.R.* 60 (1946), 17–18) has studied this usage and suggests semantic and rhythmical principles which dictated his choice.

reliquus, etc., 'what is left is a mass of weaklings and cowards'; cf. Hor. *Ep.* 1. 2, 27 *numerus sumus*; *H.* 4. 15, 3; *A.* 14. 27, 3. For the thought cf. Livy 21. 40, 10.

ignavorum et timentium. [Till's emendation of *E*'s *dementium* is more probable than the marginal reading of *E*[2] *et metuentium,* since *metu* occurs in § 3 below and whereas the combination of *ignavus* with a derivative of *metus* would be unique (but cf. Virg. *Aen.* 11, 732), it is frequently paired with *timidus* (cf. Livy 21. 44, 8, etc.).]

34, 3. quod . . . invenistis, 'as to the fact that you have found them'. Such a use of *quod* is perhaps found here alone in Tacitus (in other apparently similar examples the *quod* clause is the subject of a following verb). *Quos quod* stands for *qui, quod eos,* like *quibus si* for *quae, si eis* in *H.* 3. 36, 1.

non restiterunt, etc., 'they have not made a stand, but have been caught'. We have here, perhaps, a reminiscence of Livy 21. 40, 6 'nec nunc illi quia audent sed quia necesse est pugnaturi sunt'.

novissimae res, 'their extremity': cf. *novissimum casum* (*H.* 2. 48, 2; *A.* 12. 33, 1, etc.). The word is coupled with *extremus* in *G.* 24, 2. As regards *E*'s reading in this clause, it is plain that *corpora* and *aciem* can hardly both stand. Ritter's *torpor* for *corpora* is clearly right (cf. Livy 9. 2, 10 'stupor omnium animos ac velut torpor . . . membra tenet'; for the construction cf. Livy 3. 47, 6 'stupor omnes defixit') and the combination of *novissimae res* and *torpor* as subject is no harder than *recessus ipse ac sinus famae* (c. 30, 3). The passage perhaps recalls Livy 22. 53, 6 'quod malum cum torpidos defixisset'.

in his vestigiis, 'on the ground on which they stand', cf. *H.* 4. 60, 2; Livy 22. 49, 4 *mori in vestigio.*

ederetis, final subjunctive, 'were destined to produce': *G.* 29, 1, cf. 'pars . . . imperii fierent'. *Edere victoriam,* only here, is written on the analogy of *edere pugnam* (Livy 4. 20, 10, etc.), *edere caedem* (Livy 5. 13, 11), etc.

transigite, 'have done with' (cf. *G.* 19, 2 *semel transigitur*), an extension of the classical *transigere cum aliquo.*

imponite, etc., 'crown with one great day', analogous to *finem imponere.*

quinquaginta, a stretch of rhetoric: only forty-two years at most (A.D. 43–84) had intervened since the invasion of Claudius. Round numbers are common in such contexts (cf. Livy 1. 29, 6).

adprobate, 'prove'. So with acc. and inf. in *H.* 1. 3, 2.

exercitui: emphatic, to want of spirit in the *soldiers.* The allusion is to Trebellius' and Bolanus' *inertia* (c. 16, 3–4) and to Paulinus' severity (c. 16, 2).

35, 1. alacritas, cf. c. 33, 1 n.

ad arma discursum, cf. Livy 5. 36, 5; 25. 37, 11; 27. 41, 8, etc. As is to be expected Tacitus throughout uses military language that had been developed by earlier historians, especially by Sallust and Livy.

35, 2. instinctos, 'inspired': cf. c. 16, 1. *Ruentes* 'eager to charge'; so used by itself of charging the enemy in c. 37, 3; *H.* 4. 78, 1.

ita disposuit: on the troops present see Introd., p. 78. The 8,000 foot will have consisted of the four Batavian and two Tungrian milliary cohorts mentioned in c. 36, 1 (see note) and two other cohorts, probably raised in Britain itself. The 3,000 horse are distinct from the four *alae* of c. 37, 1 and probably comprised another six *alae quingenariae.*

firmarent, 'should make a strong centre' (cf. c. 14, 3, and note); so Livy 22. 46, 3 'media acie peditibus firmata'; 23. 29, 4, etc.

adfunderentur, 'were spread out on', as in Sen. *N.Q.* 1 *praef.* 10: *circumfundere* or *circumfundi* is so used of horse in *A.* 3. 46, 3, etc.

pro vallo, 'in front of' (cf. *A.* 2. 80, 3, etc.), not 'along' or 'upon' (as in *H.* 1. 36, 3; 2. 26, 2).

victoriae, dative, 'in the event of victory', parallel to *si pellerentur. Decus* and *auxilium* are in apposition to the whole preceding clause.

citra, 'stopping short of (i.e. without) shedding Roman blood'. Cf. c. 1, 3; *citra sanguinem* in Sen. *de Clem.* 1. 25, 1.

That these tactics were the practice of the age is amply shown on Trajan's Column where at least three scenes of battle depict auxiliaries in action and legionaries in reserve (see Plate v). Tacitus is the first literary source to describe the new method (*H.* 5. 16, 1; cf. Claud. *B.G.* 579 ff.). The concept of conserving Roman lives was a commonplace (*A.* 12. 17, 2; *G.* 33, 1).

bellandi defines *decus*: cf. 'effugium . . . prorumpendi' (*A.* 2. 47, 1), and *A.* 3. 63, 3.

pellerentur. The subject (*auxilia*) is supplied from the sense.

35, 3. in speciem, etc., 'for show and to strike terror'. The words are joined in *A.* 2. 6, 2.

in aequo, sc. *esset* or *consisteret.*

conexi velut insurgerent. *E*'s *convexi* is impossible: it cannot mean 'sloping' and could certainly not be applied to people. *Conexi* 'packed together' (over the sloping hill, *per adclive iugum*) is a minimal change. *Insurgerent* 'towered up' is used only here for *exsurgere* which Tacitus employs without qualification in this sense at *H.* 2. 14, 2; 4. 23, 1. It is a stronger word than *exsurgere* and, therefore, is qualified by *velut.*

media campi, 'the intervening space of plain': see on *silvarum profunda* (c. 25, 1). For the whole phrase cf. Livy 24. 21, 9 'omnia strepitu vario complentur', and for *media campi* cf. Livy 4. 18, 3; Ovid, *F.* 3, 219, etc.

covinnarius eques. British charioteers were called *essedarii* by Caesar (*B.G.* 4. 24, 1) and the vehicle *essedum.* The difference in name may be only a tribal variation (*covinnus,* like *essedum,* is a Celtic word = *co*+**vignos* 'wagon'; cf. Ir. *fen*) since the *essedum* was a particularly Belgic machine (Virg. *G.* 3, 204; Σ Lucan 1, 426), but it may denote a different form of construction in that the *essedum* had a special front seat for the driver. The Celtic chariot was a two-wheeled vehicle with a guard round the sides and the front (as depicted on a sarcophagus from Chiusi: see Plate vi), low enough for the fighter to run easily over it along the yoke-pole. It was drawn by two horses sometimes in traces but more commonly yoked to a pole. Fragments of such chariots have been found at Llyn Cerrig, Stanwick, and elsewhere in Britain. See I. M.

Stead, *Antiquity* 39 (1965), 259 ff. The model became popular for a carriage at Rome (Martial 12. 24, 1), perhaps made fashionable by Maecenas (Prop. 2. 1, 76).

35, 4. ne in frontem simul et, etc. In support of the omission of the MSS. *simul* after *ne* Wölfflin notes that *simul . . . simul et* is not Tacitean, and *simul . . . simul* is used with simple cases, as in c. 25, 1; 36, 1; 41, 4.

diductis, etc., 'extending his line'. Cf. Frontin. *Strat.* 2. 3, 12.

porrectior, 'too thin'. A simple comparative would here be a mere truism.

promptior : often with *in* or *ad*. With *firmus* the construction changes to a simple case (probably abl., for the usual *adversus*, rather than dat., since the dat. and gerund(ive) after *firmus* has a quite different force from that required here). Such variations are common: cf. c. 22, 4, etc. The words describe Agricola's general character: 'disposed to optimism, and resolute in face of difficulties'.

pedes ante vexilla. Similar examples were set by Caesar (*B.G.* 1. 25, 1) and Catiline (Sall. *Cat.* 59, 1); see c. 18, 2, note. The *vexilla* are the flags of the auxiliaries. Strictly the term denotes the regimental standards (as they probably were) of *alae*, the troop standards of the mounted section of certain infantry regiments, *cohortes equitatae*, and the ensigns of all detachments; but the word is sometimes loosely used by Tacitus.

36, 1. gladiis . . . excutere, 'parry with their swords or keep off with their shields'. Cf. Livy 38. 21, 3, 'scuta, ut missilia . . . vitarent', and Veget. 1, 4, 'obliquis ictibus tela deflectere'. It seems best to take *constantia* and *arte* as modal ablatives, *gladiis* and *caetris* as instrumental, to which the infinitives *vitare* and *excutere* answer chiastically. The weapons may be compared to the Highland targe and claymore.

[The meaning of *excutere* is difficult. In this context it should naturally mean 'to shake off' (of spears that had hit the shields; cf. Virg. *Aen.* 10, 777 (*hasta*) *volans clipeo est excussa*) but this would leave no verb to describe the defensive effect of the swords. We cannot find any precise parallel for *excutere* in the sense 'to parry' but we have preferred to assume this meaning rather than suppose that Tacitus has chosen two verbs to be understood only with *caetris*. Andresen and others take *gladiis* and *caetris* as ablatives of quality, like *legionariis armis* in *A*. 3. 43, 2.]

Caetra, originally a Spanish word, denotes a small leather-covered shield, not necessarily of circular shape (Herodian 3. 14, 8). They were smaller than the Roman legionary or auxiliary shield. British *caetrae* seem to have been rectangular in shape (*c*. 60 cm. by 35 cm.). A finely preserved example, of alderwood with an oak handle and leather facing, was found at Clonoura, Co. Tipperary (*Journ. Royal Soc. Ant. of Ireland*, 1962, p. 152 and Plate xvii) and they are depicted on the arms of defeated Caledonii on a monument of Antonine date from Bridgeness (see Plate VII).

quattuor Batavorum cohortes. On the Batavian and Tungrian cohorts in Agricola's army see Introd., p. 78. The Batavians lived in the island formed by the bifurcation of the lower Rhine (see *G*. 29, etc.); the Tungri were a tribe settled in the district of Tongres near Liège (cf. *G*. 2, 3). The centre of Agricola's line was composed of eight cohorts (c. 35, 2 and note) and it may be presumed that the other two cohorts, which Agricola held back, were recently recruited British troops about whose steadiness against their compatriots Agricola may have had doubts (E. Birley, *Roman Britain and the Roman Army*, pp. 21–22).

rem ad mucrones ac manus. *Mucrones* is used for the more usual *gladios* to fix attention on the distinction between the Roman and the British sword.

inhabile, 'awkward'.

parva . . . gerentibus. It is just possible, perhaps, that both explanations (*parva . . . gerentibus* and *nam . . . tolerabant*) are genuine, as they do not altogether repeat each other. The smallness of their shields and great size of their swords were disadvantages, and the pointlessness of the latter an additional disadvantage at close quarters. But the first clause is open to grave suspicion, as it repeats what has been already mentioned ('brevibus caetris et ingentibus gladiis') and takes *sine mucrone* too far away from *ad mucrones*. Livy speaks (22. 46, 5) of the Gaulish swords as 'praelongi et sine mucronibus', and contrasts them with the Spanish, and they are similarly described by Polybius (2. 33, 3; 3. 114, 3, with Walbank's notes). This iron sword, 90 cm. in length compared with the Roman gladius of 55–60 cm., too unwieldy for thrusting and therefore made without a point, is well known from the middle La Tène period (La Tène II/III). It is less common in Britain (Piggott, *Proc. Prehist. Soc.* 16 (1950), 1–28: La Tène III = Piggott's Group V), where British swords were already influenced by Roman design, but well preserved examples, with a blade-length of 80 cm. and with a rounded tip, have been found in the River Witham (now in Lincoln Museum) and from Gelliniog-wen (now in the National Museum of Wales). No swords of this date have yet been recovered from Caledonian country but being remote from Roman influence they are likely to have retained the La Tène dimensions.

complexum armorum, 'a grapple', crossing swords hand to hand. The expression occurs only here, but *complexus* is often used of a close-locked struggle (Curt. 9. 7, 22; Pliny, *N.H.* 9, 91; Sil. 14, 552; [Quint.] *Decl.* 4, 22). In battle generally the Britons seem to have relied on their greater agility and rapidity of movement as against Roman soldiers (see Caes. *B.G.* 5. 16).

in arto. This correction of *in aperto* is required by sense and context, and a scribe might easily confound *arto* with *apto.* Cf. Livy 28. 33, 9, and the description of the Germans in *A.* 2. 21, 1.

tolerabant: predicated of ships in *A.* 2. 6, 2, as *pati* of the sea in *H.* 5. 6, 2. Here the swords are boldly personified, or the swordsmen rather than the swords are thought of.

36, 2. miscere ictus, 'to rain blows indiscriminately', cf. *A.* 2. 14, 3 *deserent ictus*; Virg. *Aen.* 12, 720 *vulnera miscent*; 2, 329 *incendia miscet*; Val. Fl. 6, 631. For another use of *miscere* cf. c. 15, 2 and note.

ferire umbonibus, as in Livy 5. 47, 4.

fodere, the regular word for stabbing (Livy 8. 10, 6; Curt. 4. 15, 31; *A.* 2. 21, 1, etc.). *Foedare* (*E*) would imply that Agricola, like Caesar at Pharsalus (Florus 2, 13) ordered his soldiers to disfigure the handsome faces of the British (cf. *H.* 3. 77, 3). Till, Forni, and Koestermann retain *foedare* in the more neutral sense of 'to wound' but the passages which are quoted to support this meaning (Plaut. *Amph.* 244 ff.; Virg. *Aen.* 3, 241; 2, 286, etc.) do not justify it. The same corruption occurs in a manuscript at Livy 8. 10, 6 (*foedantes* for *fodientes*; see Drakenborch's note).

stratis, as in *H.* 2. 43, 1, etc.

adstiterant: often in the military sense of taking position.

proximos quosque. The superlative with *quisque* or *quaeque* in Republican Latin is usually found only in the singular (but cf. Cic. *de Am.* 34; *Off.* 2, 75) but the plural becomes increasingly common later (cf., e.g., Livy 1. 9, 8; *A.* 14. 31, 1, etc.).

semineces, as in Livy 23. 15, 8; 29. 2, 15; Ovid and Virgil. The point is that the Roman attack was so effective that in their eagerness to clinch victory they pressed on leaving many of the British unscathed or only wounded.

36, 3. interim equitum turmae. With *E*'s text it would be necessary to understand the *equites* to be Caledonians and *hostes* in the next sentence to be the Romans. But no Caledonian horse appear to be present (if indeed any existed, see Introd., p. 79), and *turmae*, rarely used of other than Roman cavalry, apparently refers here to those on the wings (c. 35, 2), who must have repelled the chariots in the plain (between the opposing forces) before the infantry could close, and presumably had been further engaged with them while the battle was going on, though Tacitus does not trouble about such a detail. [To make the words *fugere covinnarii* a parenthesis (Persson) would be very abrupt. Urlichs' suggestion, that the *enim* which the MSS. have before *pugnae* (below) has dropped out here and was wrongly reinserted there by a later corrector, is attractive. Most editors acceps Wex's alternative suggestion that *ut* 'when' should be supplemented before *fugere* but this would seem to require a previous mention of the *covinnarii*. Cf. the displacement of *nam* in c. 37, 4.]

recentem terrorem: not 'fresh terror', but 'terror at their recent coming'. The same phrase in *A.* 14. 23, 1 means terror recently struck, the effect of which is still fresh (cf. Cic. *Tusc.* 3, 75). Here it appears to mean the terror caused for the moment by the attack of the cavalry before they became wedged, rather than that which had just been caused by the repulse of the chariots.

haerebant: so with abl. in *A.* 1. 65, 4. The enemy did not give way as they expected.

minimeque equestris ea [enim] pugnae facies. With the minor correction of *equestres* to *equestris* and the deletion of *enim* which cannot be construed and may be interpolated from above (see note), *E*'s text makes good sense. The situation is as follows: after a preliminary barrage of long-distance fire (c. 36, 1), the Roman cavalry cleared the *covinnarii* from the intervening field and allowed the six cohorts of foot to close with the Caledonians and to press them back up the hill. At this juncture the Roman cavalry returned to join the infantry attack on the slopes held by the Caledonians but became wedged in an interlocking mass of men. As a result it ceased to look like a cavalry action which is essentially an affair of movement and fluidity and became a confused and stationary mêlée. So Livy 22. 47, 1 'minime equestris more pugnae', of the situation at Cannae when the cavalry were restricted by the terrain and were hedged in by the infantry; Sall. *Jug.* 59, 3 'non uti equestri proelio solet', describing a scrimmage of cavalry and infantry. For *pugnae facies* cf. *H.* 2. 42, 2. The emendation *aequa nostris*, adopted by Anderson and others, is unnecessary. *Enim* is often interpolated, as, e.g., by manuscripts at Livy 1. 50, 4; 53, 7, etc.

aegre in gradu stantes. *E*'s *aegradiu aut stante* is clearly corrupt. A subject must be provided for *impellerentur* which should refer to the Roman infantry who were being pushed forwards by the weight of the horses in their rear. Most editors adopt *aegre clivo instantes* (Triller) 'those who were with difficulty pressing up the hill' or *aegre clivo adstantes* (Andresen) 'those who were with difficulty maintaining their position on the hill', but Tacitus does not use *instare* in this sense (cf. Virg. *Aen.* 11, 529) and *adstare* would require *in clivo* and is improbable after *in aequo adstiterant* in § 2. Lipsius conjectured *in gradu stantes* (cf. Ovid, *M.* 9, 43; Sen. *Const. Sap.* 2. 16, 2,) but *E*'s reading is best accounted for by *aegre in gradu stantes* (haplography would reduce it to *aegradu stantes*). The Roman infantry found it difficult to maintain their ground against the Caledonians who had the advantage of the slope; at the same time they were pushed on from behind, so that they found themselves sandwiched between two opposing forces. [Other interpretations such as that by L. A. Mackay, *Studies in Honour of G. Norwood* (Phoenix, Suppl. vol. i), pp. 224–8, are too fantastic to merit consideration.]

exterriti, etc.: taken from Sall. *H.* fr. 1. 139 M. 'equi sine rectoribus exterriti aut saucii consternantur'. They are loose horses from the Caledonian chariots.

transversos aut obvios, 'in flank or front', of the Romans. The next words show that, though the Romans are called *vincentes,* their progress was very difficult, and that the British reserve was thereby induced to advance.

37, 1. Britanni : those in the rear (c. 35, 3).

vacui, 'unengaged', explained by *pugnae expertes*: in *H*. 4. 17, 5 it is opposed to *occupati*. Cf. also Sall. *H*. fr. 2. 35 M. 'at Sertorius vacuus hieme copias augere'.

vincentium, 'the conquering side' (not yet victorious); cf. *H*. 4. 78, 2, etc.

coeperant, i.e. they had begun to do so and would have done so, had not . . . . See on c. 13, 2.

subita belli, 'emergencies'. So in Livy 6. 32, 5, etc.

37, 2. consilium : that of attacking in the rear. This cavalry attack in the rear was the turning-point in the Roman victory. It is commemorated by a rare type of reverse on the coins issued to commemorate Domitian's imperatorial salutation on this occasion. See note on c. 39, 1 and Plate iv.

aversam : on the opposite side, i.e. in the rear.

tum vero, etc. The description is evidently imitated from Sall. *Jug*. 101, 11 'tum spectaculum horribile in campis patentibus: sequi fugere, occidi capi; equi atque viri adflicti, ac multi vulneribus acceptis neque fugere posse neque quietem pati, niti modo ac statim concidere; postremo omnia, qua visus erat, constrata telis armis cadaveribus, et inter ea humus infecta sanguine'.

grande . . . spectaculum. Tacitus' interest is throughout psychological rather than military. Like Sallust (cf. *spectaculum horribile* quoted above) and Livy, he exploits the emotional potentialities of a battle-description.

37, 3. prout cuique ingenium, 'as each was inclined' (to flee or face death), as in Sall. *Jug*. 93, 7; Livy 8. 21, 1. *Hostium* depends both on *catervae* and on *quidam*, which are contrasted, like *armatorum* and *inermes*.

terga praestare : for the familiar *dare, praebere*. So Juvenal 15, 75; cf. also Lucan 5, 770 ff. *Praestare* was much used as a choice synonym for *praebere* (Löfstedt, *Peregrinatio*, p. 204). *Ruere* 'charge' (cf. c. 34, 2). The sense of *contra* is here implied.

aliquando, etc., suggested by Virg. *Aen*. 2, 367 'quondam etiam victis redit in praecordia virtus'. These are distinguished from those who flung away their lives.

37, 4. nam postquam, etc. After *adpropinquaverunt E* has *nam*. Andresen's transference of the conjunction to the head of the sentence seems the best correction. *Nam* can hardly be retained by attaching the *postquam* clause to the preceding sentence, as Leuze suggested.

adpropinquaverunt: the subject is *Britanni*, supplied from *victis*.

gnari. This correction of *ignari* is supported by *H*. 2. 13, 1; 85, 2; 5. 6, 4. The *i* probably arose from the preceding *m*, cf. c. 19, 4. Tacitus seems to have had in mind Livy 22. 31, 4 'cum a frequentibus palantes ab locorum gnaris ignari circumvenirentur'.

frequens, 'busy': cf. *A*. 13. 35, 4, etc.

indaginis modo, more commonly *velut indagine*, as in *A*. 13. 42, 4. *Indago* is the process of hedging round the cover of wild animals by toils or by a line or circle of men to prevent their escape and then rounding them up (or the means by which they are hedged round): cf. Virg. *Aen*. 4, 121 'saltusque indagine cingunt'. The comparison is used by Hirtius (Caes.), *B.G*. 8. 18, 1; Livy 7. 37, 14; Lucan 6, 42; Sil. 13, 141; Stat. *Theb*. 12, 451; Florus 4. 12, 48. Here the light infantry 'forming as it were a hunters' cordon' are aided in their drive by dismounted horsemen for the thicker parts of the forests and mounted men for the thinner.

et, sicubi artiora . . . persultare. The sentence as a whole is made a little obscure by the straining after conciseness, but it is not necessary to suppose a zeugma (and supply with *cohortes* a verb of motion like *progredi*), since *persultare* has the sense of 'scour' in *H*. 3. 49, 1 and *A*. 11. 9, 1, and does not appear to be confined to horsemen. 'He ordered strong light-armed infantry, like a cordon, to scour the woods and where they were thicker dismounted cavalry and where thinner mounted cavalry to do the same.' For *artior* 'thicker' (of woods) cf. Caes. *B.G*. 7. 18, 3.

vulnus: so used metaphorically in c. 29, 1; 45, 4. *Accipere vulnus* is common in this sense in Justin (cf. 1. 8, 10; 2. 11, 19; 42. 4, 10) and may have been suggested to Tacitus here by its literal use in Sall. *Jug*. 101, 11 (quoted in § 2 above).

37, 5. rursus, i.e. after their check.

agminibus, modal abl.: cf. *A*. 4, 51, 1 'catervis decurrentes'.

rari, adversative asyndeton for *sed rari*: see note on c. 22, 4.

vitabundi in vicem, 'avoiding each other': see on c. 6, 1.

avia petiere, cf. c. 19, 4 'remota et avia'. This use of *avia* 'pathless areas' is first found in Virg. *Aen.* 2, 736 (cf. *A.* 15. 11, 2), *avia petere* in Sil. 4, 177 (cf. *A.* 2. 68, 1).

nox et satietas : so *nox* is coupled with *laetitia* (*H.* 4. 14, 2), with *lascivia* (*A.* 13. 15, 2), but here there is not a hendiadys; the two motives are distinct as in *A.* 1. 68, 5 'donec ira et dies permansit'.

37, 6. ad decem milia. This is no doubt a mere guess, but looks moderate as compared with such guesses elsewhere: cf. *A.* 14. 37, 2. The auxiliaries, however, who won the battle numbered only from 13,000 to 14,000.

nostrorum . . . cecidere. As a general rule, Tacitus omits the number of Romans slain, and appears to have professed to follow Sallust in doing so (Oros. 7. 10, 4). The only exceptions besides this passage are found in *A.* 14. 37, 2; *H.* 2. 17, 2.

quis (for *quibus*) occurs only here in the minor works. It is more common in the *Hist.* (23 *quis*: 71 *quibus*), outnumbers *quibus* in the first twelve books of the *Annals* (54:45) but declines sharply in the last four (7:50).

Aulus Atticus : the only subordinate officer of Agricola mentioned in this treatise. Atticus was perhaps a family friend of Agricola and Tacitus. The *praenomen* Aulus occurs among the Cornelii and oftener among the Iulii. Several Iulii Attici are known, and Columella tells us that one of them wrote on viticulture, 'cuius velut discipulus Iulius Graecinus (Agricola's father, c. 4, 1) duo volumina similium praeceptorum de vineis', etc. (1. 1, 14). Cf. Introd., p. 2. This is not the Domitius Atticus known at Newstead (J. Curle, *A Roman frontier post and its people* (1911), p. 174, pl. xxxi) since the Domitii do not carry the praenomen Aulus.

[**iuvenili.** Livy uses both the *-ilis* and the *-alis* form of the adj. but *iuvenalis* is chiefly linked with *ludus* (1. 57, 11; 5. 22, 5, etc.) perhaps under the influence of the games known as Juvenalia. Tacitus elsewhere has the form *-alis* only of those games themselves (*A.* 15. 23, 1; 16. 21, 1, etc.) and Guarnieri's correction should be accepted.]

ferocia equi : the army mounts were high-mettled stallions, as can be seen on Trajan's Column (Taf. LXXI, ed. Cichorius).

38. From the military narrative Tacitus passes to an evocative vision of the desolate aftermath of the battle. It should be compared with the similar picture of the field of Bedriacum (*H.* 2. 70) or with the eerie scene of Varus' disaster (*A.* 1. 61). The confused disorder is conveyed by the short, abrupt sentences (see note on § 1 *trahere*) and the pathos is suggested by a heightened use of language (see notes on § 2 *secreti, fumantia, incerta, spargi*).

38, 1. gaudio praedaque : combination of abstract and concrete, as in c. 25, 1 *copiis et laetitia*, etc.

palantes, not 'wandering about' (*vagi*) but 'dispersing', as always in Tacitus.

trahere, etc.: the accumulation of ten historical infinitives is remarkable. In Sall. *Jug.* 66, 1 there are eleven.

per iram : cf. c. 29, 1, and note.

miscere . . . consilia aliqua, 'take counsel of some sort together'. Cf. *H.* 2. 7, 2 *mixtis consiliis. Aliqua* (needlessly taken by some to be an interpolation) seems contemptuous. *Consilia* is again supplied with *separare* (= *separatim capere*), i.e. then each took thought for himself.

pignorum : without such a genit. as *amoris*, as in poets (Propertius and Ovid) and Livy. Cf. *G.* 7, 2, etc.

tamquam misererentur, 'as if in pity', to prevent their captivity. On the feeling of Germans in this respect cf. *G.* 8, 1. For the thought cf. Sen. *de Ira* 1. 16, 3 'interim optimum misericordiae genus est occidere'. It is not necessary to suppose the pity to be a pretence.

38, 2. faciem : cf. c. 36, 3. *Aperire faciem* is a new phrase: it is perhaps modelled on Livy's *lux aperuit bellum* (3. 15, 9; cf. 27. 2, 10). Cf. *aciem dies aperuit* (*H.* 4. 29, 3) and c. 22, 1.

vastum . . . silentium, 'the silence of desolation' (so in *A.* 4. 50, 4; *H.* 3. 13, 2) from Livy 10. 34, 6.

secreti colles, 'lonely hills' (because deserted by their inhabitants), a sense implicit in the ordinary meaning of *secretus*, 'secluded', 'retired', i.e. remote from human society (cf. *G.* 40, 4 *secreto lacu*). Ovid has *secretos montes* (*M.* 11, 765).

fumantia . . . tecta, as in Ovid, *M.* 13, 421; cf. Livy 3. 68, 2.

incerta fugae vestigia : from Lucan 8, 4.

spargi, 'spread over a wider area'. Cicero writes *bellum dispersum* (*de Leg. Man.* 35) but the simple verb is first used in this phrase by Lucan (2, 682; 3, 64) after the model of *spargam arma* in Virg. *Aen.* 7, 551.

Borestorum : a wholly unknown people, but they presumably lay on the sea-coast and since the retreat to winter quarters had not yet begun, it would not be unreasonable to seek the Boresti on the Moray coast, perhaps in Forres. The name Forres cannot, however, be connected, since it is the Gaelic *Farais*, a loan-word from the French.

deducit, that is from the high ground to the coast where he linked up with the fleet.

38, 3. circumvehi. This implies a circumnavigation of Britain (in the reverse direction to the Usipi) and therefore a return to a southern base for winter.

praecipit : so with infin. for *ut* in c. 46, 3, a construction common with other such verbs. On the voyage cf. c. 10, 4.

vires, 'forces'. With these a landing was made on the Orkneys (c. 10, 4).

peditem atque equites. Tacitus in combining these words never elsewhere varies the number, except in adversative clauses (as *G.* 6, 1) or different syntactical relations (as *A.* 13. 40, 2).

novarum gentium, presumably between Moray and Aberdeen.

in hibernis, the ordinary quarters of the troops on the Forth–Clyde isthmus and perhaps also further south, e.g. at Trimontium (Newstead on the Tweed), etc. There is no evidence of permanent structures north of the isthmus at this date. The legionary fortress of Inchtuthil was still under construction when it was abandoned *c.* A.D. 87, and the legions, as opposed to the auxiliaries, may well have withdrawn to York and Chester. Cf. Introd., pp. 67ff.

38, 4. secunda, qualifies both the concrete *tempestate* and the abstract *fama*.

Trucculensem. Both the name and the locality are unknown and the text is probably corrupt. If it was a southern port (see note on *proximo* below) *Rutupiensem* or *Rutupensem*, as first suggested by Lipsius, would suit well. It is an easy change palaeographically. Richborough was the base of the *classis*

Britannica at this period (c. 25, 1 n.), *Profecta* is to be understood with *unde*, the sense being *quo, inde profecta . . ., redierat*. Cf. *unde* taken with a participle only, in *A*. 15. 44, 1, etc.: there is a somewhat similar brachylogy in *H*. 4. 29, 2 'unde clamor acciderat, circumagere corpora'. For *praelegere* 'to skirt' cf. *A*. 2. 79, 1; 6. 1, 1. The simple verb *lecto* (evidently a conjecture) in the margin of *E* is not used in this sense by Tacitus. *E*'s *omnis* (sc. *classis*), emphasizing that the whole fleet returned intact (cf. Caes. *B.G*. 4. 36, 3; 5. 23, 6: in Caesar, as in Tacitus, the season is the autumn when losses at sea were normally expected so that the return of the whole fleet was a circumstance to be remarked upon), is to be preferred to the correction *omni* (with *latere*) which gives a weaker sense and an unparalleled word-order. *Proximo* should mean 'adjacent' (cf. *A*. 4. 5, 1 'Italiam . . . proximumque Galliae litus' but it is difficult to see which shore of Britain is meant or what it is adjacent to. It can hardly be the shore adjacent to the army's camp on the Moray Firth. It might be the shore adjacent or nearest to Rome, that is the south coast, which the Romans thought was parallel and close to the French coast (cf. c. 11, 2: see Fig. 1): if so, it would make *Rutupiensem* probable. For a different view see A. R. Burn, *C.R*. 18 (1968), 316; *Tacitus* (ed. T. A. Dorey), 59–60.

tenuit, 'reached', a nautical term common in Livy.

unde . . . praelecto . . . redierat. The voyage must be supposed to have taken place between the date of the command issued (§ 3) and the winter, i.e. during the indefinite time occupied by his *lentum iter*. The season of the year is confirmed by *hiems adpetebat* (c. 10, 4), and the time required need not have been long. The descent on the Orkneys and sighting of Thule (c. 10, 4) would not entail a great divergence from the direct route. The total distance from Moray to Richborough might be as much as 1,500 miles but with favourable weather that could have been accomplished in a month.

39–42. *The Recall and Last Years of Agricola*

Agricola was recalled in the winter of A.D. 84–85. He died nine years later on 23 August A.D. 93 without having been employed again in the public service. The circumstances of his last years are clothed by Tacitus with such a cloak of innuendo that grave doubt has been cast upon the veracity of these chapters. It is important to distinguish the actual facts from the surrounding tissue of rumours and inferences. He says that Domitian rewarded Agricola's success but at the same time tried to belittle it (c. 40, 1; 40, 3): this is confirmed by the character of the imperatorial salutation (see note on c. 40, 1 *triumphalia*). He says that Agricola was forced to decline the province of Asia and was refused the *salarium*: there is no reason to disbelieve this (see note on c. 42, 2). For the rest he does not commit himself to the truth of the story about the offer of the province of Syria: it was merely a topical belief (c. 40, 2 *credidere plerique*). He does not say that Agricola should have been appointed to Pannonia or Dacia: it was a popular agitation (c. 41, 3–4 'ore vulgi . . . sermonibus aures verberatas'). He does not allege that Agricola was poisoned: he only attests a rumour to that effect (c. 43, 2 *constans rumor*) which in the climate of the year A.D. 93 was natural enough. Such gossip and rumours may well have circulated: we cannot dispose of them but we can see how Tacitus has used them to blacken Domitian. Discreditable motives are throughout attributed to Domitian (cf. c. 39, 1–3; 40, 2; 40, 3; 41, 1; 41, 3–4; 43, 1–2). It is, for instance, implied that jealousy inspired Domitian to recall Agricola and not to employ him further. In fact Agricola had already served for an unprecedented time in Britain and could not in the normal course of events have anticipated another command. Nor would his experience as a specialist in British affairs have suited him for a Danube command. The decision to limit the Roman commitment in Britain was sensible strategy when the Danubian frontier was a source of anxiety and when Britain was likely to involve a further drain on manpower. We do not know how fair Tacitus' estimate of Domitian's motives in general was. Tacitus saw Domitian as a tyrant and his view of human nature led him to assume that Domitian must have reacted and behaved like a tyrant. Jealousy of a successful general (c. 39, 2), patient accumulation of grievances (c. 39, 2), the façade of success (c. 39, 1) are characteristic of tyrants and are, therefore, assumed for Domitian too. But these are not facts of the same kind as the objective details of Agricola's life. There may be doubts about Tacitus' picture of Domitian but not about his biography of Agricola.

39, 1. hunc. Cf. c. 18, 1 and note for this use of *hic* to introduce a new section.

epistulis. Probably only one dispatch concerning each campaign was sent, at its conclusion, though there will have been a regular series of dispatches on other business throughout the year; but Tacitus frequently uses this plural (as *litterae* is always used) of a single letter. For *iactantia* cf. note on c. 25, 1.

auctum is used in the sense of 'exaggerated' with *cuncta* (*A.* 2. 82, 1) and other words implying statements; so here with *rerum cursum* in the pregnant sense of 'the news of this course of events'. The MSS. *actum* (cf. note on c. 19, 4) in this context could only mean 'performed'.

ut erat Domitiano moris: for 'Domitianus, ut ei moris erat', an attraction apparently due to straining after conciseness. For *moris est* cf. c. 33, 1; c. 42, 4. *Ut erat Domitianus* (text of *E*) is probably a deliberate simplification of a difficult piece of syntax.

fronte, 'outwardly'. Cf. *H.* 2. 65, 1; the antithesis is very common (cf., e.g., Caecil. *com.* 79 'fronte hilaro, corde tristi').

inerat conscientia, 'he felt conscious'. So *H.* 4. 41, 1 'quis flagitii conscientia inerat'.

derisui, first found in Phaedr. 1. 11, 2, occurs nowhere else in Tacitus. Caesar uses *irrisui*, Livy *risui*.

falsum . . . triumphum. Domitian triumphed twice for successes in Germany (cf. Suet. *Dom.* 6). The first occasion, which is that here mentioned and came late in the year, the *congiarium* being paid early in A.D. 84 (*Inscr. It.* xiii, 1, p. 102), followed the expedition against the Chatti in A.D. 83, when the frontier was advanced and secured in the Taunus district. The war of A.D. 83 was celebrated by the conferment on Domitian of the title *Germanicus* early in the following year (Kraay, *Am. Num. Soc., Museum Notes IX*, p. 112). That the war was a sham (cf. *G.* 37, 5) is maintained by other writers hostile to Domitian, Dio 67. 4, 1 (μηδ' ἑορακώς που πόλεμον), and Pliny, *Pan.* 16, 3 ('mimicos currus, falsae simulacra victoriae'); but there is no doubt about the substantial results attained, as excavation has shown. Frontinus, who may have served in the expedition, says: 'victis hostibus cognomen Germanici meruit' and speaks of Domitian's justice to the Germans (*Strat.* 2. 11, 7), and again (2. 3, 23) of his directing a battle. He also refers to the newly instituted system of frontier defence (see note on c. 41, 2 *limite imperii*). Domitian's ludicrous exploitation of his success (he regularly appeared in the Senate in triumphal dress and renamed September Germanicus) would have been enough to nauseate even unprejudiced judges.

per commercia, 'commercial transactions': cf. c. 28, 3; *G.* 24, 2.

quorum, etc.: a similar story had been told of Gaius (Suet. *Cal.* 47). Cf. c. 4, 1; 13, 2; Introd., p. 20 note. [The collective sing. ***crinis*** (*E*) is only used of a single individual.] **The blond** German hair (c. 11, 2 and note) was so much admired at Rome that it was often cut from the heads of captives to make ladies' wigs (Ovid, *Am.* 1. 14, 45–46; *Ars* 3, 165; Mart. 14. 27). A crude soap containing some dye-stuff (*sapo*; Pliny, *N.H.* 28, 191; Mart. 14. 26) was used to achieve the same colour artificially and this is no doubt what Domitian is alleged to have employed on his slaves.

at nunc (cf. c. 1, 4). This depends on *inerat conscientia.* The adjectives form the true predicate: 'the victory which was now extolled was real and great'.

39, 2. id . . . formidolosum, developed from Sall. *Cat.* 7, 2 'regibus . . . semper aliena virtus formidulosa est'.

privati, 'a subject', as in *H.* 1. 49, 4, etc.

frustra, etc., 'In vain had public eloquence and the grace of civilian professions been silenced if someone other than himself were now to seize military fame. Qualities of other kinds could be more easily overlooked, but good generalship was the Emperor's virtue' (that is, stamped its owner as fit for the purple and, therefore, a dangerous rival). Cf. *studiis civilibus,* used of a jurist in *A.* 3. 75, 1. By *civiles artes* (cf. Cic. *Brut.* 155; Livy 10. 15, 12) political (senatorial) activities in particular are meant. The suppression of 'omnis bona ars' (see c. 2, 2) probably does not here refer to the expulsion of philosophers, which took place in A.D. 93, but to the general repression of Domitian's rule as a whole (c. 3, 2; cf. Suet. *Dom.* 2). So Pliny says (*Ep.* 8. 14, 2) 'priorum temporum servitus, ut aliarum optimarum artium, sic etiam iuris senatorii oblivionem quandam et ignorantiam induxit'. Cf. also *Pan.* 66, 76. It is a characteristic trait of a tyrannical emperor in Tacitus to be jealous of his successful generals (cf. Claudius, *A.* 11. 19, 3). Tacitus' view is doubtless also coloured by the events of the last years of Domitian's reign. So Pliny (*Pan.* 14, 5) describes Domitian as 'alienis virtutibus invidus'.

For *utcumque* cf. *A.* 2. 14, 3; for *dissimulari, A.* 4. 19, 4.

39, 3. exercitus, 'agitated', a rare but classical use (e.g. Cic.

Planc. 78); for *exercitus curis* cf. Ovid, *M.* 15, 768; Virg. *Aen.* 5, 779.

quodque, etc., 'which was a sign of his cruel purpose', in apposition to the following words, in which the stress is laid on *secreto,* as if Tacitus had written 'et secreto suo (quod . . . indicium erat) satiatus', 'after taking his fill of, indulging to the full in, his usual seclusion'. His periods of retirement and brooding are spoken of in Pliny, *Pan.* 48 ('illa immanissima belua . . . velut quodam specu inclusa . . . . Non adire quisquam . . . tenebras semper secretumque captantem', etc.), and his seclusion in his Alban villa (c. 45, 1) in many places. *Secreto* 'seclusion' as in *A.* 4. 41, 2, etc. Cf. c. 22, 4.

in praesens, as in *A.* 1. 4, 1, etc. [*In praesentia* (*E*) is not found in Tacitus.]

reponere, 'to store up'. Similarly Tacitus speaks of Tiberius 'odia in longum iaciens, quae reconderet auctaque promeret' (*A.* 1. 69, 5) and Nero's 'dissimulatum ad praesens et mox redditum odium' (*A.* 16. 5, 3). It is another characteristic trait of a tyrant.

impetus . . . languesceret, 'the first burst . . . should die down': cf. *A.* 4. 21, 1 'impetus offensionis languerat'.

nam, etc., i.e. for he was still in command and was therefore to be feared. The words form a transition to the next chapter, in which Tacitus cannot bring himself to say plainly that Agricola was recalled. Nor does his fierce resentment allow him to judge fairly. Domitian had in fact nothing to fear from a middle-class official *peritus obsequi* (c. 8, 1). Tacitus veils the fact that his hero had governed for an unusually long period, and that soldiers were urgently needed on the Danube frontier (c. 41, 2, note), whither one legion (*II Adiutrix*) and doubtless a proportionate number of auxiliary troops were transferred soon after Agricola's recall. His resentment still burned six or seven years later when he wrote 'perdomita Britannia et statim missa'.

40, 1. **triumphalia ornamenta** : called also *triumphalia insignia,* the right to be styled *triumphalis* and to wear the *toga picta, tunica palmata,* etc. In order to ensure the subordination of generals to the Emperor, Augustus had abolished the full triumph except for members of the imperial family. The last person apart from the imperial house to celebrate a triumph was L. Cornelius Balbus in 19 B.C. In its place (*pro triumpho*)

Augustus substituted the grant of *triumphalia ornamenta* to generals who distinguished themselves under imperial auspices. See A. E. Gordon, *Univ. of Calif. Publications in Class. Arch.* 2 (1952), 308 ff.

inlustris statuae: also called *statua triumphalis* (*H.* 1. 79, 5) or *laureata* (*A.* 4. 23, 1), erected in the *forum Augustum* (*A.* 15. 72, 1; *CIL.* vi. 1386). The honour, though distinct from the *ornamenta*, usually accompanied them.

quidquid. *Supplicationes* and other rites would here be meant. We should expect *quidquid aliud*, but *aliud* is often omitted, e.g. *Dial.* 35, 5; *H.* 2. 6, 2, etc.

multo verborum honore. These words refer to the fulsome terms of the decree (cf. Vell. Pat. 2. 104, 2) and not to an inscription setting out Agricola's achievements. Triumphal statues under the Empire were not accompanied by *elogia* (Degrassi, *Inscr. Ital.* xiii, 3, p. 7).

decerni . . . iubet. Such honours were decreed by the senate, but usually on the initiative of the princeps. (Cf. *CIL.* vi. 1386 'a[u]c[tore] imp. Caes(are) . . . senatus ornament(a) triumphal(ia) decr(evit) statuamque in foro Aug(usto) ponendam censuit'.)

addique . . . opinionem, 'and the suggestion to be added'; cf. *praebere opinionem, adferre opinionem*, etc. Probably the decree was so worded as to hint at further honours in contemplation, and the fact that Syria was vacant suggested that this government was meant.

Atili Rufi. A military diploma shows that T. Atilius Rufus was in A.D. 80 legate of Pannonia (*CIL.* xvi. 26). His presence in Syria is attested by a milestone of A.D. 83 (*AE*, 1925, 95). The younger Pliny may have served as a military tribune there under him (*Ep.* 8. 14, 7). Nothing is known of his antecedents. He must have been consul *c.* A.D. 75, or earlier. We do not know his successor in Syria.

maioribus, 'senior men': as in *H.* 1. 48, 1, etc. *Maior* never means 'more eminent' in Tacitus and it could hardly be claimed that Atilius was a man of eminence. Agricola was 44 at this date and Syria was usually governed by men over 50: for exceptions see Syme, *Tacitus*, 2, pp. 631–2.

40, 2. credidere plerique: placed emphatically at the beginning of the sentence, as in c. 9, 2 and elsewhere.

libertum: it is hardly credible that Domitian should have felt that Agricola might need the inducement of Syria to leave his province or that the freedman should have turned back on the Channel coast ('in ipso freto Oceani') on sighting Agricola's vessel. But the rumour, to which Tacitus deliberately does not commit himself, could have arisen from the mention of Syria in the senate and the knowledge that a freedman did go on a mission to Britain at this time. Freedmen were employed on delicate imperial business (cf. Polyclitus going 'ad spectandum Britanniae statum' for Nero in *A*. 14. 39, 1) and Domitian might well have wanted first-hand information, perhaps from the procurator, when a change-over in the government of Britain was in process.

ministeriis: for *ministris*, as in *A*. 13. 27, 1. This shows that he was a personal freedman of Domitian's and not an official from one of the civil service departments (*a rationibus*, etc.).

dabatur, 'was to be offered'. It is to be inferred that an order of recall had been already sent, and that this offer was only to be made in case he seemed disinclined to obey it.

sive . . . sive refers in sense to the whole story, *credidere*, etc. From the wording it is obvious that Tacitus was not told this by Agricola.

ex, 'in accordance with'. Cf. *G*. 3, 3; *H*. 1. 82, 2, etc.

40, 3. successori: the name of the successor is unknown.

celebritate, 'by publicity and a crowd coming to meet him'. Such a reception of an eminent citizen at his homecoming is described in the case of Cn. Piso (*A*. 3. 9, 2–3).

officio, 'the attentions': c. 18, 5, and *A*. 2. 42, 2. Cf. Suet. *Tib*. 12 'vitans praeternavigantium officia'.

ut praeceptum: Agricola's orders were only to call upon Domitian; he chose to do so unobtrusively by night.

brevi osculo, 'a hasty kiss'. So in *A*. 13. 18, 3: the custom of greeting with a kiss the emperor's more intimate and more distinguished friends appears to have been introduced by Augustus. Cf. Suet. *Tib*. 34; *Otho* 6, etc.

inmixtus est, probably passive, 'was lost amid the courtier crowd'. Cf. c. 28, 1; *H*. 3. 74, 1.

40, 4. grave inter otiosos = *molestum otiosis*, 'unpopular among civilians', who envied and disliked it. *Otiosus* 'peaceful' as in *Dial.* 40, 2. The thought is a commonplace, exemplified by the fate of Lamachus in Aristophanes' *Acharnians.*

hausit, 'took his fill of' (cf. c. 4, 3 and note). No such phrase as *haurire otium* is elsewhere found, but Wölfflin (*Philol.* 26 (1867), 153) noted that the metaphor *temperaret* ('blend') is kept up, and that *libertatem haurire* (*H.* 4. 5, 2; Livy 39, 26, 7) is a near parallel. For *penitus* cf. *Dial.* 30, 3. [The corruption (*auxit E*) is common (Brink, *Thes. Ling. Lat.* 'haurire' col. 2567, 45).]

cultu, probably 'dress', as it usually means in association with *modicus* (Pliny, *Ep.* 1. 22, 4; *Pan.* 83, 7; *CIL.* vi. 1527 (Laud. Turiae) and *sermo* (*G.* 43, 1; 46, 1), implying that Agricola did not avail himself of the flamboyant costume of a *triumphator.* But in Curt. 9. 8, 23 'modico civilique cultu' it must mean 'way of life', as editors have understood it here.

facilis, 'affable': cf. *A.* 3. 8, 2, and c. 9, 3 *facilitas.*

uno aut altero, 'one or at most two'. Cf. c. 12, 2; 15, 2; *G.* 6, 1 *uni alterive.*

per ambitionem, 'by their ostentatious display'. Cf. 'funerum nulla ambitio' (*G.* 27, 1). *Aestimare* has usually an abl. of the standard (with or without *ex*), to which the construction with *per* (cf. Suet. *Cal.* 60) is equivalent.

viso aspectoque : having not merely seen him, but observed his mode of life and demeanour.

quaererent famam, 'asked about his celebrity', asked how he could have won such fame. Not 'sought for his fame', which would be a very obscure way of saying 'sought vainly for its explanation'; nor 'missed his fame', i.e. saw no sign of it, which would seem to require *requirerent.* Quite different is the use of *famam quaerere* in c. 9, 4.

interpretarentur, 'understood', 'interpreted rightly', as often. Somewhat different is *H.* 1. 14, 2. It is natural to supply *famam*: many sought for the reason of his reputation, few understood it (i.e. because Agricola did not advertise by behaviour or by word what he had achieved). Anderson and other editors supply 'his modest demeanour' from the earlier part of the sentence: few people understood the reason of his unobtrusive behaviour (e.g. that it sprang from an unambitious character).

41, 1. eos dies : those following his return and preceding the winter of A.D. 85 (note on § 2).

absens : repeated for emphasis. The facts may be exaggerated since prosecutions in absence are unusual (*A*. 15. 59–60). Agricola may have been informed against but no action taken.

laudantes, 'panegyrists': cf. *peccantes* (c. 4, 2), etc. Whether they were insidious enemies or indiscreet friends, their praise would be equally pernicious in result.

41, 2. et, 'and indeed' the times forced his name into notoriety.

rei publicae tempora : the wars spoken of below.

sileri, 'to pass unmentioned'. Cf. Pliny, *Ep*. 8. 22, 4. The verb is used with accus. of the thing in Cicero.

sinerent, i.e. in spite of his own endeavour.

tot exercitus in Moesia, etc. These disasters are enumerated in chronological order. A broken account of the campaigns is to be found in Dio 67. 7–8 and 10, and allusions in Suet. *Dom.* 6, Statius, Martial, etc. For a discussion see Syme, *CAH* xi. pp. 168 ff. The Dacians, probably in the late summer of A.D. 85, invaded Moesia, and defeated and killed the *legatus*, Oppius Sabinus. Domitian took the field early in 86, and drove them back across the Danube, but in his absence the praetorian prefect Cornelius Fuscus was killed and his army cut to pieces. One disaster, probably the second, was commemorated on the altar of Adamklissi in the Dobrudja. After these disasters, probably the greatest since that of Quintilius Varus, two years passed, in the latter of which (A.D. 88) ensued the revolt (*bellum civile*) of Antonius Saturninus on the Rhine, called in some inscriptions *bellum Germanicum* (*ILS* 1006, 2127, 2710). In 88 Tettius Julianus restored Roman prestige by a considerable victory over the Dacians at Tapae, but in 89 Domitian himself, after suppressing Saturninus and inflicting chastisement on his allies, the Chatti (see on c. 39, 1), attacked the Marcomani and Quadi (German tribes, Suebi, in Bohemia and Moravia) from Pannonia for not sending aid against the Dacians, and was defeated by them with the help of their neighbours the Sarmatian Iazyges (*ILS* 9200), whereupon he made a timely peace with the Dacians (Dio 67. 7, 2). (The precise nature of this peace and, in particular, the question whether the Romans withdrew south of the Danube are disputed and cannot be resolved without careful archaeological investigation. See Syme, *J.R.S.* 49 (1959), 26 ff.; D. M. Pippidi, *Hommages à A. Grenier* (1962), pp. 1265 ff.). In the early spring of A.D. 92 the Iazyges, in alliance with the

Marcomani and Quadi, invaded Pannonia and annihilated a legion (Suet. *Dom.* 6). Domitian was at the seat of war probably from May 92 to Jan. 93, and on his return consecrated a laurel without claiming a triumph (Suet., loc. cit.). This war is called in inscriptions *bellum Suebicum et Sarmaticum* (*ILS* 1017, 2719). The allusions in the text to Moesia and Dacia are explained by the disasters of Sabinus and Fuscus, those to Germany and Pannonia by the disastrous incidents of the Suebo–Sarmatic campaigns.

temeritate aut per ignaviam. For the variation of construction cf. c. 46, 3, etc. The terms are regularly paired by Livy (e.g. 2. 65, 4; 6. 24, 5; 9. 5, 6).

militares viri: the phrase is used of experienced or professional soldiers of the officer-class (*H.* 2. 75, 1; 3. 73, 2; *A.* 4. 42, 2; 15. 10, 1; 15. 20, 3; 15. 67, 3: so also in Livy). Tacitus intends a contrast between the defeat of field-armies under incompetent generals and the annihilation of garrisons with their commanders ('tot militares viri cum tot cohortibus' virtually forms a simple subject to balance 'tot exercitus'; for the anaphora cf. Livy, 25. 24, 12. *E*'s *totis cohortibus* is less effective than the marginal correction *tot*). *Expugnati*, implying that they were in possession of forts, is often used of people (*H.* 3. 19, 1; Caes. *B.G.* 7. 10, 1; Livy 23. 30, 2, etc.). The lack of archaeological investigation makes it impossible to know where these forts were.

nec iam de limite imperii et ripa, etc. 'No longer was it the frontier-line of the empire and river-bank that were imperilled'. The *ripa* is that of the Danube. The losses of which Tacitus is speaking were incurred on the middle and lower Danube, and *et ripa* is, therefore, to be taken as explanatory of *limite*. *Limes*, which in the earlier first century had denoted a military line of penetration, protected by forts ('limitem scindere, aperire, agere'), is here passing into its later use of defensive frontier-lines or frontier-line in general.

possessione, 'maintenance' of whole provinces. Wex compares Cic. *Acad.* 2, 132 'non de terminis sed de tota possessione contentio'.

41, 3. continuarentur, 'followed continuously upon'.

omnis annus, 'every year', as *omnis aetas*, etc.

insigniretur. This use of the verb is a new development of the expression 'annus insignis (erat)', common in Livy (2. 60, 4; 4. 29, 8, etc.): cf. *A.* 16. 13, 1; Suet. *Cal.* 31.

ore vulgi. It is unlikely that such popular agitation was widespread: the lower classes are never interested in foreign policy (*H.* 1. 4, 3; 4. 38, 2). Tacitus exaggerated but there may have been some demand for Agricola. Domitian, however, recognized that the Danube provinces required local knowledge and a sure control of the army. It is therefore not surprising that L. Funisulanus Vettonianus was moved from Pannonia to be governor of Moesia Superior in A.D. 85/86 (*ILS* 1005).

expertum bellis. Cf. *Bell. Alex.* 61, 1; Livy 29. 24, 12.

cum inertia . . . eorum. The MSS. *eorum* should perhaps be retained referring back to the *duces* of § 2. The editors of the Bipontine edition read *aliorum* (cf. *H.* 3. 3, 1) but the change is not needed and *aliorum* occurs again in § 4.

41, 4. aures verberatas: a common figure (cf., e.g., Plaut. *Amph.* 333; Sen. *ad Marc.* 19, 6; *Ep.* 56, 4; Lucan 7, 25, etc.).

dum: best taken as only temporal ('while'): cf. *H.* 1. 1, 1; *A.* 13. 3, 1. It has also in Tacitus a causal force ('inasmuch as'), but only with the present.

libertorum, sc. *Caesaris.*

amore et fide, sc. *in Domitianum* and with *malignitate et livore* sc. *in Agricolam.* The ablatives are causal, and *exstimulabant,* 'spurred on', applies to both pairs. For *malignitas et livor* cf. Suet. *Cal.* 34, 1; Pliny, *Pan.* 58, 4.

deterioribus: neuter (cf. c. 16, 3), *pronus in* being used when persons are spoken of (*H.* 1. 13, 4, etc.). A similar dative with *facilis* (*A.* 2. 27, 2), *promptus* (*A.* 2. 78, 1), etc., is noted by Andresen.

simul . . . simul: cf. c. 25, 1; c. 36, 1; c. 35, 4 note. A similar antithesis occurs in *H.* 4. 34, 3 'non minus vitiis hostium quam virtute suorum fretus'; cf. also Cic. *Leg. Man.* 67 'cum suis virtutibus tum etiam alienis vitiis magnum'. *Vitiis* refers to the *malignitas et livor,* not to the *inertia* and *formido* of other generals.

in ipsam gloriam, the very prominence that he anxiously sought to avoid, as leading to ruin. Under an emperor like Domitian *gloria* was a precipice: cf. § 1 above *gloria viri*; c. 5, 3; c. 40, 3–4; 42, 3 (*famam fatumque*). A special mention of the perils accompanying glory might be expected, like 'unde gloria egregiis viris et pericula gliscebant' (*A.* 15. 23, 4), but the preceding narrative has emphasized the perils.

42. The veracity of the following episode has been much attacked. It is argued that Agricola's desire for *otium* (cf. c. 40, 4) would have made him anxious to decline any further command and that he did not even trouble to ask for the *salarium*. Both these actions, Traub (*C.P.* 49 (1954), 255–7) alleges, have been distorted by Tacitus to Domitian's discredit. Dorey (*Greece & Rome* 7 (1960), 66 ff.), on the other hand, believes that Agricola was in poor health and had to be dissuaded, for his own good, by Domitian from assuming another heavy burden and that Tacitus has spitefully misrepresented the situation. These reinterpretations run so counter to the facts as given by Tacitus that they should not be maintained. Although provinces could be declined on score of age or infirmity, Agricola was relatively young and, so far as is known,[1] active. He was forced to play a humiliating game because he did not know what Domitian's real attitude was. He tried to remain non-committal until he was brutally enlightened by Domitian's agents. See K. von Fritz, *C.P.* 52 (1957), 73–97.

42, 1. quo proconsulatum Africae, etc. The governorships of Asia and Africa, the most important of the senatorial provinces, were awarded every year to the two senior consulars who had not held either, the lot determining which was to have which. Sometimes one or other was given *extra sortem* (*A.* 3. 32, 2), or a candidate was prohibited by the *princeps* (*A.* 6. 40, 2), or declined it, as did Salvius Liberalis under Trajan (*ILS* 1011). From a survey of known instances during this period it appears that the turn of Agricola would have fallen between twelve and fifteen years after his consulship. 'Occiso Civica nuper' suggests that it was about A.D. 90, since Civica was proconsul in A.D. 88/89 and was put to death for suspected complicity in the revolt of Saturninus.

sortiretur = *sortiri debebat.*

Civica : His full name C. Vettulenus Civica Cerialis is given in a military diploma, which names him as *legatus* of Moesia in A.D. 82 (*ILS* 1995). Suet. (*Dom.* 10) states that he was put to death by Domitian during his proconsulate of Asia. His

[1] But it should be noted that several consulars in their fifties or younger died during this period (cf. Domitius Lucanus, T. Aurelius Fulvus, M. Otacilius Catulus: see Syme, *Tacitus*, p. 69 n. 6).

family may have been Sabine in origin. He was suffect consul about A.D. 76 and his brother Sextus about A.D. 73. He appears to have been the deceased proconsul whose place was temporarily taken by C. Minicius Italus, 'proc. provinciae Asiae quam mandatu principis vice defuncti procos. rexit' (*ILS* 1374).

consilium, 'counsel given', i.e. a warning, as in *H.* 2. 89, 1, etc.

exemplum, 'a precedent'. So *vestra exempla* (*A.* 3. 50, 2), etc.

ultro, 'of their own accord' (without waiting for him to say anything about it). Cf. c. 19, 4, etc.

occultius, 'covertly', opposed to *non obscuri*, 'in plain words'. So *occultus* is used of a person who conceals his thoughts, *H.* 2. 38, 1.

quietem et otium : as in c. 6, 3.

adprobanda, 'commending' (to Domitian): cf. c. 5, 1.

42, 2. paratus simulatione, 'well equipped with hypocrisy', i.e. with pretended ignorance of any pressure put upon Agricola. Cf. *Dial.* 33, 6 'instructum his artibus animum paratiorem ad', etc.; Cic. *ad Att.* 9. 13, 4 'paratum peditatu, equitatu', etc.; Suet. *Cl.* 42, 'sermone paratus'.

in adrogantiam compositus, 'set in an authoritarian demeanour', i.e. appearing stern and impartial and allowing himself to be entreated to do what he really wished to do. Cf. 'in securitatem compositus' (*A.* 3. 44, 4), etc.

excusantis. We should expect the addition of *se* (as in *A.* 3. 35, 2; *ILS* 1011), or of an accus. of the thing pleaded in excuse, or apologized for. The ellipse of *se* is, however, common with *excusare*; cf. *A.* 1. 10, 7, etc.

agi sibi, etc. Cf. 'actae . . . gratiae consuetudine servitii' (*H.* 2. 71, 2); 'Seneca, qui finis omnium cum dominante sermonum, grates agit' (*A.* 14. 56, 3), and Seneca's anecdotes of others (*de Ira*, 2. 33, 2; *de Tranq.* 14, 4).

nec erubuit, etc., 'did not blush for the odiousness of the concession', for granting as a favour what was really the gratification of his own dislike. The abl. is causal, as in *G.* 28, 4. For a similar sarcasm cf. *H.* 1. 21, 1 *exilii honorem*.

salarium, etc. This substantive is not found earlier than Seneca and the elder Pliny. Its use dates from the Augustan regulations by which all provincial governors had fixed pay on a scale proportioned to their rank (Dio 52. 23, 1; 53. 15, 4).

It is stated in Dio 79. 22, 5, that in A.D. 217 the salary offered to a proconsul of Africa in lieu of the office was a million sesterces. See Sherwin-White, *Letters of Pliny*, pp. 149–50.

offensus: so with accus. and inf. apparently only in Suet. *Aug.* 89; *Tib.* 34; Phaedr. 4. 11, 6. It is analogous to *dolens*, *aegre ferens*.

ex conscientia, 'from conscience'. **quod vetuerat**: by his agents.

42, 3. proprium humani ingenii. For similar remarks on human nature, cf. *H.* 1. 55, 1; 2. 20, 1; 38, 1.

odisse quem laeseris. In this commonplace sentiment Tacitus seems to have followed Seneca, who says, 'pertinaciores nos facit iniquitas irae' (*de Ira* 3. 29, 2), and 'quos laeserunt, et oderunt' (sc. 'magna fortuna insolentes', ibid. 2. 33, 1). Cf. also *A.* 1. 33, 1.

vero, pointing the contrast: Domitian would feel this far more than other men. The sense is 'inrevocabilior erat, ac tamen'. *Praeceps* is so used with *in* in *H.* 1. 24, 1; *praeceps in iram* in Livy 23. 7, 12.

obscurior, 'more given to concealing his feelings'. So of persons in *A.* 4. 1, 2; 6. 24, 3.

inrevocabilior, a rare word (originally poetical but in prose from the time of Livy), is here alone used by Tacitus, who has elsewhere *implacabilis*. It is common in this sense in Seneca (cf. *de Ira* 3. 7, 4, etc.).

leniebatur. It is to be noticed that, though Agricola himself received no further distinction during these years, his son-in-law Tacitus was praetor and *quindecemvir sacris faciundis* in A.D. 88, and probably received about A.D. 90 a legionary command and, thereafter, a province. See Introd., p. 8.

famam fatumque, 'renown and ruin'. The two are closely joined in idea, the one being regarded as the sure precursor of the other (cf. c. 41, 4, and note), and the connexion being strengthened by the alliteration. The words are similarly coupled by Virg. *Aen.* 7, 79; 8, 731.

42, 4. sciant, etc. This sentence, concluding the narrative of Agricola's life and leading on to the epilogue, should not be seen as an embarrassed defence of Agricola's and Tacitus' personal collaboration with the régime of Domitian. It is in line with all Tacitus' political thinking about the futility of ostentatious independence under the principate. The contrast with the immediately following account of Agricola's death must be deliberate and reflect Tacitus' view that Agricola no less than more controversial heroes of the time displayed *virtus*.

inlicita, 'forbidden', that is, not approved by the Emperor (*A*. 6. 8, 3) rather than contrary to the constitution (cf. *A*. 3. 27, 1 *inlicitos honores*). The attractions of the forbidden are proverbial (cf., e.g., Sen. *H.O.* 357).

vigor, especially in a military sense: cf. c. 41, 3; *H*. 1. 87, 2, etc. It is linked with *industria* also by Vitruv. 5. 5, 6 and Pliny, *Ep*. 1. 14, 7.

eo laudis excedere, 'attain to a height of honour'. *Excedere* in this sense, though used nowhere else by Tacitus, is common in Valerius Maximus, e.g. 'ad clarissimum gloriae lumen excessit' (3. 4, ext. 1; cf. 5. 6, 4, etc.) and is used by Seneca (e.g. *Ep*. 85, 12).

quo plerique per abrupta. *Per abrupta*, 'by perilous courses', a metaphorical use of the phrase, which means literally 'over steep ground' (Florus 1. 23, 10; Sidon. *Ep*. 3. 12, 2, etc., cf. *A*. 2. 55, 3). Cf. the very similar passage in *A*. 4. 20, 3 'an . . . liceat inter abruptam contumaciam et deforme obsequium pergere iter ambitione ac periculis vacuum'.

Excesserunt or *excedunt* is to be understood with *quo per abrupta*, and Tacitus then passes on to a new main verb (*sed . . . inclaruerunt*). This form of syntactical variation is common in Tacitus; cf. *H*. 2. 5, 2 'consuluere, primum per amicos, dein Titus prava certamina . . . aboleverat', etc. Editors have supplied *nisi* or *enisi* but the addition is unnecessary. Tacitus admits that the Stoic martyrs won praise but adds that they achieved no good by it. The reference is to Thrasea Paetus and, above all, to Helvidius Priscus (see notes on c. 2, 1). Similar sentiments are expressed by Sen. *Ep*. 113, 32 'qui virtutem suam publicari vult non virtuti laborat sed gloriae' and Martial (1. 8, 5–6, citing Thrasea) 'nolo virum facili redimit qui sanguine famam: hunc volo, laudari qui sine morte potest'.

ambitiosa, 'ostentatious': cf. c. 29, 1, and note.

inclaruerunt: cf. *A*. 12. 37, 3; also in the elder Pliny. Cf. Val. Max. 5. 4, ext. 3 'specioso exitu vitae inclaruistis'.

43–46. *Epilogue*

The epilogue consists of an account of Agricola's last hours (c. 43) and a consolation in the conventional manner (cc. 44–46; see note). Tacitus had a taste for such final scenes, as can be seen from his treatment of the deaths of Rubellius (*A*. 14. 58, 1), Britannicus (*A*. 13. 17, 2) and, above all, Germanicus (*A*. 2. 71, 1–4) who mirrors many of Agricola's qualities (see note on c. 9, 3). In this he may have been influenced by a prevailing literary fashion, the 'exitus virorum illustrium' (see Introd., p. 13). With the end of the narrative proper, his writing undergoes a marked change. From an historical style, modelled on Sallust and Livy, he passes to a more oratorical manner in keeping with the subject-matter. In particular these chapters owe a clear debt to Cicero's description of the death of the orator Crassus in the *prooemium* of the third book of the *de Oratore*.

43, 1. Finis vitae, etc. Cf. Cic. *de Orat*. 3, 8 'fuit hoc (the death of Crassus) luctuosum suis, acerbum patriae, grave bonis omnibus'. Notice the *tricolon auctum* (cf. c. 31, 2 n.).

extraneis, 'outside his circle': so used in contrast to one of the family in *A*. 4. 11, 1. Here it is an intermediate term between *amici* and *ignoti* (complete strangers). For this use of *ignotus* 'someone who does not know him' cf. *A*. 2. 71, 3; Cic. *Planc*. 42; Nep. *Ages*. 8, 1.

vulgus . . . populus. *Et* is probably explanatory. *Vulgus* usually involves the idea of social and intellectual inferiority; *populus* is a more honourable term with a political reference. But the two words are thus grouped as virtually synonymous in *H*. 1. 89, 1; *Dial*. 7, 4 (followed by a singular verb). So *populus* and *plebs* in *H*. 1. 35, 1.

aliud agens: 'otherwise occupied', often used in the sense of 'indifferent' or 'heedless'. Cf., e.g., *Dial*. 32, 1. It would thus answer to the 'communium curarum expers populus' of *H*. 1. 89, 1, the 'vulgus vacuum curis' of *H*. 2. 90, 2.

circulos, 'social gatherings': *A*. 3. 54, 1 'in conviviis et circulis'; Quint. 12. 10, 74 'illi per fora atque aggerem circuli'.

locuti sunt. *Obitum Agricolae* is supplied from the sense: *loqui* has the meaning of *in ore habere* in *H*. 1. 50, 2, etc.

oblitus. It is better to treat the repeated *est* of the MSS. as dittography than to alter it to *et* or *set*, since *augebat* is emphatic at the beginning of the sentence as in Cic. *Brut.* 2. For the shape of the sentence cf. c. 3, 3 'laudatus erit aut excusatus'.

43, 2. rumor. The existence of such rumours has been denied, particularly since they are a recurring feature in Hellenistic historiography. It cannot, however, be proved that they are fictitious and in a suspicious atmosphere such as surrounded Domitian's last years gossip does notoriously circulate. Tacitus evidently did not himself believe them (see, however, c. 44, 4 *festinatae*) but perhaps hoped that his readers would. They are discussed by B. Walker, *The Annals of Tacitus*, p. 142.

interceptum (sc. *esse*): used with *veneno* in *A.* 3. 12, 4 (of Germanicus), and often of other kinds of treacherous death.

nobis nihil comperti adfirmare ausim. The MSS. text may be translated 'I would not venture to assert that we have any ascertained evidence'. For this use of *nihil* cf. Cic. *ad Att.* 2. 4, 3 'nihil certi polliceor' 'I do not promise anything for sure' rather than 'I promise nothing sure'. Not recognizing this use, editors have supposed that *ut* (Wex) or *quod* (Walch) has fallen out after *comperti* but no change is needed. *Nobis* probably means Tacitus himself: the change from plural to singular is common (e.g. *H.* 2. 77, 1; 4. 5, 1). For the whole phrase cf. Livy 3. 23, 7 certum adfirmare ... non ausim'. For *nihil comperti* cf. Cic. *pro Clu.* 131 'nihil cogniti, nihil comperti'. Suetonius does not mention Agricola among Domitian's victims; Dio (66. 20, 3) gives the fact as undoubted (ἐσφάγη).

ceterum : passing on to known facts which might give some support to the rumour.

per omnem valetudinem, 'throughout his illness'. Cf. c. 45, 4.

principatus . . . visentis = *principum . . . visentium.* The point is that princes, who regularly pay such visits of inquiry through messengers, do not usually pay them so often.

primi . . . intimi. Cf. *A.* 6. 38, 2 'praecipuos libertorum' (4. 10, 2, etc.) and *A.* 4. 3, 4 'medicus . . . frequens secretis' (12. 67, 2, etc.). Trusted doctors were an essential part of the imperial establishment.

sive, etc., 'whether that action meant real concern or prying' (i.e. to see if all was going on as they wished). For *cura* cf. § 1; for *inquisitio* c. 2, 3. *Cura* has been strained to mean 'pretended interest', but Tacitus here, as often, puts the possibility of its reality, thereby securing the appearance of impartiality, while conveying the impression of its falsity. Compare the account of Piso's messengers during the illness of Germanicus (*A*. 2. 69, 3): 'incusabantur ut valetudinis adversa rimantes'. For the use of *illud* cf. *A*. 1. 49, 2; 4. 19, 3; and note on c. 21, 2.

43, 3. momenta, 'turning points', the stages of his sinking (cf. *A*. 5. 4, 1 'brevibus momentis summa verti').

cursores : couriers posted at intervals, probably to his Alban villa (c. 45, 1).

constabat and *constat* are uniformly impersonal in Tacitus. The imperfect is adapted to the time of the reports (when Tacitus was away from Rome), as in § 4; c. 38, 1.

nullo credente, 'and none believed that news brought with such dispatch could be unwelcome'.

speciem, etc. Although *animus* is often contrasted with *vultus* as inner feelings against outward appearance (cf. Livy 40. 5, 14 'animo magis quam vultu ea crimina accipiebat'; *H*. 4. 31, 2, etc.), *animus* is also used of the air radiating from a person, which need not be sincere, as opposed to his facial expression (cf. Livy 2. 5, 8 'vultus spectaculo esset, eminente animo patrio'; *Bell. Afr.* 10 '(nil) auxili nisi in imperatoris vultu . . . animum enim altum prae se gerebat'). So here Domitian displayed a pretence of grief both by his manner and by his countenance. There is no need to change the text to *animi vultu* (Baehrens) or *habitu vultuque* (Ernesti; cf. *H*. 1. 14, 2, etc.)

securus . . . odii, 'untroubled in his hatred'. He no longer had to fear that his dislike for Agricola would lead to repercussions and could therefore even afford to display grief. Cf. *A*. 3. 28, 2 'potentiae securus', etc.

et qui, 'and being a man who': cf. *H*. 2. 25, 2 'et cui . . . placerent'.

43, 4. coheredem . . . Domitianum scripsit. It became a common practice under bad emperors thus to sacrifice a part of the property to save the rest for the relatives. Gaius exacted a share (Suet. *Cal.* 38), and some famous instances are given under Nero in which testators made, or were recommended to make, the emperor joint heir (*A*. 14. 31, 1; 16. 11, 1). Tiberius refused such legacies (*A*. 2. 48, 1), as Domitian did at first (Suet. *Dom.* 9), though afterwards he seized them eagerly (ibid. 21), so that Pliny speaks of him as 'unus omnium, nunc quia scriptus, nunc quia non scriptus, heres' (*Pan.* 43, 1).

piissimae. This superlative, condemned by Cicero when used by Antony (*Phil.* 13, 43), is frequent in and after Seneca. It is particularly common in epitaphs (*ILS* 4954, etc.) and the whole phrase 'optimae . . . filiae' is taken from the formal language of conventional praise. See also next note.

honore iudicioque, 'honour and approval'. Both words are commonly used in connexion with wills; for *honor* cf. Cic. *Quinct.* 14; Val. Max. 7. 7 pr.; Pliny, *Ep.* 7. 24, 8: for *iudicium* cf. Pliny, *Ep.* 10. 94, 2. They are linked in *ILS* 8394 (Laud. Murdiae) 'pecuniam legavit ut ius dotis honore iudici augeretur' but a hendiadys should not be assumed here.

44–46. *Consolatio*

The work closes with a Consolatio in the traditional manner, which balances in length and dignity the Preface. The Consolatio had a long history going back to Homer but it became systematized by successive philosophical and rhetorical schools. One of the most influential examples was the περὶ πένθους of the Academic philosopher Crantor (*c.* 300 B.C.). It is therefore natural to find recurring topics and much of Tacitus' material, as illustrated below, is conventional. In particular he is indebted to Cicero's *de Oratore*, but instructive comparisons can also be made with Seneca (e.g. *Consol. ad Marciam*; *Ep.* 99) and with the famous letter of Sulpicius to Cicero on the death of his daughter (*ad Fam.* 4. 5). But Tacitus introduces a more personal note and in the lyrical phrases of the final chapter (see notes on c. 46, 1) expresses an intensity of feeling not found in earlier literature. The subject is studied, but without reference to the *Agricola*, by R. Kassel, *Untersuchungen zur griechischen und römischen Konsolationliteratur* (1958), with earlier bibliography.

44, 1. Natus erat, etc. The date of Agricola's birth was 13 June A.D. 40, which suits that of his praetorship (c. 6, 4 note and App. 1) and is quite reconcilable with that of his father's death (note on c. 4, 1). Gaius was consul for the third time during the first fortnight of Jan. A.D. 40, and he was sole consul, his designated colleague having died before 1 Jan. (Suet. *Cal.* 17). There is no reason to suspect the number *ter*, as given by the MSS. Normal Latin would say *tertium*, which is doubtless what Tacitus wrote. *Ter consul* strictly means 'one who has been consul three times', and though Martial says *bis* (for *iterum*) *consule* (10. 48, 20), the MSS. *ter* is no doubt merely an expansion of *iii*. Gellius devotes a whole chapter (10. 1) to discussing whether *tertium* or *tertio consul* is correct Latin for 'consul for the third time'; but, as he does not mention *ter* as a possibility, it was evidently not a prose construction.

The date of his death was 23 Aug. A.D. 93. *Sexto* of the MSS. should therefore be *quarto*, the error being due to a misreading of *iv* as *vi*. Cf. c. 33, 2, where *vii* was misread as *viii*. In some cases errors of chronology appear to be due to Tacitus himself (e.g. *A*. 14. 64, 1), but he would not err here.

decimum kalendas. Tacitus often omits *ante* in such expressions: cf. *A*. 6. 25, 3.

Collega Priscinoque. For the correct style of these *consules ordinarii* of A.D. 93 see *ILS* 9059.

44, 2. Agricola's appearance and character. It was usual to begin a Consolatio with a description of the deceased in laudatory terms: cf. [Plut.] *Cons. ad Apoll.* 101 E; Kassel, p. 52.

habitum, 'personal appearance' (cf. c. 11, 1 and note).

decentior quam sublimior, 'handsome rather than imposing'. *Decens* is used in this sense in both prose and poetry from the time of Ovid and Horace on and *decentior* serves also as comparative of *decorus*.

nihil impetus, etc. 'nothing violent in his face, kindliness of expression abounded'. *Impetus* makes far better sense than *metus*, the marginal variant, which can mean something causing fear (or to be feared) but is generally used of alarming circumstances. In later writers *impetus animi* is used technically of an impetuous temperament (cf. Firmic. *Mat.* 6. 22, 4) and contrasted with *gratia ingenii*. For *gratia oris* cf. Quint. 6 pr. 7 'quid ille gratiae in vultu ostenderit'. For *superesse* in this sense cf. c. 45, 5; *G*. 6, 1, etc.

bonum, etc., 'you would readily have believed him to be a good man, and gladly to be a great one': 'gladly', because of his placid face and winning expression.

44, 3. The timeliness (εὐκαιρία) of Agricola's death. When a person died prematurely it was conventional to claim that he died at the height of his achievements, of his prosperity, and of his happiness, without witnessing the disappointments that were to follow his death and surrounded by his contented family. Cf. Cic. *Tusc. Disp.* 1, 109; Sen. *Cons. ad Marc.* 20, 4; [Plut.] *Cons. ad Apoll.* 111 A ff.; Kassel, pp. 87 ff. See details below.

ipse, 'he himself', as distinct from his *habitus.*

medio in spatio integrae aetatis, 'in mid career of his prime'. Cf. Cic. *de Orat.* 3, 7 'O . . . inanes nostras contentiones quae medio in spatio saepe franguntur'; Sen. *Ep.* 93, 1 'in medio cursu raptus est'. *Integra aetas* is used of Tiberius at a considerably less advanced age (Suet. *Tib.* 10), *integra iuventa* of Agrippina at about thirty-three (*A.* 12. 2, 3).

quantum ad, 'as far as concerned', = *quantum attinet ad.* So used *G.* 21, 2; *H.* 5. 10, 2.

quippe, etc. The thought is as follows. 'Agricola's glory was as complete as if he had lived the longest life. For he had realized to the full the true blessings, which lie in virtues, and what else could fortune add to one who had the rank of consular and *vir triumphalis*? He did not enjoy excessive riches, but possessed a handsome fortune. And he was happy, too, in that he died neither childless nor a widower, without loss of position or reputation or relatives and friends, and dying escaped the evil to come.'

vera bona, according to the Stoic creed, are defined in *H.* 4. 5, 2: 'virtue is the only good, nothing is evil but what is base; power, rank, and all other things outside the mind (i.e. *bona fortunae, bona externa*) are neither good nor evil'. Tacitus naturally begins with the highest good (cf. *filia*, etc., below). It is to be noted that for rhetorical reasons Tacitus finds it convenient to adopt Stoic language here but in fact his concept of *virtus* is Roman rather than Stoic.

impleverat, 'realized to the full', as in *H.* 1. 16, 4 'impletum est omne consilium'; Pliny, *Ep.* 2. 1, 2 'summum fastigium privati hominis impleret'.

adstruere, 'to add': cf. Vell. Pat. 2. 55, 2 'vix quicquam gloriae eius adstruxit'; Pliny, *N.H.* 9, 119; Pliny, *Ep.* 3. 2, 5 'dignitati eius aliquid adstruere': this meaning is first attested in Ovid. The thought is a commonplace in such contexts, going back at least to Isocr. *Evag.* 28.

44, 4. opibus, etc.: specifies another aspect of Agricola's *felicitas* in life, answering the rhetorical question *quid aliud?*

For, as Aristotle said, material prosperity is an essential condition of complete εὐδαιμονία. As such it regularly figures in Consolations. Wealth is part of a man's good fortune (Men. περὶ ἐπιδεικτ. 9, 292 ff. Walz) but to be proud of excessive wealth is degrading ([Plut.] *Cons. ad Apoll.* 103 E). For *nimiis opibus* cf. Livy 33. 46, 3. The possession of wealth and children is similarly linked in the eulogy of L. Metellus (Pliny, *N.H.* 7, 140).

speciosae contigerant. It seems necessary to omit *non* with the marginal variant and regard it as wrongly repeated from the previous clause. *Speciosae* need not mean more than 'handsome' (cf. [Quint.] *Decl.* 9, 16 'cum solum tam speciosae fortunae crederem fructum posse prodesse') and is here contrasted with *nimiis*. The corrupt intrusion of *non* may be very old. Dio (66. 20, 3) says that he died in poverty and, since Dio's source is the *Agricola*, it is likely that he read *non* in his text, which would account for his statement. There is no evidence that Agricola was poor. He was careful in money matters (c. 6, 4) and, although his father's estate was looted, it was not confiscated (c. 7, 1).

If *non* before *contigerant* is retained, the contrast is weak and irrelevant, and the logical connexion with *non gaudebat* is lost. *Contigit* is usually used of a stroke of good fortune (*A.* 13. 37, 5) which would exclude the meaning 'extravagant' for *speciosae*. An ample competence was not considered in those days a matter for shame or condolence.

filia . . . superstitibus. There is no punctuation in *E*, which begins a new sentence after *gaudebat*. *A* and *B* put a stop after *superstitibus*, which is impossible. The words mean that it was an element in Agricola's happiness that he did not outlive those who were dearest to him. The sentence is compressed in Tacitus' manner: he was blest by fortune in that (1) he died with his position and surroundings unimpaired, and (2) he escaped the evil days to come. *Filia . . . superstitibus* might have been expected to follow *beatus*, on which stress is laid (as e.g. on *tristis* in c. 43, 3), but the position gives the words a special emphasis corresponding to the store which Agricola set on this element of his happiness and to the ancient sentiment that it was a misfortune to outlive children and spouse, and receive the last rites from alien hands (cf. Sulp. *ad Fam.* 4, 5, 5; Val. Max. 2. 5, 8; Vell. Pat. 1. 11, 7; Virg. *Aen.* 11, 159; Quint. 6 pr. 4 'duos enixa filios felix decessit'; Pliny, *Ep.* 1. 12, 11 'decessit superstitibus suis, florente republica'; Kassel, p. 96).

[*potest . . . beatus* separates the personal *filia . . . superstitibus* from the three abstracts *incolumi . . . amicitiis*, which form a characteristic *tricolon auctum*; cf. note on c. 31, 2.] In death, as in life, Agricola was *felix*.

dignitate : so used for *dignitate senatoria* in *A*. 3. 17, 4. For *incolumi dignitate* cf. Cic. *Lig*. 19.

44, 5. nam sicut ei non licuit. Here there is clearly an antithesis (as in *Dial*. 11, 2; *H*. 5. 7, 2, etc.), between *sicut* and *ita*. As on the one hand he missed a great happiness, so on the other he escaped great misery. The attempt of Boetticher and others to explain the manuscript text by supposing such an unprecedented ellipse as that of supplying *solacium tulisset* from *solacium tulit* may be dismissed. To omit or bracket *quod* or alter it to *quodam* or *quondam* does not help, for *ominabatur durare* can hardly stand for *ominabatur se duraturum*. And even the latter would be wrong. What Agricola could be said *ominari* was that Trajan would be emperor, not that he would live to see him such. It is necessary to suppose a lacuna, which is best filled by Dahl's *non licuit*. It is probably to be connected with a similar lacuna in c. 45, 1 (see note).

durare. So with *ad* in *Dial*. 17, 6; *A*. 3. 16, 1.

quod, etc., 'an event which he used to presage by prophecy and prayer, i.e. to foretell and long for'. Some such sense as that of *optabat* is supplied with *votis*. Cf. Virg. *Aen*. 7, 273 'si quid veri mens augurat, opto'. Trajan early attracted notice: he had inherited considerable fame from his father, who was a distinguished *legatus legionis* in the Jewish war (Jos. *B.J*. 3. 7, 31), must have become consul soon after (probably in A.D. 70; see Syme, *Tacitus*, pp. 30–31), was legate of Syria from 73/74 to 76/77, where he won the *triumphalia*, and afterwards proconsul of Asia. He was a staunch supporter of Vespasian. He is spoken of as dead in Plin. *Pan*. 89. Trajan himself had earned distinction by moving his troops rapidly from Spain to assist in the suppression of Antonius Saturninus in A.D. 89 (c. 39, 2; 41, 2, notes), and became consul in A.D. 91 (two years before Agricola's death). Presages of him at this time, deriving their force *ex eventu*, are mentioned in Plin. *Pan*. 5; Dio 67. 12, 1. The forecast *apud nostras aures* was presumably made before A.D. 89/90, when Tacitus left Rome for a provincial appointment (c. 45, 5), probably in connexion with the suppression of the *coup d'état* by Saturninus.

apud . . . aures: personification, as in many places: cf. *A.* 1. 31, 5. By *nostras* Tacitus means his own.

festinatae. The word seems to suggest foul play, 'precipitated' not 'premature': cf. *A.* 1. 6, 2; 4. 28, 2; Juv. 4, 96, etc.

solacium tulit. Agricola is the subject of *tulit* and *evasisse* is epexegetic of *solacium*, 'he took this consolation with him, that he had escaped . . .', as in Ovid, *M.* 5, 191–2 'magna feres tacitas solacia mortis ad umbras a tanto cecidisse viro': cf. Lucan 8, 314–16; Stat. *Silv.* 2. 5. 24–27. Furneaux–Anderson understood *tulit* as *accepit* ('he received this consolation'; cf. *H.* 4. 85, 1), Gerber–Greef as *adtulit* (sc. *nobis*: 'he brought us this consolation') but the passage of Ovid supports the simple interpretation. [*Grave* 'effective' (*E*: cf. Cic. *ad Fam.* 6. 3, 4 *consolatio gravior*) is choicer than the marginal reading of E^2 *grande* which became the popular synonym for *magnum* in late Latin. A similar confusion occurs in the manuscripts at Pliny, *N.H.* 16, 220; *Pan. in Mess.* 96.]

spiramenta, 'breathing spaces', i.e. 'pauses' (so used apparently only here and in Ammianus), rhetorically synonymous with *intervalla.* For the use of *per* see on c. 29, 1.

continuo, an adjective, as everywhere else in Tacitus.

uno ictu: a figure perhaps suggested by the famous wish of Gaius (Sen. *de Ira* 3. 19, 2) 'ut populus Romanus unam cervicem haberet, ut scelera sua . . . in unum ictum . . . cogeret'.

exhausit, 'drained of its blood': cf. *A.* 12. 10, 1 *caedibus exhaustos*; and *A.* 13. 42, 4 *Italiam . . . hauriri.*

45, 1. Non vidit, etc., cf. *de Orat.* 3. 8 'non vidit flagrantem bello Italiam, non ardentem invidia senatum, non sceleris nefarii principes civitatis reos, non luctum filiae, non exilium generi', etc.; Sulp. *ad Fam.* 4. 5, 5; Sen. *Suas.* 6, 6; Sen. *Cons. ad Marc.* 20, 5; [Plut.] *Cons. ad Apoll.* 110 F; Kassel, p. 82.

obsessam curiam, recalling the act of Nero at the trial of Thrasea (*A.* 16. 27, 1).

eadem strage: so in the denunciation of Regulus in *H.* 4. 42, 4 'cum . . . innoxios pueros, inlustres senes, conspicuas feminas eadem ruina prosterneres'.

consularium. In addition to Rusticus and Helvidius, Suetonius (*Dom.* 10, 11, 15) lists as distinguished victims Domitian's relatives T. Flavius Sabinus (*cos.* 82) and T. Flavius Clemens (*cos.* 95), Civica Cerialis (cf. c. 42, 1), Acilius Glabrio (*cos.* 91), Ser. Cornelius Salvidienus Orfitus, a friend of Nerva (*cos. ann. inc.*; cf. Phil. *Vit. Apoll.* 7. 8, 33), L. Salvius Otho Cocceianus, nephew of the emperor Otho and a connexion of Nerva (*cos. ann. inc.*), Mettius Pompusianus (*cos.* under Vespasian; cf. Dio 67. 12, 3), L. Aelius Lamia (*cos. suff.* 80; cf. Juv. 4, 154), and Sallustius Lucullus, a latter-day governor of Britain.

feminarum exilia: cf. *H.* 1. 3, 1. Pliny tells us of Gratilla, perhaps wife of Arulenus Rusticus (c. 2, 1), Arria, widow of Thrasea, and her daughter Fannia, wife of Helvidius (*Ep.* 3. 11, 6; cf. 7. 19, 4; 9. 13, 5). Domitilla, the widow of T. Flavius Clemens, was also banished (Suet. *Dom.* 15, 1).

exilia et fugas: often paired, as in Cic. *de Orat.* 3, 9; *fuga* is a more general term, covering *relegatio* or other less severe forms of banishment.

Carus Mettius: the famous delator, the accuser of Senecio (see on c. 2, 1), of Fannia (Pliny, *Ep.* 7. 19, 5) and of many others. See Pliny, *Ep.* 1. 5, 3; 7. 27, 14; Mart. 12. 25, 5; Juv. 1, 36 (where the Scholiast gives some further particulars). The *una victoria* may be the case of the Vestal Cornelia (Pliny, *Ep.* 4. 11, 6).

censebatur, 'was estimated', 'appraised', a post-Augustan use. The abl. is that of value; as yet his power was counted by one victory only.

arcem. Domitian had constructed a huge villa in the Alban Hills on the site of the ancient Alba Longa (mod. Castel Gandolfo). Traces of the three great terraces on which it was constructed still survive (Lugli, *Bull. Comm. Arch. Comm. di Roma*, 1918, 29 ff.). It was called *arx* (Dio 67. 1, 2) not because it was a stronghold in which Domitian lurked but because it incorporated the citadel of the ancient city (Juv. 4, 60; Livy 7. 24, 8; *CIL.* vi. 2172). *Intra* is emphatic; his voice was not yet heard beyond it, not in the senate.

Messalini: L. Valerius Catullus Messalinus, ordinary consul with Domitian in A.D. 73, a famous blind accuser, eloquently described in Pliny, *Ep.* 4. 22, 5, and in Juv. 4, 113–22, who calls him 'mortifer', and 'grande et conspicuum nostro quoque tempore monstrum'. He died apparently before Domitian (Pliny, loc. cit.). His family was perhaps connected with the poet. On him see Syme, *Tacitus*, pp. 5–6.

Massa Baebius: mentioned in A.D. 70, as then a procurator of Africa, 'iam tunc optimo cuique exitiosus, et inter causas malorum, quae mox tulimus, saepius rediturus' (*H.* 4. 50, 2). The Schol. on Juv. 1, 35, makes him, as also Carus and Latinus, to have been among the freedmen buffoons of Nero's court.

etiam tum reus. In A.D. 93 Pliny was deputed by the senate, with Senecio, to prosecute him for misconduct as proconsul of Hispania Baetica, and gives an account of the proceedings to Tacitus for insertion in the *Histories* (*Ep.* 7. 33; cf. also 3. 4, 4; 6. 29, 8). He was condemned and his property was confiscated, but he turned upon Senecio with a charge of *impietas.* This passage shows that he was on trial at the time of Agricola's death, and that it was believed that he would be crushed, but that later he became formidable again. *Etiam tum* 'still as before'. In A.D. 93 Massa was still a mere *reus*; he had not yet become the dangerous *accusator* that he was to be. The prosecution of Senecio led to attacks on his friends and allies, Rusticus, Mauricus, Helvidius, Fannia, etc., and to the expulsion of the philosophers (see note on c. 2, 2: Suet. *Dom.* 10, 3; Dio 67. 13, 3; Sherwin-White, *Letters of Pliny*, pp. 444 f.).

nostrae : those of senators. Tacitus treats as the act of the whole order, including himself (cf. Introd., p. 16), what may have been that of one person only: Publicius Certus was especially noted for having laid hands on Helvidius, and was attacked by Pliny after the death of Domitian (*Ep.* 9. 13). He says 'nullum (scelus) atrocius videbatur quam quod in senatu senator senatori, praetorius consulari, reo iudex manus intulisset' (§ 2). But the emphatic violence of the language suggests that Tacitus was himself present at some of these scenes and felt a personal sense of guilt. He probably returned to Rome shortly after Agricola's death.

Helvidium : son of the Helvidius of c. 2, 1, and stepson of Fannia (Pliny, *Ep.* 9. 13, 3). He was consul before A.D. 87. He was indicted for a supposed allusion to Domitian in a play about Paris and Oenone written by him and was put to death (Suet. *Dom.* 10). Pliny mentions his wife Anteia, and wrote a treatise in vindication of him (*Ep.* 7. 30, 4; 9. 13, 1).

Maurici Rusticique. On the death of Arulenus Rusticus, brother of Mauricus, see on c. 2, 1. Junius Mauricus was banished. He was prominent in the senate by A.D. 68 (Plut. *Galba* 8) and is mentioned in A.D. 70 as asking for the publication of the *commentarii principales* (*H.* 4. 40, 3)—an appropriate request if his father had been keeper of the senatorial minutes (cf. note on c. 2, 1). His friend Pliny speaks of his relegation (3. 11, 3), and of his high character (4. 22, 3), and addresses some letters to him. He returned after Domitian's death and became an intimate of Nerva and Trajan.

All three belonged to a hereditary group which preferred outspoken opposition to Domitian's régime to the policy of quiet collaboration which was pursued by men like Agricola, Tacitus, and Pliny.

visus. *Visus* means either 'the appearance of Mauricus and Rusticus' or 'the look which they gave us'. The latter is probably correct (cf. Pliny, *N.H.* 8, 80 'luporum visus noxius esse creditur'), since a positive look makes a better balance with Senecio's blood. In either event a verb seems to have dropped out. It is impossible to assume a zeugma with *perfudit*. The context demands a sense such as 'put us to shame' or 'wounded us' but there is no means of deciding what the missing verb was: Reitzenstein's *adflixit* or Anderson's *dehonestavit* are among the best proposals. We have also considered *polluit*. [The loss may be connected with the loss of *non licuit* or a similar phrase in c. 44, 5. On the hypothesis that the archetype of *E* had approximately 13 letters to the line, the two passages would be some 44 lines apart. The manuscript may have suffered damage which affected both sides of a page and resulted in the two losses. The marginal reading *Mauricum . . . divisimus*, 'parted (the brothers) M. and R.', cf. *H.* 4, 14, 3 'fratres a fratribus dividantur' and Ovid, *Trist.* 1. 3, 73, is a very unhappy conjecture, giving an insipid and unsuitable meaning in this context. It is also excluded by the anaphora which demands that *nos* should be acc. in both places.]

sanguine . . . perfudit, as in Virg. *G.* 2, 510; Catull. 64, 399.

45, 2. Nero tamen, i.e. cruel though he was, yet he refrained from beholding the outrages he commanded. Nero was not present at the trial of Thrasea and Soranus (*A.* 16. 27, 1). Cf. also *A.* 13. 17, 3.

pars: cf. Ovid, *M.* 9, 292 'pars est meminisse doloris'.

videre et aspici: cf. c. 40, 4. So *A.* 3. 45, 1 'viderent modo . . . et aspicerentur'.

subscriberentur, 'were noted down', to be laid to our charge. Cf. Suet. *Aug.* 27 'Pinarium . . . cum contionante se . . . subscribere quaedam animadvertisset'; also Suet., *Cal.* 29; Quint.

12. 8, 8. Elsewhere the word is often used of signing an accusation (cf. *A*. 1. 74, 1). Persons were accused *ob lacrimas* under Tiberius (*A*. 6. 10, 1).

cum denotandis . . . sufficeret, etc., 'when that savage face, crimsoned with the flush by which he made himself proof against all token of shame, marked out without wincing (lit. 'was equal to marking out', sc. to his agents) so many pale cheeks'. *Sufficeret* is generally translated 'was enough to': a look was sufficient to mark them out to informers to note. But *sufficere* with the dative of a gerundive or of a noun with active sense (*labori, obsequio, bello,* etc.) means 'to be equal to' and this interpretation seems to fit better both *tot* and *pudorem*. For the use of *denotare* cf. *denotantibus vobis ora ac metum* (*A*. 3. 53, 1); *notat et designat oculis ad caedem* (Cic. *Cat*. 1, 2).

tot hominum palloribus. The abstract plural (cf. Lucr. 4, 336) is used rhetorically for 'pale faces', faces which by their paleness betrayed sympathy. The appearance of Domitian's intimates is similarly described by Juv. 4, 74 'in quorum facie miserae magnaeque sedebat pallor amicitiae'.

rubor. Domitian's countenance was naturally flushed (Suet. *Dom*. 18; Phil. *Vit. Apoll*. 7, 28). In his youth, before his character was known, says Tacitus, his frequent blush was taken as a sign of modesty (*H*. 4. 40, 1). Here his natural complexion is rhetorically said to have been used as a screen against shame, because it prevented him from betraying shame. Pliny speaks like Tacitus, 'superbia in fronte, ira in oculis, femineus pallor in corpore, in ore impudentia multo rubore suffusa' (*Pan*. 48).

45, 3. Tu vero, etc.: again a reminiscence of *de Orat*. 3, 12 'ego vero te, Crasse, cum vitae flore tum mortis opportunitate divino consilio et ornatum et exstinctum esse arbitror'. For the thought cf. also Aesch. *Pers*. 712; Xen. *Ages*. 10, 4; Virg. *Aen*. 11, 159, etc.

constans et libens, 'bravely and cheerfully'.

fatum: normally used of natural, in contrast to violent, death (as in *A*. 2. 42, 3; 71, 1; 6. 10, 3 *fato obiit*), though in c. 42, 3 with the latter meaning. Here it makes against the suspicion of poison.

tamquam expresses the judgement of those who heard his words.

pro virili portione, also in *H.* 3. 20, 2, for the more usual *pro virili parte,* 'as far as a man could'. This use of *portio* is not found before the imperial period (cf. Quint. 10. 7, 28, etc.).

innocentiam . . . donares, 'you would make him a present of acquittal', dispel the suspicion of foul play by speaking of your illness as natural. He seems to imply that the present was undeserved, and this is the impression conveyed by his whole narrative.

45, 4. adsidere valetudini, 'to sit by his illness', for *aegrotanti.* To die with one's family round one's bedside was counted a blessing ([Dionys.] *Rhet.* 5, p. 30; [Plut.] *Cons. ad Apoll.* 117 B) just as, for Andromache and others afterwards, it was a source of grief not to have been present to hear the last words of a dying man (*Iliad* 24, 743–5; [Plut.] *Cons. ad Apoll.,* loc. cit.).

satiari : cf. c. 39, 3. With *complexu* in Livy 38. 37, 8.

excepissemus, 'we should have caught up'. Cf. *A.* 6. 24, 1.

figeremus = *infigeremus.* Cf. *A.* 1. 65, 5. This use of *figo* with the abl. is rare but classical (Lepidus in Cic. *ad Fam.* 10. 34a, 3; Livy 1. 26, 10; 22. 20, 2; Suet. *Cl.* 17, 3: and poets, e.g. Virg. *Aen.* 4, 15, etc.).

45, 5. noster . . . nostrum, emphatic 'our special'.

condicione : causal abl., 'owing to the circumstance' (a post-Republican use; cf. Val. Max. 3. 8 ext. 4, etc.). *Absentiae* is a defining genitive.

ante quadriennium, for *quadriennio ante.* Cf. Nep. *Dat.* 11 *ante aliquot dies.* On the absence of Tacitus from A.D. 89/90 to 93 see Introd., p. 9.

amissus est : *es* should not be read, as the apostrophe is dropped after *donares* (cf. *filiae eius*) and resumed in *optime parentum.* Cf. c. 32, 3 for a similar *variatio.*

superfuere, 'were in abundance', see on c. 44, 2. The presence of his wife ensured all outward marks of respect; but there was still an unsatisfied longing for others dear to him. Andresen notes the use of *aliquid* to express the vague longing of a dying man.

compositus. *Componere* means 'to lay out' a body at death (*H.* 1. 47, 2; Catull. 68, 98; Hor. *Sat.* 1. 9, 28, etc.) and is the appropriate conclusion to this description of Agricola's death.

Comploratus (E^{2m}: probably a conjecture) is not found in Tacitus, is repetitive with *lacrimis*, and anticipates the final moment related in the second half of the sentence.

46, 1. We should not mourn Agricola's death but should rather be inspired by the memory of his life. This stately conclusion is woven out of commonplaces which are illustrated in the notes but it expresses an unusually moving sense of bereavement. R. G. Austin (*C.R.* 53 (1939), 116–17) has called attention to the prevalence of lyric rhythms in it (e.g. *si quis piorum* – – ⏑ – –, *ut sapientibus* – ⏑ ⏑ – ⏑ –, etc.) which intensify its expressive character. It may be compared as a whole with Cic. *Arch.* 30 and Sen. *Ep.* 99. The younger Pliny was deeply affected by it and made use of it in his account of his uncle's death (*Ep.* 2. 1, 10).

Si quis, etc. The conditional clause tells us nothing about Tacitus' personal convictions about survival after death. It is a purely conventional way of speaking which is found in Consolations (Sulp. *ad Fam.* 4. 5, 6; Sen. *Ep.* 63, 16; [Plut.] *Cons. in Apoll.* 120 B), in epitaphs (e.g. *C.E.* 433. 7–8 'si liceat saltem post tam crudelia damna / sedibus aeternis sensus refovere piorum . . .'; 1031), and in other literature (e.g. Eur. *Alc.* 744; Isocr. *Evag.* 1; Hyperid. 6, 43; Catull. 96, 1; Ovid, *Am.* 3. 9, 59; Antipater, *Anth. Pal.* 7. 23, 6; etc.). See Lattimore, *Themes in Greek and Roman Epitaphs*, p. 61; Kassel, p. 97.

sapientibus, i.e. philosophers, especially Plato and the Stoics.

placide quiescas. A regular prayer (cf. Tib. 2. 4, 49; Virg. *E.* 10, 33), often found in Consolations (e.g. [Plut.] *Cons. ad Apoll.* 120 B) and abbreviated to the formula *b(ene) q(uiescas)* in epitaphs (Lattimore, *Themes*, p. 72).

nosque domum. *Et* has been added by some editors, on the ground that Tacitus could not belong to the *domus* of Agricola; and *-que et* is a favourite combination of Tacitus, especially when a pronoun precedes, as in c. 18, 4. But Tacitus ranks himself as a member of the family in c. 45, 4 and 5. Agricola's wife would be the sole representative of his *domus*, strictly understood.

muliebribus. Unrestrained grief was regarded by the ancients as unmanly (Archil. fr. 7, 10 D.; Eur. *Orest.* 1022; Plato, *Rep.* 3, 388 a–b; cf. the sentiment ascribed to the Germans 'feminis lugere honestum est, viris meminisse' in *G.* 27, 1) and is, therefore, deprecated by the writers of Consolations ([Plut.] *Cons.*

ad Apoll. 102 D, 113 A; Sen. *Cons. ad Marc.* 1, 1; *Ep.* 99, 24, etc.).

voces. This is a familiar injunction on epitaphs (cf., e.g., *C.E.* 59, 12 'desinite luctu, questu lacrumas fundere'; Lattimore, *Themes*, p. 217 ff.) and is often ascribed to dying men (cf. *A.* 15. 62, 1 (Seneca); *A.* 2. 71, 3 (Germanicus)).

quas . . . fas est. The removal of such virtues to a higher sphere is an event for which we must neither feel nor manifest sorrow. *Fas est* is a strong phrase, with overtones of religious sanction, for *licet* or the like. It has been suggested that it is used here because of the ancient belief that excessive lamentation disturbed the peace of the dead (Rohde, *Psyche*, p. 206 n. 2) but it is in keeping with the heightened language of the whole passage.

46, 2. admiratione, etc. *E* evidently omitted *te* and read *temporalibus*. *Te* is needed and is restored in the right position by E^2 since it is a characteristic of *te, se* etc. to be placed as near the second place in the sentence or clause as possible (Kühner–Stegmann 2. 593). (Reitzenstein placed it after *potius*, Gudeman, Till, and others after *similitudine*.) *Temporalibus* 'short-lived' has been defended (as by Brotier) by making Tacitus allude to the ephemeral nature of any kind of laudation, but *temporalis*, though found in Seneca, is not used elsewhere by Tacitus and the idea is foreign to such Consolations which regard praise and imitation as being the proper and lasting rewards of the virtuous dead (Thuc. 2. 43, 2; Hyperid. 6, 42; [Plut.] *Cons. ad Apoll.* 114 D; Kassel, p. 90). Muretus saw that *temporalibus* is likely to have resulted from a dittography of *te potius* (*te potius* → tēporibus → tēporalibus) and should be deleted. In triple polysyndeton of nouns coupled by *et . . . et*, Tacitus never has the arrangement 'noun *et* adj. noun *et* noun' whereas the collocation 'noun *et* noun *et* noun with qualification', is found (see Gerber–Greef, Lexicon, p. 374 f. 'et'). [The commonly accepted correction *immortalibus* (Acidalius; cf. Cic. *Phil.* 4, 4; 10, 7, etc.) robs *similitudine* of its contrast and climax and is not palaeographically plausible. Nor is it clear what precisely it would mean in this context: neither 'lifelong' (cf. Plancus in Cic. *ad Fam.* 10. 11, 1) nor, more strictly, 'immortal' is appropriate. The reading of E^{2m} *temporibus* has been made the basis of different reconstructions of this passage by Lenchantin and Forni, but it gives an absurd sense and destroys the characteristically rhetorical *tricolon auctum* (see c. 31, 1 n. and below)—*admiratione, laudibus, similitudine*. A novel interpretation is given by R. Verdière, *Latomus* 19 (1960), 792–6.]

si. Tacitus often uses a qualifying conditional clause to build up a *tricolon auctum*; cf. *G.* 40, 4 'vehiculum et vestes et, si credere velis, numen ipsum'; *Dial.* 12, 4, etc.

suppeditet, 'suffices', as in *H.* 1. 1, 4.

similitudine colamus is a very probable correction of *militum decoramus*. The corruption could have arisen from the loss of *si-* (after *si* above) and a copyist's transposition of *d* and *n*, producing (*si*)*militunidecolamus*, which was then 'corrected' to *militum decoramus*. The correction is supported against Heinsius's *aemulatu* by *si natura suppeditet*. We can all strive to emulate a great man but actually to resemble him presupposes that we are granted the power to do so. *Colamus* 'let us honour him' is the natural word in the context (cf. Statius quoted below; Mart. 7. 23, 4; Sen. *Tro.* 300) and restores a typically Ciceronian clausula (–⏑⏑⏑– ×; cf. *esse videatur*). *Decoremus*, proposed by Fulvio Orsino, gives a less good clausula and is palaeographically harder (see above). It is defended by reference to Ennius 17 V. 'nemo me lacrumis decoret'; cf. Virg. *Aen.* 11, 25.

46, 3. id: explained by the infinitive following, as in c. 39, 2. From the fact that only Agricola's daughter and wife are mentioned (cf. cc. 44, 45) it appears that Tacitus had no children of his own in A.D. 98.

revolvant, 'ponder over'; so (without *in animo* or a pronoun), in *A.* 3. 18, 4; 4. 21, 1. This use appears in Virgil and Ovid, but apparently first in prose in Sen. *Ep.* 80, 3 (with *mecum*).

formamque ac figuram, 'the form and shape', rhetorical synonyms, thus coupled in Cic. *Tusc.* 1, 37; *de Or.* 3, 179; Plin. *Pan.* 55.

corporis . . . imaginibus. For this custom cf. Plin. *Ep.* 2. 1, 12 'Verginium iam vanis imaginibus . . . audio, adloquor, teneo' (= *complector* in Tacitus), and Stat. *Silv.* 2. 7, 124 ff.:

> Haec (Lucan's widow) te non . . .
> falsi numinis induit figura,
> ipsum sed colit et frequentat ipsum
> imis altius insitum medullis
> ac solacia vana subministrat
> vultus, qui simili notatus auro
> stratis praenitet,

where *falsi numinis* alludes to the common practice of representing the dead in the form of a divinity (cf. Suet. *Cal.* 7).

complectantur, 'cling to', 'cherish' (sc. *animo*).

non quia, etc., 'not that I would forbid', subjunctive of rejected reason.

imbecilla ac mortalia. The transience of physical memorials, such as statues, etc., in comparison with the durability of character was a commonplace as old as Pindar (*Nem.* 5, 1). Cf. especially Isocr. *Evag.* 73–75; Xen. *Ages.* 11, 14; Cic. *Arch.* 30; Sen. *Cons. ad Marc.* 24, 5. The words of Tiberius in *A.* 4. 38, 1 are very similar: '. . . haec mihi in animis vestris templa, hae pulcherrimae effigies et mansurae'.

forma mentis, cf. *forma animi* above. Here it seems to be used almost in a Platonic sense.

quam tenere, etc., 'which you can preserve and reproduce not by the material and artistic skill of another, but only in your own character'. For *per* cf. notes on c. 3, 2; 29, 1.

46, 4. quidquid, etc. The thought that a man's fine qualities will be immortal through memory was conventional in Consolations; cf. Sen. *Ep.* 99, 4; *Cons. ad Marc.* 1, 3; Livy 39. 40, 7; Vell. Pat. 2. 66, 5. Notice the strong use of the simple verb *miror* = *admiror* as in Livy 5. 26, 8, etc.

ex, 'belonging to': we should say 'in'.

mansurumque est, 'and is destined to abide', stronger than *manebit*.

in animis, etc., 'in the hearts of men through the endless course of ages, by the fame of his achievements'. *Fama rerum* has this sense in *H.* 4. 39, 3 'claros rerum fama', and Livy 25. 38, 8 'vivunt vigentque fama rerum gestarum'. Halm's *in fama* is neither acceptable nor necessary. *Fama rerum* is instr. or causal ablative, and the meaning is made clear by the next clause. The first *in* is local, the second temporal: *in aetern. temporum* = *in aeternum* or *aeternum* (*A.* 3. 26, 2; 12. 28, 2). The double use of *in*, followed by an abl., is characteristic of Tacitus' striving after variety.

nam, etc., explains the preceding statement: Agricola's glory will not be forgotten, because his achievements are placed on record. The threefold alliteration is noteworthy.

inglorios et ignobiles, as in Cic. *Tusc.* 3, 57 which Skutsch (*Kl. Schr.* 302) suggested was an echo of Ennius.

obruet. *E* can be inferred to have read 'multos veterum velut inglorios et ignobilis oblivio obruet' but Decembrio, who saw the Hersfeldensis in 1455, states that he read 'multos veluti inglorios et ignobiles oblivio obruet'. There can be no doubt that the manuscript which Decembrio saw was *E* (see Introd., p. 83) and, therefore, Decembrio must have copied the text inaccurately as he did elsewhere (R. P. Robinson, *The Germania of Tacitus*, p. 11) and there is no authority for deleting *veterum*. Editors, feeling that Tacitus should be making the point that many of the heroes of old have been forgotten whereas Agricola will be remembered, have accepted Haupt's conjecture *obruit* (perfect) but the point is as effective if Tacitus says that although many of the heroes of old are still remembered and so may seem assured of perpetual remembrance, they will in fact be forgotten whereas Agricola's memory will survive for all time because in Tacitus he has found a *vates sacer* which they lack. For *oblivio obruere* cf. Cic. *Brut.* 60.

The sentiment resembles that of Hor. *Odes* 4. 9, 25–28. Tacitus ends on the same note with which he began, 'clarorum virorum facta moresque posteris tradere'. And the fact is that without this book Agricola's name would have been known to us only from a brief reference in two passages of Dio (39. 50, 4 and 66. 20, 1–2) and from two inscriptions of A.D. 79; one on leaden waterpipes found at Chester (*ILS* 8704*a*): *Imp*(*eratore*) *Vesp*(*asiano*) *viiii* *T*(*ito*) *imp*(*eratore*) *vii* *co*(*n*)*s*(*ulibus*), *Cn. Iulio Agricola leg*(*ato*) *Aug*(*usti*) *pr*(*o*)*pr*(*aetore*); the other on a building inscription from the forum at Verulamium; containing the fragmentary allusion: [*Cn. Iulio A*]*gric*[*ola leg. Aug. pro*] *pr.* (*Ant. Journal* 36 (1956), 10). See Plate I. Another inscribed leaden-pipe has recently been discovered at Chester.

narratus et traditus superstes erit. The expression of this feeling in contemporary sculpture is exemplified by the Arch of Titus and the Column of Trajan, both narrating in stone the *fama rerum* of the commemorated dead.

APPENDIX 1

THE CHRONOLOGY OF AGRICOLA'S CAREER

AGRICOLA was born on 13 June, A.D. 40, shortly before his father's execution by the Emperor Gaius.[1] He was praetor in A.D. 68, the last year of Nero's reign. His tribunate, therefore, would normally fall in 66 and his quaestorship in 64, one year's interval being required between the tenure of those various offices. As quaestor in the province of Asia he served at first under the proconsul Salvius Titianus, and later, perhaps, under L. Antistius Vetus, though Tacitus does not mention the second stage (see n. on c. 6, 2).

Agricola thus held the quaestorship in his twenty-fourth year, and the praetorship in his twenty-eighth year, one year and two years respectively before the prescribed age for those offices. He had, in fact, availed himself of the *Lex Papia Poppaea*, which granted candidates a remission of one year for each child born to them. Agricola's first son, born in 63 or late in 62, lived long enough to enable his father to claim the privilege of the law (c. 6, 2); and his daughter, born in 64, made possible the remission of yet another year. Two years were thus gained in all, despite the death of the son meantime.

In A.D. 70 he was appointed to command the Twentieth Legion, probably stationed at Wroxeter, and returned to Rome either in 73 or, with his chief Cerialis, before May 74. On return he was elevated to the patriciate and was thereafter appointed governor of Aquitania, a post which he held less than three years. During his governorship he was designated consul and held the office in 77. The precise months in which this consulship was held remain unknown, and the matter has been much discussed, since the date of Agricola's arrival in Britain has seemed to depend upon it. The question may be summarized as follows.

[1] c. 44, 1.

Agricola was *consul suffectus*, the *consules ordinarii* being Vespasian and Titus, the latter soon giving place to his brother Domitian. There is, however, no evidence either for the date when Agricola assumed the consulate or for the number of months that he held it. Tacitus writes that 'during his consulship Agricola betrothed his daughter to me, and after it he gave her in marriage and was immediately (*statim*) appointed to Britain, following the additional honour of a pontificate' (c. 9, 6). Agricola's arrival is noted as follows: *Hunc Britanniae statum . . . media iam aestate transgressus . . . invenit, cum . . . milites velut omissa expeditione ad securitatem . . . verterentur* (c. 18, 1). He arrived, then, late in the season, and this lateness has been explained by supposing that he went to Britain in the year of his consulship.[1] Undoubtedly, the first impression conveyed by the narration of the preliminaries in c. 9 is that all followed hard upon each other.

Considering the question, then, in these terms, it must first be asked whether Agricola could have held a consulship and arrived in Britain by midsummer of the same year. The Roman summer was normally considered to run from mid May to mid August; and, in the absence of definition, Tacitus may be taken to have used the term in its ordinary sense. This view wins support from his usage in *Hist.* 5. 23, where the equinox (22–23 Sept.) marks the turn (*flexus*) of autumn, and in *Ann.* 11. 31, 4, where *adultus autumnus* is the month ending not later than 15 Oct. *Media aestas* would thus cover from mid June to mid July.

In the matter of the consulate, by the time of Vespasian, in order to secure a steady stream of consulars for the appropriate offices, the suffect consulship, as held by Agricola, had been reduced to as little as two months. If, therefore, Agricola had held a consulship in March–April, he could have reached Britain, a journey of about 25 days at normal rate, by the beginning of June. But when allowance is made for the wedding and for a pontificate, the date of midsummer (mid June to mid July) begins to look more difficult. Moreover, the very possi-

[1] Most recently by Büchner, *Tacitus und Ausklang* (Studien zur römischen Literatur, Band iv, 1964), pp. 99 ff. Older literature in Anderson's edition of the *Agricola*, App. 1, pp. 166 ff.

bility of arrival by midsummer depends entirely upon the assumption that Agricola's consulate came early in the year, a point on which there can be no certainty, pending further discoveries of *Fasti*.

No less uncertain is the argument based upon the mutiny of the Usipi, who had been recently levied in Germany and brought to Britain for initial training. Mommsen concluded that these Usipi had been annexed and conscripted by Domitian in 83, and that this fixed the date of the summer in question.[1] But the territory of the Usipi, who were by this time neighbours of the Mattiaci and inhabitants of the Lahn basin, lay so close to the Empire that conscription, as among the Frisii, may already have operated among them. Much discussion has been devoted to the matter, but no certain argument is to be founded upon it.

A much stronger argument, and one in favour of A.D. 84 as the year of Mons Graupius, is implicit in the description of its relation to the Chattan war and the triumph which followed its end. The impact of the news from Caledonia, where the victory was won *exacta iam aestate* (c. 38, 2), was contrasted with the *conscientia derisui fuisse nuper falsum e Germania triumphum*. This triumph took place late in 83, and it is the plentiful coins of 84 that bear the new title *Germanicus* for the first time and the fifth imperatorial acclamation. The inference must therefore be that the news of Mons Graupius came almost a year later, not at the same time as the German victory, whose triumph it would then have preceded.[2]

Coins of the seventh imperatorial acclamation, of the year 84, are not only uncommon, but their commemorative reverse has a most unusual design. It represents a cavalry trooper riding down a barbarian, while another foe lies dead on the field. This is a battle composition of the kind which appears in Imperial sculpture at Adamklissi, but is hardly seen isolated again in Roman art. It is, however, highly relevant to the victory of Mons Graupius, which was won by cavalry (c. 35, 2; 36, 3; 37, 1) in the manner depicted on the coin—*sequi, vulnerare capere atque eosdem oblatis aliis trucidare* (c. 37, 2). The

[1] *Provinces* I. p. 150 and note.

[2] So Syme, *CAH* xi (1936), p. 164 = *Tacitus*, I, p. 22.

acclamation and commemorative issue of coins is the Emperor's share of the honours conferred upon Agricola, and this point is decisive for the dating of the governorship to A.D. 78–84.[1]

A subsidiary point of some interest may be appended.[2] An inscription (*ILS* 1025) from Tivoli records a substantial draft from the Ninth Legion as taking part in the *expeditio Germanica* under a *tribunus laticlavius*, who was decorated by Domitian in what must thus be the Chattan war. The seconding of this draft would very well explain the weakness of that legion (c. 26, 1) in Agricola's sixth campaign, which would then be dated to 83. The inscription of Velius Rufus (*ILS* 9200), which has been assigned to the same year, mentions other detachments from all British legions, under a *primus pilus*, but the dating is uncertain and cannot be used for the purposes of this argument.

[1] C. M. Kraay, *American Numismatic Society, Museum Notes* ix, 109 ff.

[2] R. K. McElderry, *J.R.S.* 10 (1920), 68 ff.

APPENDIX 2

THE MUTINY OF THE USIPI

FOUR accounts at least existed in antiquity of this famous mutiny, which gave rise to an accidental circumnavigation of Britain. The best known and most detailed is that of Tacitus (*Agr.* 28). But it had been preceded by a version which was drawn upon by Pompullus, a rival poet mocked by Martial (*Epigr.* 6. 61, 1–4) about A.D. 90. The latest account is by Cassius Dio (66, 20), and this, differing in certain respects from that by Tacitus, may also draw upon the unknown source used by Pompullus (but see notes on c. 28, 1; 44, 4).

The differences between Tacitus and Dio are interesting. Tacitus specifies the unit as composed of conscripted Usipi, irritated by initial training; Dio mentions soldiers only, but adds that they killed their centurions and a tribune, thereby implying a milliary cohort, where Tacitus mentions a centurion and legionary instructors. Tacitus next recounts that three warships were commandeered, their pilots being murdered or otherwise disappearing, while Dio mentions only ships, but then states that the men sailed along the west coast, as waves and wind carried them, whereas Tacitus refers to no explicit course. The accounts then diverge. Tacitus describes numerous opposed landings for water and necessities on British territory. Dio notes that the mutineers escaped detection when they touched at Roman garrisoned positions on the east coast, and then closes the story: but Tacitus, omitting any reference to landings in Roman-occupied Britain, ends his narrative much more dramatically by relating a resort to cannibalism, shipwreck on the Suebic shore, and the enslavement of the survivors, some of whom passed by repeated sale to the Roman Rhineland, where their story made them famous. As for the date of the episode, Tacitus specifies the penultimate summer of Agricola's campaigns, namely, A.D. 83, while Dio puts it into connexion with Titus and dates it to A.D. 79.

The discrepant chronology must surely be due to compression by Dio of other events in the British war, which he mentions but fails to describe: Tacitus can hardly be wrong on the date. Dio's tribune may be a false inference, while the undetected landings of Dio might lie behind the Tacitean phrase *nondum vulgato rumore, ut miraculum praevehebantur* which would be intended to cover any slur arising from the fact that the mutineers escaped unintercepted. Dio's statement that the Usipi sailed eastwards indicates that the resultant circumnavigation was clockwise, and this would agree equally well with his account of later unintercepted contacts on the other coast (ἐκ τοῦ ἐπὶ θάτερα) and with the reference by Tacitus to the many landings in British territory, that is, anywhere between the Mull of Kintyre and Aberdeen. It would also fit with the final shipwreck in Germany or Denmark (*peninsula Suebica*) as recounted by Tacitus. The different versions are, in short, different summaries of a single story out of which Dio and Tacitus have each selected what interested them. The facts as recounted are not contradictory, but complementary, and both contributions are useful, though Tacitus, as might be expected, is more informative. Neither author discloses the ultimate fate of the survivors; whether recognized or self-declared, they were presumably thoroughly interrogated and sentenced to death. The complete version of the facts must certainly have emerged at the official inquiry following recognition of the slaves as ex-recruits.

APPENDIX 3

AGRICOLA AND THE FORTH-CLYDE ISTHMUS

Inventus in ipsa Britannia terminus. namque Clota et Bodotria, diversi maris aestibus per inmensum revectae, angusto terrarum spatio dirimuntur: quod tum praesidiis firmabatur (c. 23).

TACITUS makes it clear that this frontier was a temporary one, established in the fourth campaign (c. 23) and relinquished in the sixth (c. 25, 1). Traces of it have been recovered by the spade, sometimes certain and sometimes shadowy. Much more spadework is needed if any detailed picture is to be recovered. A review of the present position is here offered.

The island of Great Britain is reduced to narrow width at two points in its length, between Tyne and Solway and between Forth and Clyde. The first neck, traversed by Hadrian's Wall, was for long the north-west frontier of the Roman Empire. But the second and more northerly neck is twice as narrow as the first, and offers an almost straight passage from tideway to tideway with only a slight climb in between.

In A.D. 140–2 Antoninus Pius chose the second neck as the site for a frontier barrier held for over forty years. The new Wall, described as *murus caespiticius*, was a rampart built of turf sods laid in regular courses. The forts along it were about two miles apart, and defended mostly by earth ramparts, though sometimes by a stone wall. A deep, wide ditch ran throughout in front of the Wall and a military road behind it. The course occupies the south rim of the gap between Forth and Clyde and is skilfully planned for immediate tactical advantage and northward outlook.[1]

The siting of the Antonine Wall is highly relevant to

[1] The Wall is studied in detail by Sir George Macdonald, *The Roman Wall in Scotland*[2] (O.U.P., 1934). More recent research is summarized by K. A. Steer, *J.R.S.* 50 (1960), 87 ff.

Agricola's line of posts, since in general there is no doubt that any posts intended to command the Isthmus must have taken the line chosen for the Wall, while in particular pre-existent Agricolan posts could well have decided the line which the Wall-builders selected. It is thus not surprising that some sites on the Antonine line should have yielded vestiges of Agricola. These are as follows:

(i) *Old Kilpatrick*, the western terminal fort of the Antonine Wall. Excavations produced Flavian and late first-century Samian ware and mortaria.[1]

(ii) *Cadder*. Excavations yielded slight ceramic evidence for an Agricolan occupation, possibly not on the same site, though the excavator thought that the ditch-system of the 3-acre fort was Agricolan in origin.[2] Haverfield had already suggested that an altar from Cadder erected by the prefect L. Tanicius Verus was of Agricolan date, identifying him with a homonymous centurion in Egypt (*CIL*. iii. 34), who kept visiting the Singing Statue of Memnon in A.D. 80–81. Tanicius is indeed a rare *nomen*; but, as Professor E. Birley has observed,[3] the careers of the men forbid their identity, the prefect probably being a lineal descendant of the centurion.

(iii) *Bar Hill*, the central fort of the Antonine Wall. An earlier rectangular fortlet, 180 × 140 feet in size, lies below the second-century work, but only its ditches were recovered, enclosing under two-thirds of an acre, with an attached annexe not fully traced but matching Croy Hill (see iv, below). The fortlet ditch had become overgrown with trees before being filled in to make room for the Antonine fort. This creates the strongest presumption that the work was Agricolan.[4] The plan suggests an intermediate patrol-post with signal-beacons in the annexe. See Fig. 17.

(iv) *Croy Hill*, the next fort east of Bar Hill. An earlier fortlet 200 × 140 feet in size, with an annexe, was again found below the Antonine fort, though evidence for a gap between the two occupations was not recovered. The plan of these

[1] Macdonald 340 f.
[2] Macdonald 310 f.
[3] *Roman Britain and the Roman Army*,[2] p. 91.
[4] Macdonald 272 f. See, however, Steer 90.

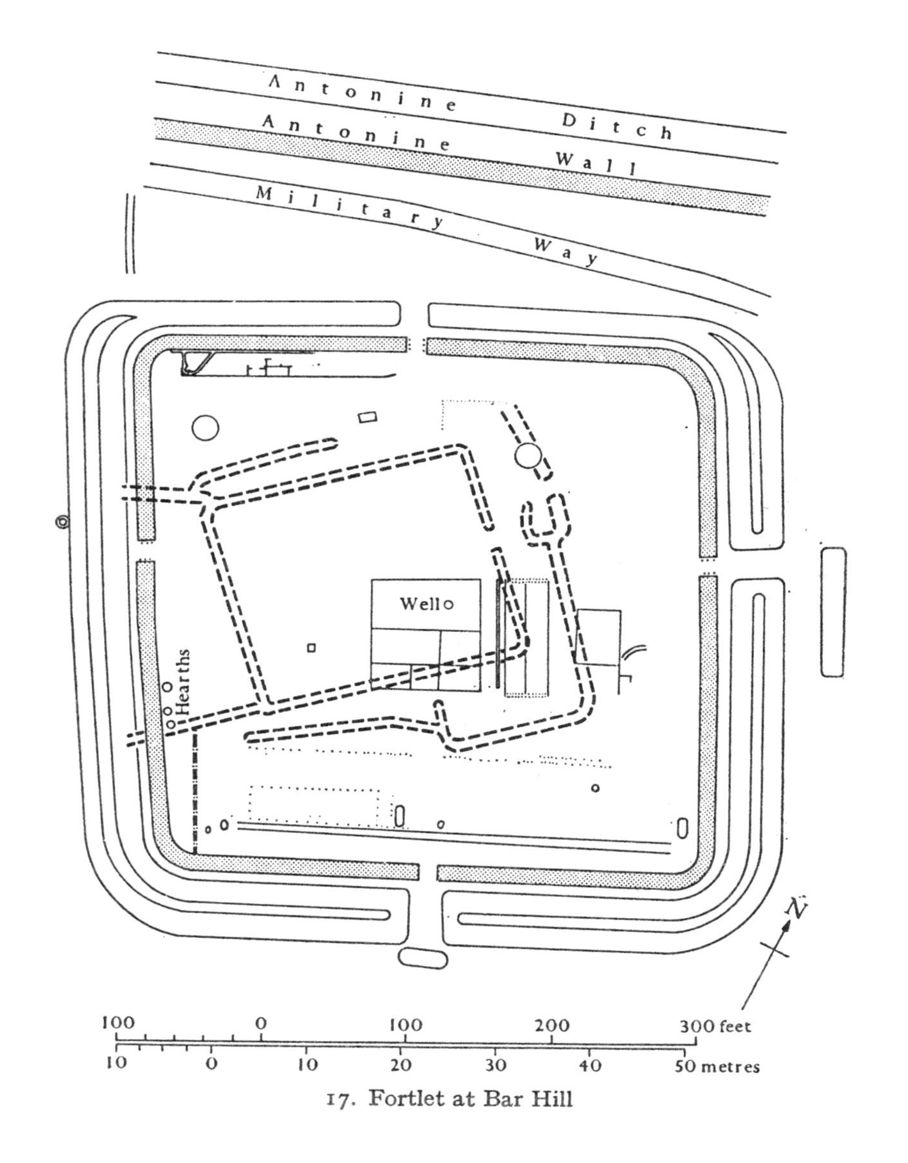

17. Fortlet at Bar Hill

structures is so similar that they must belong to the same series.[1] See Fig. 18.

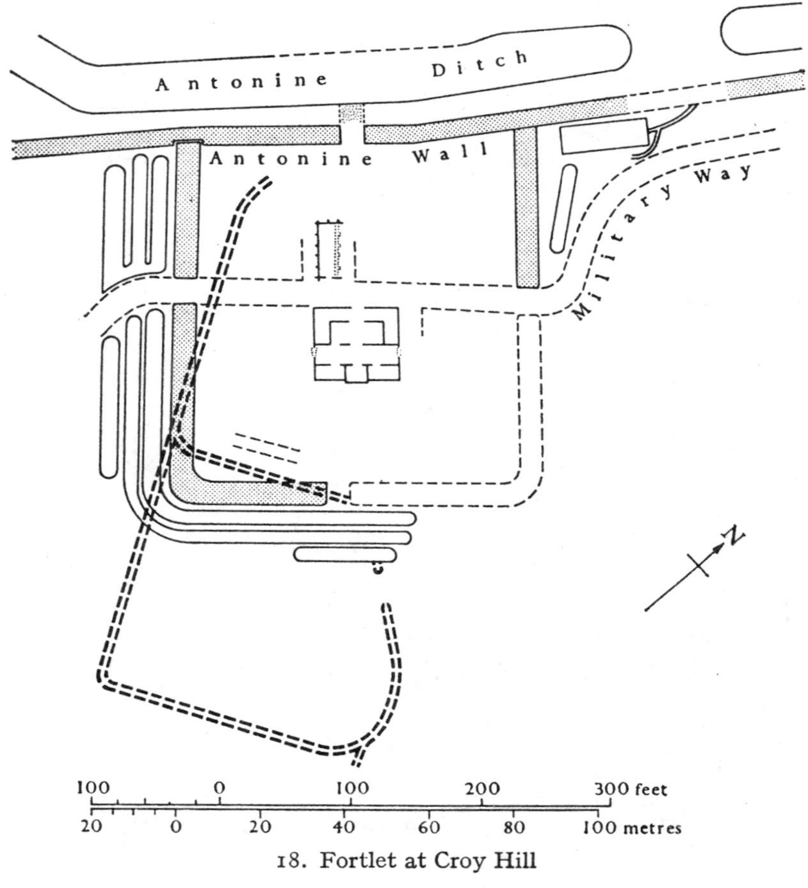

18. Fortlet at Croy Hill

(v) *Castlecary*, near Bonnybridge. A little decorated and plain Samian ware found here belongs to the age of Agricola.[2]

[1] Macdonald 267 f.; Steer 90.

[2] Macdonald 250 ff., 467; *Royal Comm. on Anc. and Hist. Mon. of Scotland, Stirlingshire* I (1963), pp. 102–3.

The site is strategically important, since it occupies a position where the line is accessible from the south.

(vi) *Camelon*, three-quarters of a mile north of the Antonine Wall at Falkirk. Excavations of 1899 yielded Flavian pottery and evidence for an extensive Flavian site.[1] If the temporary frontier now took to the south bank of the Carron and so to the Forth, Camelon could have formed part of it: but if it continued with the Wall along the higher ground, Camelon must be concerned with the Carron bridge-head on the north road, as it was when the Wall came.

By contrast with the six sites enumerated, two putative Agricolan sites on the Isthmus must on existing evidence be dismissed. At *Rough Castle* a field of concealed obstacle-pits, or *lilia*, covering the approach to the Antonine fort, has been thought to antedate it.[2] There is no evidence for this; and recent excavation has shown the site to be devoid of pre-Antonine remains.[3] At *Mumrills* recent investigation has shown that ditches thought in 1929 to be Agricolan in fact belong to the defences of the Antonine fort, while the site as a whole does not yield pre-Antonine relics.[4]

The results which emerge are that an Agricolan occupation of six sites from Camelon westwards to Old Kilpatrick seems proved, whatever may have been the arrangement further east. These sites fall into two classes. Larger, if ill-defined, sites seem to exist at Camelon, Castlecary, Cadder, and Old Kilpatrick. Smaller structures certainly occur at Croy Hill and Bar Hill, and these are clearly definable as signalling posts, with annexes for beacons. It is evident that the known sites do not form an alternating system of large and small sites, but it will be observed that the two small signalling forts in adjacent positions occupy high ground, away from any obvious line of penetration. They are not signalling-towers alone, but are somewhat larger versions of the fortlet at

[1] *Proc. Soc. Ant. Scot.* 35 (1900–1), 329 ff.; *Royal Comm., Stirlingshire*, pp. 107 ff. For recent excavations of Flavian material see *J.R.S.* 50 (1960), 213.

[2] Macdonald 234 ff.

[3] *Royal Comm., Stirlingshire*, pp. 100 ff.

[4] Steer 89; *Proc. Soc. Ant. Scot.* 94 (1960–1), 87–90.

Martinhoe,[1] which housed a *centuria* for beacon signalling, though a tower for signals of another kind may also have crowned their gateway. The remains on the Isthmus, however, are in general insufficient to permit as yet any coherent picture of Agricola's temporary disposition.[2]

[1] *Antiquity* 39 (1965), 255.

[2] Doubt must also attach to the material from Old Kilpatrick. Sir George Macdonald identified much of the pottery as Flavian but Mr. B. R. Hartley, who has re-examined all the Samian ware from Scottish and North British sites recently, states that none of the Old Kilpatrick finds are likely to be as early as Agricola.

APPENDIX 4

RAW METAL SUPPLIES IN ROMAN BRITAIN

Fert Britannia aurum et argentum et alia metalla, pretium victoriae (c. 12, 6).

AMONG the raw metals of Britain,[1] Tacitus specifies only the nobler by name: the rest are contained within the vague term *metalla*. Nor do his words, *pretium victoriae*, imply that Claudius invaded Britain in order to win these minerals, although by Roman law they became imperial property.[2]

1. GOLD

In south-west Wales, a gold-mine existed at Dolaucothi, Pumsaint (Carmarthenshire), close to the Roman road from Brecon to Llanio (Cardiganshire). Opencast workings, drifts, and well-cut adits following an auriferous quartz reef extend for about half a mile along the hillside on the left bank of the Cothi. No other spot in Britain displays such a potential for revealing Roman mining-technology at its best, although

[1] This survey excludes coal, for which see G. Webster, *Antiq. Journ.* 35 (1955), 195–217; Adams, Bradburn, and Boon, *Geol. Mag.* 102 (1965), 469–73.

[2] Solinus, *Collectan. Rerum Memor.* 22, 11 speaks of *metallorum larga variaque copia quibus Brittaniae solum pollet venis locupletibus*; cf. *Paneg.* 5, 11, *tot metallorum fluens rivis*. For general accounts see O. Davies, *Roman Mines in Europe* (1935), pp. 140–64; R. G. Collingwood, *An Economic Survey of Ancient Rome* (ed. T. Frank), 3 (1937), pp. 34–47; I. A. Richmond, *Roman Britain*[2] (1963), pp. 149–59. R. F. Tylecote's *Metallurgy in Archaeology* (1962) is indispensable for information on extraction and other processes. R. J. Forbes, *Studies in Ancient Technology*, 7–9 (1963–4) has much important material with especial reference to documentary sources. For detail of sites consult W. Bonser, *A Romano-British Bibliography* (1964), §§ 57–67.

allowance must be made possibly for medieval, and certainly for modern, exploitation (Smyth, *Mem. Geol. Surv.* i (1846), 480–4, plan and section; O. Davies, *Roman Mines in Europe* (1935), pp. 154–5; *Arch. Camb.* 91 (1936), 51–57; Nash-Williams, *B.B.C.S.* 14 (1952), 79–84). Finds include part of a de-watering wheel of well-known Roman type (Boon and Williams, *J.R.S.* 56 (1966), 122 ff.; cf. Vitruvius 10, 4). For washing milled ore (selected ore assays around 0·0006 per cent., but the vanished parts of the lode were almost certainly very much richer), a leat brought an estimated maximum of 2½–3½ million gallons a day from a dam in the Cothi 7 miles upstream (G. D. B Jones and others, *B.B.C.S.* 19 (1962), 71–84 for the course as far as the workings; the leat thereafter discharged south-eastward away from them). A bath-house about a quarter of a mile away from the mine is probably connected with a fort, of which traces may be claimed in first–second century material found at Pumsaint itself. Other finds include a hoard of second–third century gold ornaments (British and Carmarthen County Museums) and two coin-hoards, one found in 1965, closing with Carausius.[1]

There is no reason to suspect pre-Roman mining at Dolaucothi, and the gold mentioned by Strabo as a British export (4, 199) may have come from alluvial or placer deposits such as those of the Mawddach estuary (Mer.).

2. SILVER AND LEAD

Tacitus specifically mentions silver, as did Strabo before him (4, 199). The metal was extracted by cupellation from the lead ore which the British province had in abundance (cf. Pliny, *N.H.* 34, 164), although the silver-content of British ores is low.[2] Over 80 pigs of lead have been recorded from

[1] Boon, *Num. Chron.* forthcoming.

[2] The average of modern assays of British ore is about 0·013 per cent. silver, contrasting markedly with the Laurium (Athens) 1·8 per cent. It should, however, be remembered that the silver-content is very variable indeed, not only from leadfield to leadfield, but from lode to lode (e.g. Mendip ore has assayed as much as 0·4 per cent.) and also in each lode, being naturally richest near the surface of the ground. For cupellation see Tylecote, op. cit. or *Brit. Journ. Hist. Sci.* 2, 1 (5) (1964), 30–32. It seems possible that early imperial

Britain, mostly with imperial names or those of lessees.[1] Some (e.g. *CIL.* vii. 1215–16) are marked with the inscription EX ARG(*entariis*), which suggests that the mines were known as silver mines: the wording cannot mean 'from the silver-works' since it is clear that some pigs so marked cannot have been cast from desilvered lead.[2] Since very few workings can be accepted as Roman, the pigs are the best evidence for the Roman exploitation of lead in Britain. The main areas of production are as follows.

(i) *Somerset.* On the limestone Mendip plateau accessible deposits were already known to the Britons of Glastonbury Lake-village. The earliest Roman pig is of A.D. 49 (*CIL.* vii. 1201) and others run to the years A.D. 164–9 (*CIL.* vii. 1211; *EE.* 3, 121*e*; 4, p. 206). The mining-settlement of Charterhouse has yielded an inscription of Severus or Caracalla (*RIB* 185) and smelting-sites are known near Priddy (*J.R.S.* 43 (1953), 123).

(ii) *Shropshire.* The richest parts of this field lie around Shelve and Minsterley. While the evidence of old workings remains ambiguous, proof of Roman activity is supplied by four pigs of Hadrian's reign (*CIL.* vii. 1209 *a* (*b*), *c*, *e*, *f*).[3]

(iii) *Flintshire.* Here the medieval mining-district of Tegeingl preserves the tribal name Deceangli recorded by Tacitus or his copyists as Deceangi.[4] Traces of supposed Roman mining are doubtful, but badly preserved smelting-furnaces have been excavated at Pentre (Flint) (*Publ. Flints. Hist. Soc.* 10 (1924)), and about 26 pigs of lead have been recovered, two of A.D. 74 with the inscription DECEANGL(*icum*) (*CIL.* vii. 1204; *EE.* 7, 1121; cf. 9, p. 642) and about 20 in a wreck (?) in the

exploitation, depending on superficial assay, proved disappointing, and that mines were made available to concessionaires as a result. It is significant that Cicero's brother, on Caesar's staff in Britain in 54 B.C., reported that there was not a grain of silver in the island (Cic. *ad Att.* 4. 16, 7).

[1] Tylecote, *Metallurgy in Archaeology*, tables 33–34 for best list; add *J.R.S.* 53 (1963), 162.

[2] J. A. Smythe, *Trans. Newc. Soc.* 20 (1939–40), 139–45.

[3] Whittick (*Trans. Shrops. A. & N.H. Soc.* 46 (1932), 129–35) shows that 1209*a–b* are probably one and the same piece, found with wooden shovels (*Trans. Newc. Soc.* 12 (1931–2), 72–73 and pl. 8).

[4] *A.* 12. 32, 1, *inde Cangos* = *in Deceangos.*

Mersey estuary at Runcorn. One pig bears the name of the lessee C. Nipius Ascanius (*J.R.S.* 41 (1951), 142), who cold-stamped a Mendip pig of A.D. 59 found at Stockbridge (Hants) (*CIL.* vii. 1203 = *EE.* 7, 1120).

(iv) *Other Welsh leadfields.* There is no proof that old workings in west Wales are Roman, but lead found at the Severn valley fort of Caersws (Mont.) suggests that interest was taken in the resources of this region. In south Wales a mine on Cefn-pwll-dû, Draethen (Glam.) has recently been proved Roman from finds of first–second century pottery, some coated with stalagmite and associated with broken ore and charcoal suggestive of fire-setting, as well as a coin of *c.* A.D. 270–4; while a flourishing mining-settlement has been discovered at Lower Machen (Mon.) near by (*Arch. Camb.* 94 (1939), 108–10). The stump of a pig found at Caerwent (*J.R.S.* 38 (1948), 101) bearing the stamp of the Second Legion, which was stationed at Caerleon, suggests that the south-east Wales leadfield was under military management at some period.

(v) *Derbyshire.* The limestone hills of the Peak supplied much lead. One ingot carries Hadrian's name; eleven (*CIL.* vii. 1214–16; *EE.* 9, 1265–6; *J.R.S.* 31 (1941), 146 [five]; 48 (1958), 152) bear those of private lessees linked individually or in a company (*socii*) with the place-name *Lutudarum* (cf. *Archaeologia* 93 (1949), 38). No certainly Roman workings are known.

(vi) *Yorkshire.* Two pigs from Hayshaw Moor, eight miles from Ripley, bear the name of Domitian and the toponym BRIG(*anticum*), and belong to A.D. 81 (*CIL.* vii. 1207).

(vii) *Northumberland. Scoriae* and Roman remains are said to have been found near Alston, and Roman lead sealings bearing the word METAL(*la*) and the name of C(*ohors*) II NER(*viorum*), the third-century garrison of the adjacent fort of Whitley Castle, are known (*Cumb. & Westm. Trans.* 36 (1936), 109).

About two dozen silver ingots, of double-axe or bar shape, have been found in the British Isles, half of them in Ireland, where they presumably represent piratical loot. Eight carry inscriptions declaring an official source (e.g. *CIL.* vii. 1196 EX OF FL HONORINI; 1198 CUR(*ator*) MISSI(*onum*)). The weight of some approximates to a Roman pound (about 327·45 g.)

and the most probable explanation of their existence is that they were soldiers' donatives (cf. Amm. Marc. 20. 4, 8). They are all late-fourth century.[1]

3. COPPER

The copper resources of north-west Wales were extensively exploited by the Romans. Two dozen bun-shaped ingots, eight with stamped inscriptions, are the principal evidence in the usual default of well-attested workings (except at the Gt. Ormes Head; cf. Davies, *Arch. Camb.* 100 (1945), 61–66). Fifteen ingots have been found in Anglesey, six in Caernarvonshire, and three in Denbighshire.[2] Two from the Parys mine, Amlwch (Anglesey), suggest interest in the rich deposits so vastly mined in the eighteenth–nineteenth centuries. No imperial names appear on the ingots, and the industry was evidently in the hands of individuals or companies (cf. SOCIOR(rum) ROMAE, counterstamped NAT SOL [a name] of *CIL.* vii. 1200, from Aberffraw, Anglesey). A Roman mine in Llanymynech Hill (Mont.) sought copper, and lower-lying lead ore was exploited by later miners. In addition to occupation-material, the Ogof ('cave') has recently yielded a hoard of 33 *denarii* closing *c.* A.D. 148–9.[3] No copper ingots are known in the Border region.

4. TIN

Caesar mentions British tin, mistakenly declaring that the metal was found *in mediterraneis regionibus* (*B.G.* 5, 12); but it may well be that his wars not only cut, but destroyed, the well-established trade route from Bolerion (W. Cornwall) via the Rhône delta to the Mediterranean (Diod. 5. 22, 4; 5. 38, 5; Strabo (from Posidonius) 3, 147), since Strabo does

[1] K. S. Painter, *Journ. Brit. Arch. Assn.*[3] 28 (1965), 12–13 for the best list; discussion, 3–7.

[2] Lists with authenticated provenances in Roy. Comm. Anc. and Hist. Mons. (Wales): *Anglesey* (1937), *Caernarvonshire* 1 (1956) and 2 (1960); Denbighshire: *J.R.S.* 44 (1954), 106; *Arch. Camb.* 113 (1958), 69–70; Gardner and Savory, *Dinorben* (1964), 226.

[3] Boon, *Num. Chron.* forthcoming.

not list it among British exports. It is certain that in Flavian times the Spanish mines were the principal source (Pliny, *N.H.* 34, 156). This is borne out by the British remains, which are mostly late. Suspicion has undeservedly been cast on the Carnanton (St. Mawgan-in-Pyder) block (*EE.* 9, 1262; *Proc. Soc. Antiq. Lond.*[2] 18 (1899–1901), 117–23 and pl.; *V.C.H.* (Cornwall) 2 (1924), fig. 11 [upside down]: cf. Smythe, *Trans. Newc. Soc.* 18 (1937–8), 256–7; Collingwood, *Camb. Anc. Hist.* xii (1939), 290) with cold stamps, of a helmeted bust with shield, and DDNN (i.e. *Dominorum Nostrorum*). There are also eight blocks of pewter from the Thames at Battersea (London) with a fourth-century stamp (*EE.* 9, 1263). Roman coins found in Cornwall are mostly late, although several belong to the reign of Nero, while five milestones comprise three of the third (*RIB* 2230, 2232, 2234) and two of the fourth century (*RIB* 2231, 2233). It looks as if the area was opened up after the closure of mines in Spain and Dacia. Most of the Romano-British vessels and objects of tin and pewter are of third–fourth century date. The failure of Tacitus, therefore, to mention tin specifically seems intelligible.

5. IRON

Britain is rich in iron ores and these were often worked in Roman times, frequently on a small scale. Among the most interesting finds are traces of six shaft-furnaces at Ashwicken (*J.R.S.* 50 (1960), 227) in Norfolk and the Redesdale deposits near the fort of *Habitancum* (Risingham) which appear to have yielded blooms resmelted at Corbridge and forged into weapons and other gear.[1]

Although the province cannot be ranked among the great

[1] *Arch. Ael.*[3], 8 (1912), 207–9; Richmond, *A History of Northumberland*, 15 (1940), 80. Unlike the other metals mentioned in this appendix, iron was not liquefied in the smelting, except accidentally. The result of reduction was a spongy mass which required prolonged beating to consolidate it and to extrude foreign matter. The resultant blocks were sometimes forged together to make bars several feet in length for constructional purposes: two can be seen in Chedworth Roman villa site museum. Although examination of edged tools shows little use of steel, carbonized iron (steel) was well known in Roman times.

iron districts of the Roman Empire such as were portions of Italy, Spain, Macedonia, Noricum and Gaul, large-scale Roman workings occur in several districts,[1] and virtually all the well-known sources were exploited to some extent: principally:

(i) *The Weald,* east Sussex, north and north-west of St. Leonards: a region near enough to the Channel to explain Caesar's phrase *in maritimis (regionibus)* (*B.G.* 5, 12) as the source of British iron.[2] It would seem, however, from his comment *sed eius exigua est copia* that the workings were less developed then than they later became, for coins and pottery suggest exploitation throughout the Roman period,[3] much of the slag being used for road-metal. It has been suggested that the Chichester inscription (*RIB* 91) which mentions a COL-LE]GIUM FABROR(*um*) relates to a group of iron-workers organized in Roman fashion, during the period of the first-century client kingdom of Cogidubnus.[4]

(ii) *The Forest of Dean.* In west Gloucestershire (and in adjacent parts of Monmouthshire and Herefordshire) Roman working is proved at Lydney,[5] at Weston-under-Penyard (*Ariconium*) where extensive slag-heaps covering many acres have been recognized,[6] and elsewhere.

[1] Richmond, *Roman Britain*², pp. 157–9.

[2] The iron currency-bars mentioned by Caesar in the same passage (*utuntur aut aere aut nummo aureo aut taleis* (π: *aliis* χBMS: *anulis* LN) *ferreis ad certum pondus examinatis pro nummo*) are identified with objects resembling unfinished sword-blades, found mainly in the west and south of England. Strabo lists iron as a British export (4, 199) and it was employed as early as the later seventh century B.C. (cf. Fox, *Antiq. Journ.* 19 (1939), 386–7 for an iron copy of a native bronze sickle-type from the Llyn Fawr (Glam.) hoard). The Glastonbury finds (Gray, *Glastonbury Lake-village*² (1917), pp. 360 ff.) are still the best instances of pre-Roman iron-working in Britain.

[3] E. Straker and I. D. Margary, *Geogr. Journ.* 92 (1938), 55–60 and pl.

[4] Richmond, *Roman Britain*², p. 158.

[5] R.E.M. and T. V. Wheeler, *Lydney Report* (1932), 18–22.

[6] G. H. Jack, *Ariconium* (Woolhope Nat. Field Club, 1923).

PLATE I

a. Inscription from Verulamium *forum*, A.D. 79

b. Inscribed water-pipe from Chester, A.D. 79

PLATE II

Head of Claudius

PLATE III

Monument and Inscription of Classicianus

PLATE IV

Sestertius of Domitian, A.D. 84

PLATE V

Scene from Trajan's Column

PLATE VI

Carved sarcophagus from Chiusi

Distance slab from Bridgeness

PLATE VIII

Fendoch and the Sma' Glen

MAPS—KEY TO NUMBERING

1. Richborough, Kent
2. Canterbury, Kent
3. Rochester, Kent
4. London
5. Chelmsford, Essex
6. Fingringhoe, Essex
7. Colchester, Essex
8. Coddenham, Suff.
9. Ixworth, Suff.
10. Scole, Norf.
11. Great Chesterford, Essex
12. Cambridge
13. St. Albans, Herts.
14. Godmanchester, Hunts.
15. Irchester, Northants.
16. Towcester, Northants.
17. Alchester, Oxon.
18. Dorchester, Oxon.
19. Fishbourne, Sussex
20. Bitterne, Hants
21. Hamworthy, Dorset
22. Wimbourne, Dorset
23. Hod Hill, Dorset
24. Dorchester, Dorset
25. Ham Hill, Somerset
26. Waddon Hill, Dorset
27. Topsham, Devon
28. Exeter, Devon
29. North Tawton, Devon
30. Nanstallon, Corn.
31. Martinhoe, Devon
32. Old Burrow, Devon
33. Wiveliscombe, Somerset
34. Sea Mills, Glos.
35. Bath, Somerset
36. Mildenhall, Wilts.
37. Wanborough, Wilts.
38. Cirencester, Glos.
39. Kingsholm, Glos.
40. Gloucester
41. Dorn, Glos.
42. Chesterton, Warwicks.
43. Caves Inn, Warwicks.
44. Water Newton, Hunts.
45. Longthorpe, Northants.
46. Great Casterton, Rutland
47. Leicester
48. Mancetter, Warwicks.
49. Baginton, Warwicks.
50. Alcester, Warwicks.
51. Worcester
52. Droitwich, Worcs.
53. Metchley, Warwicks.
54. Greensforge, Staffs.
55. Eaton House, Staffs.
56. Stretton Mill, Staffs.
57. Kinvaston, Staffs.
58. Red Hill, Salop.
59. Wall, Staffs.
60. Littlechester, Derbys.
61. Broxtowe, Notts.
62. Castle Hill (Margidunum), Notts.
63. Thorpe, Notts.
64. Brough (Crococalana), Notts.
65. Ancaster, Lincs.
66. Lincoln
67. Newton on Trent, Lincs.
68. Littleborough, Notts.
69. Winteringham, Lincs.
70. Brough on Humber, Yorks.
71. York
72. Malton, Yorks.
73. Templeborough, Yorks.
74. Wroxeter, Salop

75. Wall Town, Salop.
76. Leintwardine, Herefords.
77. Kenchester, Herefords.
78. Clyro, Radnor.
79. Abergavenny, Mon.
80. Usk, Mon.
81. Coed y Caerau, Mon.
82. Caerleon, Mon.
83. Cardiff, Glam.
84. Caerphilly, Glam.
85. Gelligaer, Glam.
86. Penydarren, Glam.
87. Coelbren, Glam.
88. Neath, Glam.
89. Carmarthen
90. Llandovery, Carm.
91. Brecon
92. Caerau, Brecknocks.
93. Llanio, Cards.
94. Trawscoed, Cards.
95. Castell Collen, Radnor.
96. Nantmel, Radnor.
97. Pennal, Mer.
98. Caersws, Montgom.
99. Forden Gaer, Montgom.
100. Brithdir, Mer.
101. Caer Gai, Mer.
102. Tomen-y-Mur, Mer.
103. Pen Llystyn, Caern.
104. Caernarvon
105. Bryn-y-Gefeillau, Caern.
106. Caerhun, Caern.
107. Whitchurch, Salop
108. Chester
109. Middlewich, Ches.
110. Chesterton, Staffs.
111. Rocester, Staffs.
112. Buxton, Derbys.
113. Brough-on-Noe, Derbys.
114. Doncaster, Yorks.
115. Melandra, Derbys.
116. Manchester, Lancs.
117. Wilderspool, Lancs.
118. Wigan, Lancs.
119. Castleshaw, Yorks.
120. Slack, Yorks.
121. Castleford, Yorks.
122. Newton Kyme, Yorks.
123. Aldborough, Yorks.
124. Adel, Yorks.
125. Ilkley, Yorks.
126. Elslack, Yorks.
127. Ribchester, Lancs.
128. Kirkham, Lancs.
129. Overborough, Lancs.
130. Watercrook, Westmor.
131. Low Borrow Bridge, Westmor.
132. Bainbridge, Yorks.
133. Catterick, Yorks.
134. Bowes, Yorks.
135. Brough under Stainmore, Westmor.
136. Kirkby Thore, Westmor.
137. Brougham, Westmor.
138. Old Penrith, Cumb.
139. Caermote, Cumb.
140. Papcastle, Cumb.
141. Maryport, Cumb.
142. Kirkbride, Cumb.
143. Carlisle, Cumb.
144. Binchester, Co. Dur.
145. Ebchester, Co. Dur.
146. Corbridge, Northumb.
147. High Rochester, Northumb.
148. Learchild, Northumb.
149. Chew Green, Northumb.
150. Broomholm, Dumfriess.
151. Birrens, Dumfriess.
152. Ward Law, Dumfriess.
153. Glenlochar, Kirkcudbrights.
154. Gatehouse of Fleet, Kirkcudbrights.
155. Dalswinton, Dumfriess.
156. Tassiesholm (Milton), Dumfriess.

157. Raeburnfoot, Dumfriess.
158. Cappuck, Roxburghs.
159. Oakwood, Selkirks.
160. Newstead, Roxburghs.
161. Crawford, Lanarks.
162. Easter Happrew, Peebles.
163. Oxton, Berwicks.
164. Castledykes, Lanarks.
165. Loudoun Hill, Ayrs.
166. Inveresk, Midlothian
167. Camelon, Stirlings.
168. Castlecary, Stirlings.
169. Croy Hill, Dumbarton.
170. Bar Hill, Dumbarton.
171. Cadder, Lanarks.
172. Bochastle, Perths.
173. Ardoch, Perths.
174. Dealgin Ross, Perths.
175. Strageath, Perths.
176. Carpow, Perths.
177. Fendoch, Perths.
178. Bertha, Perths.
179. Inchtuthil, Perths.
180. Cargill, Perths.
181. Cardean, Angus
182. Stracathro, Angus
183. Dunblane, Perths.
184. Dornock, Perths.
185. Innerpefferay, Perths.
186. Broomhill, Perths.
187. Dupplin, Perths.
188. Auchtermuchty, Fife
189. Bonnytown, Fife
190. Grassy Walls, Perths.
191. Lintrose, Angus
192. Kirkbuddo, Angus
193. Battledykes-Oathlaw, Angus
194. Finavon, Forfar.
195. Battledykes-Keithock, Angus
196. Kair House, Kincar.
197. Raedykes, Kincar.
198. Normandykes, Aberdeens.
199. Kintore, Aberdeens.
200. Ythan Wells, Glenmailen, Aberdeens.
201. Muiryfold, Banffs.
202. Auchinove, Banffs.

INDEX

Emperors and classical authors are listed by their familiar English names. Roman citizens are registered under their *gentilicia* with the exception of A. Atticus and Thrasea Paetus. Places are listed under their English names with the Latin names, where known or conjectured, in brackets.

Inscriptions Discussed

OTHER TITLES IN THIS HARDBACK REPRINT PROGRAMME
FROM OXBOW BOOKS (OXFORD) AND POWELLS BOOKS (CHICAGO) www.oxbowbooks.com
www.powellschicago.com

ISBN 0–19–	Author	Title
8264011	ALEXANDER Paul J.	The Patriarch Nicephorus of Constantinople
8143567	ALFÖLDI A.	The Conversion of Constantine and Pagan Rome
9241775	ALLEN T.W	Homeri Ilias (3 volumes)
6286409	ANDERSON George K.	The Literature of the Anglo-Saxons
8219601	ARNOLD Benjamin	German Knighthood
8208618	ARNOLD T.W.	The Caliphate
8142579	ASTIN A.E.	Scipio Aemilianus
8144059	BAILEY Cyril	Lucretius: De Rerum Natura (3 volumes)
814167X	BARRETT W.S.	Euripides: Hippolytos
8228813	BARTLETT & MacKAY	Medieval Frontier Societies
8219733	BARTLETT Robert	Trial by Fire and Water
8118856	BENTLEY G.E.	William Blake's Writings (2 volumes)
8111010	BETHURUM Dorothy	Homilies of Wulfstan
8142765	BOLLING G. M.	External Evidence for Interpolation in Homer
814332X	BOLTON J.D.P.	Aristeas of Proconnesus
9240132	BOYLAN Patrick	Thoth, the Hermes of Egypt
8114222	BROOKS Kenneth R.	Andreas and the Fates of the Apostles
8214715	BUCKLER Georgina	Anna Comnena
8203543	BULL Marcus	Knightly Piety & Lay Response to the First Crusade
8216785	BUTLER Alfred J.	Arab Conquest of Egypt
8148046	CAMERON Alan	Circus Factions
8143516	CAMERON Alan	Claudian
8148054	CAMERON Alan	Porphyrius the Charioteer
8148348	CAMPBELL J.B.	The Emperor and the Roman Army 31 BC to 235
826643X	CHADWICK Henry	Priscillian of Avila
826447X	CHADWICK Henry	Boethius
8222025	COLGRAVE B. & MYNORS R.A.B.	Bede's Ecclesiastical History of the English People
8131658	COOK J.M.	The Troad
8219393	COWDREY H.E.J.	The Age of Abbot Desiderius
8241895	CROMBIE A.C.	Robert Grosseteste and the Origins of Experimental Science 1100–1700
8644043	CRUM W.E.	Coptic Dictionary
8148992	DAVIES M.	Sophocles: Trachiniae
814153X	DODDS E.R.	Plato: Gorgias
825301X	DOWNER L.	Leges Henrici Primi
814346X	DRONKE Peter	Medieval Latin and the Rise of European Love-Lyric
8142749	DUNBABIN T.J.	The Western Greeks
8154372	FAULKNER R.O.	The Ancient Egyptian Pyramid Texts
8221541	FLANAGAN Marie Therese	Irish Society, Anglo-Norman Settlers, Angevin Kingship
8143109	FRAENKEL Edward	Horace
8142781	FRASER P.M.	Ptolemaic Alexandria (3 volumes)
8201540	GOLDBERG P.J.P.	Women, Work and Life Cycle in a Medieval Economy
8140215	GOTTSCHALK H.B.	Heraclides of Pontus
8266162	HANSON R.P.C.	Saint Patrick
8581351	HARRIS C.R.S	The Heart and Vascular System in Ancient Greek Medicine
8224354	HARRISS G.L.	King, Parliament and Public Finance in Medieval England to 1369
8581114	HEATH Sir Thomas	Aristarchus of Samos
8140444	HOLLIS A.S.	Callimachus: Hecale
8212968	HOLLISTER C. Warren	Anglo-Saxon Military Institutions
9244944	HOPKIN-JAMES L.J.	The Celtic Gospels
8226470	HOULDING J.A.	Fit for Service
2115480	HENRY Blanche	British Botanical and Horticultural Literature before 1800
8219523	HOUSLEY Norman	The Italian Crusades
8223129	HURNARD Naomi	The King's Pardon for Homicide – before AD 1307
9241783	HURRY Jamieson B.	Imhotep
8140401	HUTCHINSON G.O.	Hellenistic Poetry
9240140	JOACHIM H.H.	Aristotle: On Coming-to-be and Passing-away
9240094	JONES A.H.M	Cities of the Eastern Roman Provinces
8142560	JONES A.H.M.	The Greek City
8218354	JONES Michael	Ducal Brittany 1364–1399
8271484	KNOX & PELCZYNSKI	Hegel's Political Writings
8212755	LAWRENCE C.H.	St Edmund of Abingdon
8225253	LE PATOUREL John	The Norman Empire
8212720	LENNARD Reginald	Rural England 1086–1135
8212321	LEVISON W.	England and the Continent in the 8th century
8148224	LIEBESCHUETZ J.H.W.G.	Continuity and Change in Roman Religion
8143486	LINDSAY W.M.	Early Latin Verse
8141378	LOBEL Edgar & PAGE Sir Denys	Poetarum Lesbiorum Fragmenta
9240159	LOEW E.A.	The Beneventan Script
8115881	LOOMIS Roger Sherman	Arthurian Literature in the Middle Ages
8241445	LUKASIEWICZ, Jan	Aristotle's Syllogistic
8152442	MAAS P. & TRYPANIS C.A .	Sancti Romani Melodi Cantica

8113692	MANDEVILLE Bernard	The Fable of the Bees (2 volumes)
8142684	MARSDEN E.W.	Greek and Roman Artillery—Historical
8142692	MARSDEN E.W.	Greek and Roman Artillery—Technical
8148178	MATTHEWS John	Western Aristocracies and Imperial Court AD 364–425
9240205	MAVROGORDATO John	Digenes Akrites
8223447	McFARLANE K.B.	Lancastrian Kings and Lollard Knights
8226578	McFARLANE K.B.	The Nobility of Later Medieval England
814296X	MEIGGS Russell	The Athenian Empire
8148100	MEIGGS Russell	Roman Ostia
8148402	MEIGGS Russell	Trees and Timber in the Ancient Mediterranean World
8141718	MERKELBACH R. & WEST M.L.	Fragmenta Hesiodea
8143362	MILLAR F.G.B.	Cassius Dio
8142641	MILLER J. Innes	The Spice Trade of the Roman Empire
8147813	MOORHEAD John	Theoderic in Italy
8264259	MOORMAN John	A History of the Franciscan Order
8181469	MORISON Stanley	Politics and Script
8142218	MORITZ L.A.	Grain-Mills and Flour in Classical Antiquity
8274017	MURRAY H.J.R.	History of Board Games
8274033	MURRAY H.J.R.	History of Chess
9240582	MUSURILLO H.	Acts of the Pagan Martyrs & Christian Martyrs (2 volumes)
9240213	MYRES J.L.	Herodotus The Father of History
9241791	NEWMAN W.L.	The Politics of Aristotle (4 volumes)
8219512	OBOLENSKY Dimitri	Six Byzantine Portraits
8270259	O'DONNELL J.J.	Augustine: Confessions (3 volumes)
8144385	OGILVIE R.M. & RICHMOND I.A.	Tacitus: Agricola
263268X	OSLER Sir William	Bibliotheca Osleriana
8116020	OWEN A.L.	The Famous Druids
8131445	PALMER, L.R.	The Interpretation of Mycenaean Greek Texts
8143427	PFEIFFER R.	History of Classical Scholarship (volume 1)
8143648	PFEIFFER Rudolf	History of Classical Scholarship 1300–1850
8111649	PHEIFER J.D.	Old English Glosses in the Epinal-Erfurt Glossary
8142277	PICKARD–CAMBRIDGE A.W.	Dithyramb Tragedy and Comedy
8269765	PLATER & WHITE	Grammar of the Vulgate
9256497	PLATNER S.B. & ASHBY T.	A Topographical Dictionary of Ancient Rome
8213891	PLUMMER Charles	Lives of Irish Saints (2 volumes)
820695X	POWICKE Michael	Military Obligation in Medieval England
8269684	POWICKE Sir Maurice	Stephen Langton
821460X	POWICKE Sir Maurice	The Christian Life in the Middle Ages
8225369	PRAWER Joshua	Crusader Institutions
8225571	PRAWER Joshua	The History of The Jews in the Latin Kingdom of Jerusalem
8143249	RABY F.J.E.	A History of Christian Latin Poetry
8143257	RABY F.J.E.	A History of Secular Latin Poetry in the Middle Ages (2 volumes)
8214316	RASHDALL & POWICKE	The Universities of Europe in the Middle Ages (3 volumes)
8154488	REYMOND E.A.E & BARNS J.W.B.	Four Martyrdoms from the Pierpont Morgan Coptic Codices
8148380	RICKMAN Geoffrey	The Corn Supply of Ancient Rome
8141556	ROSS Sir David	Aristotle: De Anima
8141076	ROSS Sir David	Aristotle: Metaphysics (2 volumes)
8141084	ROSS Sir David	Aristotle: Parva Naturalia
8141092	ROSS Sir David	Aristotle: Physics
9244952	ROSS Sir David	Aristotle: Prior and Posterior Analytics
8142307	ROSTOVTZEFF M.	Social and Economic History of the Hellenistic World (3 volumes)
8142315	ROSTOVTZEFF M.	Social and Economic History of the Roman Empire (2 volumes)
8264178	RUNCIMAN Sir Steven	The Eastern Schism
814833X	SALMON J.B.	Wealthy Corinth
8171587	SALZMAN L.F.	Building in England Down to 1540
8218362	SAYERS Jane E.	Papal Judges Delegate in the Province of Canterbury 1198–1254
8221657	SCHEIN Sylvia	Fideles Crucis
8148135	SHERWIN WHITE A.N.	The Roman Citizenship
825153X	SHERWIN WHITE A.N.	Roman Society and Roman Law in the New Testament
9240167	SINGER Charles	Galen: On Anatomical Procedures
8113927	SISAM, Kenneth	Studies in the History of Old English_Literature
8113668	SKEAT Walter	Langland: The Vision of William Concerning Piers the Plowman (2 volumes)
8642040	SOUTER Alexander	A Glossary of Later Latin to 600 AD
8270011	SOUTER Alexander	Earliest Latin Commentaries on the Epistles of St Paul
8222254	SOUTHERN R.W.	Eadmer: Life of St. Anselm
8251408	SQUIBB G.	The High Court of Chivalry
8212011	STEVENSON & WHITELOCK	Asser's Life of King Alfred
8212011	SWEET Henry	A Second Anglo-Saxon Reader—Archaic and Dialectical
8143443	SYME Sir Ronald	Ammianus and the Historia Augusta
8148259	SYME Sir Ronald	History in Ovid
8143273	SYME Sir Ronald	Tacitus (2 volumes)
8142714	THOMPSON E.A.	The Goths in Spain
9256500	THOMPSON Sir E.Maunde	Introduction to Greek and Latin Palaeography
8200951	THOMPSON Sally	Women Religious
924023X	WALBANK F.W.	Historical Commentary on Polybius (3 volumes)
8201745	WALKER Simon	The Lancastrian Affinity 1361–1399
8161115	WELLESZ Egon	A History of Byzantine Music and Hymnography
8140185	WEST M.L.	Greek Metre

8141696	WEST M.L.	Hesiod: Theogony
8148542	WEST M.L.	The Orphic Poems
8140053	WEST M.L.	Hesiod: Works & Days
8152663	WEST M.L.	Iambi et Elegi Graeci
9240221	WHEELWRIGHT Philip	Heraclitus
822799X	WHITBY M. & M.	The History of Theophylact Simocatta
8206186	WILLIAMSON, E.W.	Letters of Osbert of Clare
8208103	WILSON F.P.	Plague in Shakespeare's London
8247672	WOODHOUSE C.M.	Gemistos Plethon
8114877	WOOLF Rosemary	The English Religious Lyric in the Middle Ages
8119224	WRIGHT Joseph	Grammar of the Gothic Language